Grundlehren der mathematischen Wissenschaften 228

A Series of Comprehensive Studies in Mathematics

Editors

S. S. Chern   J. L. Doob   J. Douglas, jr.
A. Grothendieck   E. Heinz   F. Hirzebruch   E. Hopf
S. Mac Lane   W. Magnus   M. M. Postnikov
W. Schmidt   D. S. Scott
K. Stein   J. Tits   B. L. van der Waerden

Managing Editors

B. Eckmann   J. K. Moser

Irving E. Segal   Ray A. Kunze

Integrals
and Operators

Second Revised and Enlarged Edition

Springer-Verlag
Berlin Heidelberg New York 1978

Irving E. Segal
Massachusetts Institute of Technology, Cambridge, MA 02139, USA

Ray A. Kunze
University of California at Irvine, Irvine, CA 92664, USA

First edition © 1968 by McGraw-Hill, Inc.

AMS Subject Classifications (1970): 22 B xx, 22 C 05, 22 D xx, 28-XX, 43-XX, 46-XX, 47-XX

ISBN-13: 978-3-642-66695-7   e-ISBN-13: 978-3-642-66693-3
DOI: 10.1007/978-3-642-66693-3

Library of Congress Cataloging in Publication Data. Segal, Irving Ezra.
Integrals and operators (Grundlehren der mathematischen Wissenschaften; 228).
Bibliography: p. Includes index. 1. Integrals, Generalized. 2. Functional analysis.
I. Kunze, Ray Alden, 1928—joint author. II. Title. III. Series: Die Grund-
lehren der mathematischen Wissenschaften in Einzeldarstellungen; 228.
QA312.S4. 1978. 515'.42. 77-16682

PREFACE
TO THE SECOND EDITION

Since publication of the First Edition several excellent treatments of advanced topics in analysis have appeared. However, the concentration and penetration of these treatises naturally require much in the way of technical preliminaries and new terminology and notation. There consequently remains a need for an introduction to some of these topics which would mesh with the material of the First Edition. Such an introduction could serve to exemplify the material further, while using it to shorten and simplify its presentation.

It seemed particularly important as well as practical to treat briefly but cogently some of the central parts of operator algebra and higher operator theory, as these are presently represented in book form only with a degree of specialization rather beyond the immediate needs or interests of many readers. Semigroup and perturbation theory provide connections with the theory of partial differential equations. C^*-algebras are important in harmonic analysis and the mathematical foundations of quantum mechanics. W^*-algebras (or von Neumann rings) provide an approach to the theory of multiplicity of the spectrum and some simple but key elements of the grammar of analysis, of use in group representation theory and elsewhere. The theory of the trace for operators on Hilbert space is both important in itself and a natural extension of earlier integration-theoretic ideas.

Accordingly, four chapters have been added, one dealing with each of the subjects indicated. These form a logical extension of the standpoint of the First Edition, and at the same time convey the fundamentals of subjects which are central for aspects of higher physical mathematics, group representation theory, and growing applications to analysis on manifolds.

The opportunity has been taken to correct errors, and terminological variations, as well as some expository lapses in the First Edition, which were kindly pointed out to us by conscientious readers. It is hoped the resulting volume will be useful to students and scientists in other fields who may be interested in a cultured overview of modern analysis and its logical structure which retains continuous connections with traditional real variable theory.

PREFACE
TO THE FIRST EDITION

This book is intended as a first graduate course in contemporary real analysis. It is focused about integration theory, which we believe is appropriate. For a variety of reasons—in the interests of logic, flexibility, and curricular economy, among others—we have assumed that the reader or student is already familiar with the rudiments of modern mathematics (by this we mean the most elementary aspects of set theory, general topology, and algebra, as well as some exposure to rigorous analysis). These are not so much technical requirements—although basic concepts such as set, topological space, and uniform convergence are taken entirely for granted—as requirements of mathematical maturity and of understanding of the elementary grammar and language of modern mathematics. Assuming the adequate mathematical "aging" of the student, the book is quite self-contained. Results such as the Stone-Weierstrass theorem, the existence of a partition of unity, etc., are given full proofs rather than disposed of by reference to hypothesized preliminaries.

The aim of the book is primarily cultural, rather than vocational; the authors strive to expose the student to modern analytical thought and if

vii

possible to train him to think in such terms rather than to load him with all available information on the subject. Nevertheless, the book should represent a proper introduction to real analysis for students intending to concentrate in analysis, as well as a (possibly terminal) general course in the subject for those with other scientific interests. Indeed, thought cannot take place in a vacuum, and contrived illustrations of the theory have a way of turning out not to be as truly representative or interesting as illustrations of the actual usage of the theory for vital mathematical purposes. For this reason the book has been built around material of maximal current mathematical importance and depth, a technical mastery of which should go hand in hand with an appreciation of the general ideas.

The book is a revision, adaptation, and extension of lecture notes of courses in Integration Theory given at the University of Chicago and in Real Variable Theory at Massachusetts Institute of Technology by the first-named author. The first half of the book is suitable for a one-semester course in Lebesgue integration theory, including both abstract and classical real variable aspects. Its heart is a fresh presentation of the Daniell approach, which, combined with the use of general topological ideas, attains a high level of generality and completeness without burdening the student with heavy machinery or bulky technicalities. This material should provide a cultural experience for the student comparable to his first exposure to the calculus; indeed, the success of this theory against what appear initially as overwhelming scientific odds, and its broad applicability, render it one of the comparable intellectual achievements of mathematics. Many examples and a considerable variety of exercises, at all levels of difficulty, serve to illustrate the theory and to indicate the continuous transition between the concrete and abstract phases of integration theory. Theoretical ramifications which are secondary from the standpoint of the overall theoretical development, although frequently of considerable importance, are included among the more difficult exercises; the student is assisted with hints, and the arrangement is designed to encourage learning by students' investigations under the guidance of the instructor or by self-discovery. The exercises range in difficulty from easy ones which simply confirm an understanding of the text to relatively difficult ones, distinguished by a *, which in a controlled way, introduce the student to the beginnings of research. The * is also used to distinguish material (several sections and one chapter) which may be omitted without disturbing the main line of development.

The book as a whole is quite unified. Integration theory provides the main examples for the treatment of linear topological spaces and their duality. In the more structured situations provided by groups of transformations, new aspects of function theory arise from the consideration of invariant measures. The reducibility of commutative spectral analysis in Hilbert space to integration theory makes it natural, as well as economical, to develop spectral theory from this viewpoint. For these reasons the book may be used for the second half of a one-year course in Real Analysis, which merges naturally with the first half, provides basically new material of general importance, and yet serves at the same time to build on and provide a capstone for the student's earlier exposure to linear algebra and integration. The completion of such a course should provide the student with the key real-analytical background for work in other parts of modern mathematics; for more advanced work in analysis, whether of a more abstract or concrete variety; and for contemporary theoretical physics. We especially feel that the book takes the student rather quickly, but not too abruptly, to a good jumping-off place for the study of Fourier analysis, linear partial differential operators, the theory of group representations, operator algebras, and abstract probability theory.

Although there are now many quite competent treatments in textbook form of Lebesgue integration theory, as well as some on introductory functional analysis, we feel that none of these books achieves quite what this one is intended to do. With our treatment, it is possible to take the suitably prepared student in one year at a properly measured pace through basic contemporary real analysis, giving him the feel of the subject, a clear indication of its sweep, and an adequately detailed mastery of a number of central features. Our general viewpoint is partially in the direction of an earlier exposition by one of us on algebraic integration theory, i.e., toward the utilization of abstract integrative ideas (cf. References, p. 365); this has always been one of the long-term trends in mathematics, enforcing a type of consolidation which may be essential to prevent undue scientific complexity and bulk from imposing a crushing burden on the development of fresh ideas and methods. At the same time, as already indicated, the continuous linkage between the abstract theory and the concrete analytical situation has been everywhere insisted on—in the motivational material, in the examples, and in the exercises. We believe that this book is more likely to help cure "abstractionitis"—an unfortunate but not uncommon side effect of otherwise highly beneficial inoculations with modern mathematical ideas—than to cause it. We have made some technical innovations where they appeared useful to serve our central ideas, but have avoided them otherwise. Thus the treatment of "large" measure spaces is curtailed in the text, such primarily technical developments being outlined in the exercises; on the other hand, the uniform-space approach to the construction of invariant measures has been adopted.

As it has turned out, the book is fairly flexible from an instructional viewpoint. An independent short course, in the nature of an introduction to functional analysis and spectral theory in Hilbert space, may be given on the basis of the second half of the book, exclusive of Chapter 7, for students already familiar with the abstract Lebesgue integral.

We are much indebted to many colleagues and students for general advice and specific comments and for reducing the number of errors in the manuscript to what we hope is a superficial level. In particular, lengthy lists of corrections to draft manuscripts were supplied by Robert Kallman, Michael Weinless, and Alan Weinstein.

Irving E. Segal
Ray A. Kunze

CONTENTS

CHAPTER VI. FUNCTION SPACES   **152**

CHAPTER VII. INVARIANT INTEGRALS   **175**

CHAPTER VIII. ALGEBRAIC INTEGRATION THEORY   **206**

CHAPTER IX. SPECTRAL ANALYSIS IN HILBERT SPACE   **239**

CHAPTER X. GROUP REPRESENTATIONS AND UNBOUNDED OPERATORS   **258**

I
INTRODUCTION

1.1 GENERAL PRELIMINARIES

Before embarking on a serious study of a new subject, the intellectually prudent student will want to know why the subject is studied and what it relates to.   Let us say that he accepts on faith the assurance that integration is significant, not only as a vital tool in analysis and as the culmination of the calculus, but also as an intrinsically beautiful and complete theory, in which elements of geometry and algebra, as well as analysis, are merged. Even so, his understanding of the subject will proceed more rapidly if he has some definite, if general, knowledge of what sort of thing it is and how it is related to the subjects he is already familiar with and if he sees why it has aroused such interest.

The Riemann integral that the reader is familiar with assigns to certain functions defined on certain sets in euclidean space a number called the integral of the function over the set.   Both the function and the set must satisfy regularity conditions; it suffices, for example, if the function is continuous and the set is bounded and has a continuously differentiable boundary. In euclidean space there is defined a notion of length, area, or volume,

1

depending on whether the dimension is 1, 2, or 3, or more generally, a notion of n-dimensional volume in n-space, which plays an important part in the formation of the integral. Now a common situation in analysis, as well as in quasi-mathematical subjects such as physics and probability, is that in which one is given a kind of notion of volume, or measure, for sufficiently smooth subsets of a basic set and wishes to assign to each sufficiently well-behaved function on the set a number that will be a sort of integral. The basic set may be a Riemannian manifold, or the phase space of a dynamical system, or the space of elementary events in a probabilistic system; the notion of volume may be derived from a given Riemannian metric, or may be determined by the condition that it be unchanged with the passage of time, or from probability considerations; it may be called area, mass, charge, or probability; and the integral in question may have the interpretation of volume, pressure, potential, or expectation. But these various situations are fundamentally similar in a certain way, or to use a more precise mathematical term, *isomorphic* with regard to certain aspects. The theory of integration studies the problem of assigning an integral to a function defined on a set that is endowed with a notion of "measure," with regard to those features that are independent of the origin of measure or the interpretation of the function. The "Lebesgue" integral, then, like any other integral, is a functional on a certain class of functions, which relates to a measure on the set over which the functions are defined.

The basic difference between the Riemann and Lebesgue integral is not so much in the measure—although historically a difference in the domains of definition of the relevant measures was taken very seriously—as in the fact that the Lebesgue integral extends the Riemann integral in a certain technically advantageous fashion, and yet is itself terminal, being incapable of any further extension of the same sort. In other terms, the Lebesgue integral applies to a class of functions that is maximal in a certain simple intrinsic sense and that includes the class of functions to which the Riemann integral applies. There are many other differences, such as the fact that the natural logical extension of the Lebesgue integral is to the case of functions on an abstract set, devoid of any topology, while the Riemann integral inherently refers to functions defined on a topological space; yet in essence the Lebesgue integral, in the historically original and still most important case of integration of functions on euclidean space with respect to euclidean measure, is both in a general and a mathematical sense a completion of the Riemann integral.

The basic problems in the theory of the Lebesgue integral are generally parallel to those in the theory of the Riemann integral, and yet proceed along completely different lines. First one must define the integral and prove its existence under usefully general conditions. Next one derives properties of the integral, but these are much more extensive in the Lebesgue case and include in particular the terminal feature described in the preceding paragraph. The primary general problems that remain are those of multiple integration and differentiation. The theory of differentiation is necessarily novel since it is logical, in general, to differentiate not a point function, as in the calculus, but a set function, the derivative being a point function; the reason for this will become apparent by the time this stage of the theory is reached. When the basic problems are covered, the theory has attained a certain logical completeness, but new problems emerge from the connections of the theory with other parts of mathematics. As in most substantial and living mathematics, there is a constant tension between the abstract and the concrete in the theory of integration. Although this may be as desirable as it is inevitable, it results in the impossibility of giving a single formulation of the subject that will be adaptable to the purposes of showing its evolution from the Riemann theory, of giving its logical basis in the most intelligible and elegant fashion, or of being conveniently applicable to Fourier analysis or to abstract functional analysis. The validity of all these purposes makes it undesirable, if not impractical or misleading, to present the subject for the first time in either a wholly abstract or wholly concrete fashion. Thus the theory will be treated in this book from a mixture of several points of view.

1.2   THE IDEA OF MEASURE

Traditionally, the Lebesgue integral referred to functions defined on euclidean space, or on a subset of it, like the familiar Riemann integral. As a result of both abstract and concrete impulses, however, mathematicians began looking into the question of whether the theory could not, in some essential parts at least, be carried over to more general situations. Even before Lebesgue, the Riemann-Stieltjes integral had been treated, showing that the Riemann theory applied to functions on euclidean space, but with a rather general type of measure, important in many applications. By the latter half of the thirties, after about a decade of activity along these lines, the basic theorems in the Lebesgue theory had achieved extensions to measures on perfectly arbitrary sets, although in a few cases (notably, differentiation theory) the extension did not have the full force of the original theory when applied to the euclidean case. Concurrently, these ideas had found important applications in probability, which found in the new theory a means of making explicit the notion of "random variable" which was basic in the subject. Since that time integration theory in abstract spaces has played a part in the spectral theory of linear operators (essentially, the extension of matrix theory to an infinite number of dimensions), in analysis on topological groups (a natural and widely applicable extension of classical Fourier analysis), and in many other diverse questions.

Thus the idea of abstract measure space has technical cogency, but it is also a very natural one from an intuitive geometrical point of view. To see how it arises in this way, consider the problem of formulating the concept of "measure" on a set S as an abstraction of the notions of volume, length, mass, probability, etc., mentioned above. The most obvious approach is to define this as a function that assigns to suitable subsets E of S a number $m(E)$, representing its measure and having the properties characteristic of measure. This then raises the primary questions of (*1*) which subsets are "suitable," and (*2*) what properties are characteristic. Now one property of the familiar examples of measure is that, so to speak, the whole is the sum of its parts, in terms of measure. More specifically, if A and B are disjoint sets for which the measure is defined, then their union has measure $m(A) + m(B)$. This seems intuitively, as well as technically, to be a reasonable assumption, but as might be expected, in itself it is not a sufficient basis for the development of a useful, interesting theory. For one thing, there is too little freedom in dealing with "suitable" sets if it is known only that the union of two disjoint suitable sets is again such. It does not seem too much to require, more stringently, that the union of any finite number of suitable sets be suitable, and even to add the further requirement that the difference of suitable sets be such. If this assumption is checked against the technical situation, it still seems quite moderate, and is satisfied (or can be arranged) in the familiar cases.

A collection $\mathcal{R}$ of subsets that is closed under the formation of the union and difference of any two subsets is generally called a *ring of sets*. In other words, $\mathcal{R}$ is a ring if whenever A and B are in $\mathcal{R}$, then $A \cup B$ and $A - B$ are again in $\mathcal{R}$. It is useful to justify the use of the term "ring" in the present context. The fact is that a ring $\mathcal{R}$ of sets is actually a ring, in the conventional algebraic sense, relative to certain operations, and conversely. Specifically, these operations are the intersection $A \cap B$ as the product, and the "symmetric difference" $A \ominus B = (A - B) \cup (B - A)$ as the sum. One verifies the algebraic identities that establish associativity, distributivity, etc., in the present connection, by taking a hypothetical element of the set on one side of the identity and showing that it is a member of the set on the other side. Hence, to show that a ring of sets $\mathcal{R}$ is a ring in the standard sense relative to these operations, it suffices to show that it is closed under the operations. For this it suffices to express the intersection and symmetric difference in terms of their union and (asymmetric) differences. The latter expression is given by the definition of the symmetric difference, while a corresponding expression for the intersection is

$$A \cap B = A - (A - B).$$

It should be noted here that the empty set is the unit for addition.

Conversely, if a collection $\mathcal{R}$ is a (standard type of) ring with respect to intersection as multiplication and the symmetric difference as addition, then it is a ring of sets as defined above. For if A and B are any two sets, then

$$A \cup B = A \ominus B \ominus (A \cap B)$$

(because of the associativity of the operation $\ominus$, the expression on the right is well defined), and

$$A - B = A \ominus (A \cap B),$$

which shows that a collection is closed under unions and (asymmetric) differences in case it is closed under intersections and symmetric differences.

The system arrived at now, composed of a set S, a ring $\mathcal{R}$ of subsets of S, and a numerical-valued function m on $\mathcal{R}$ that is additive on disjoint sets,

$$m(A \cup B) = m(A) + m(B) \quad \text{if} \quad A \cap B = \emptyset,$$

is known as a *finitely additive measure space*. Usually, it is assumed in addition that m takes on only nonnegative values, and for simplicity we shall make this assumption here. Some interesting cases, such as that of a charge distribution, are thereby excluded, but it is on the whole a natural assumption, and more important, it turns out that the general case is to a large extent reducible to the case of a nonnegative measure. The general notion of a finitely additive measure space, although useful for some purposes has not (at least not yet) led to a notably rich or coherent theory, nor to results that are strong enough to be particularly useful in applications to analysis, geometry, etc. This is not altogether surprising, for one knows from a background of general mathematical experience that assumptions of a purely algebraic sort are not usually adequate for dealing with an infinite type of system. Required, in addition, are restrictions involving continuity or limits, and hence involving infinite sets of elements.

In interesting cases, such as that of euclidean space, the set S and the collection $\mathcal{R}$ of suitable subsets will not be finite, nor will $\mathcal{R}$ even have a finite set of generators; yet the only restrictions thus far imposed are on pairs of subsets. Let us now seek in a formal but intuitive way a reasonable and useful restriction of a continuity sort involving an infinite sequence E_1, $E_2, \ldots$, of suitable sets. About the weakest assumption along these lines one might think of is that, in case a sequence of sets "tends to zero" in the fairly strong sense that the sequence is monotone-decreasing,

$$E_n \supset E_{n+1}, \qquad n = 1, 2, \ldots,$$

and has an empty intersection, $\bigcap_n E_n = \emptyset$, then the corresponding measures $m(E_1)$, $m(E_2)$, $\ldots$, tend to zero as real numbers. It may be noted that the sequence of measures is easily seen to be necessarily monotone-decreasing, and so to tend to some limit, in the special but basic case in which the values of measures are always nonnegative. The foregoing continuity-type assumption is so plausible that one might be tempted to think this limit must always vanish. An example to show that this is not the case will now be sketched. This example has, incidentally, some interesting ramifications, which, however, cannot be pursued here.

Example 1.2.1 Let S be the set of all positive integers 1, 2, 3, ..., and let $\mathcal{F}$ be the collection of all finite subsets of S. Then it is easily seen that $\mathcal{F}$ is a ring; it is in fact an *ideal* in the ring $\mathcal{I} = \mathcal{I}(S)$ of all subsets of S, in the usual sense of ideal in a ring. That is, it is a subring, with the property that if A is an element of it and B is an element of the containing ring, then the product AB is again in it. For a ring of sets this means that a subring $\mathcal{R}$ is an ideal in the power set $\mathcal{I}$ in case the intersection of an element in $\mathcal{R}$ with one in $\mathcal{I}$ is always again in $\mathcal{R}$. This is clearly the case for the ring $\mathcal{F}$ because every subset of a finite set is finite. Now by transfinite induction (or the Zorn principle), it is easily seen that any proper ideal in a ring with a unit is contained in a maximal proper ideal in the ring. Hence, as S is a unit for $\mathcal{I}$, there exists a maximal ideal $\mathcal{M}$ containing $\mathcal{F}$. Now it is not difficult to show that the quotient of any ring of sets modulo a maximal ideal is isomorphic to the Galois field modulo 2 (or equivalently, to the ring of subsets of a set with exactly one element). Let m be the canonical homeomorphism of $\mathcal{I}$ onto $\mathcal{I}/\mathcal{M}$, and identify the elements of Galois field modulo 2 with the real numbers 0 and 1. Then it follows that m is a measure on S that is defined for all sets, and takes only the values 0 and 1. The sets of measure zero are precisely those in $\mathcal{M}$, so that all the finite sets have measure zero. Thus the complement E_n of the finite set $[1, 2, ..., n]$ has measure 1. On the other hand, the E_n decrease monotonely to the empty set.

Thus the following notion is materially more restricted than that of a finitely additive measure space. A *basic measure space* is a system composed of set S, a ring $\mathcal{R}$ of subsets of S, and a real nonnegative-valued function m on $\mathcal{R}$ with the properties

(*1*) $m(A \cup B) = m(A) + m(B)$ if A and B are disjoint elements of $\mathcal{R}$.

(*2*) If A_1, A_2, A_3, ... is a monotone-decreasing sequence of sets in $\mathcal{R}$ with empty intersection, then $m(A_n) \to 0$. The term *measure space* is widely used in the literature in slightly different senses; here it is defined in the modal fashion as the variation of a basic measure space in which the ring $\mathcal{R}$ is closed under countable unions (such a ring is called a *σ-ring*) and in which the values of the measure function m are permitted to be $+\infty$ as well as real.

In practice, however, it is the underlying basic measure space—and such always exists—which is crucial for analytical purposes. It turns out, a result due essentially to Daniell, that any basic measure space uniquely determines a full measure space (i.e., the concept indicated above) which is a kind of completion of the basic space. Actual constructions of measure spaces usually go no further than the construction of an underlying basic space, and rely on the general theory for the existence of the theoretically convenient completion.

Condition (*2*) above, which might have been called "continuity," is usually referred to as "countable additivity"; or rather, more precisely, (*2*) is equivalent, in the presence of (*1*), to the following condition (*2a*), which is called countable additivity, for obvious reasons:

(*2a*) If A_1, A_2, A_3, ... is a sequence of disjoint subsets of $\mathcal{R}$, and if their union $A = \bigcup_i A_i$ is in $\mathcal{R}$, then

$$m(A) = \sum_i m(A_i).$$

To see that (*2*) and (*2a*) are in fact equivalent, let (*2*) be satisfied, and let A_1, A_2, A_3, ... and A be as in the hypothesis for (*2a*).   If $B_n = A - \bigcup_{i=1}^{n} A_i$, it is clear that B_n is in $\mathcal{R}$ and that the sequence B_1, B_2, B_3, ... decreases monotonely to the empty set; so $m(B_n) \to 0$.   But by condition (*1*),

$$m(B_n) = m(A) - m\left(\bigcup_{i=1}^{n} A_i \right)$$

and

$$m\left(\bigcup_{i=1}^{n} A_i \right) = m(A_1) + \cdots + m(A_n).$$

This means that the series $\sum_i m(A_i)$ is convergent to $m(A)$ and shows that (*2*) implies (*2a*).   The converse may be obtained by a straightforward reversal of the argument.

1.3   INTEGRATION AS A TECHNIQUE IN ANALYSIS

The assumption of countable additivity is justified in part by the extensive developments that rest on it.   These developments, which include a major part of modern analysis, indicate that it is in a sense the unique generalization of the ordinary concept of volume, etc., whose examination in great detail has generally significant applications.   Although we cannot at this stage give a mathematically precise description of the advantages of countable over finite additivity, two of the main ones may be sketched.

The Lebesgue integral commutes with passage to a limit a great deal more freely than does the Riemann integral.   This is one of the most important technical features of the Lebesgue integral and both motivates in part our later approach to the definition of the integral and essentially makes it unique.

The reader has surely encountered such expressions as $\lim_{n} \int_{a}^{b} f_n(x)\, dx$, where $f_1, f_2, \ldots$ is a given sequence of functions, and has probably noted that the limit may or may not exist and that it may or may not equal $\int_{b}^{a} \lim_{n} f_n(x)\, dx$, the integrals being taken in the Riemann sense.   Even though the limit

$f(x) = \lim\limits_{n} f_n(x)$ actually exists, the limit function f may very well lack sufficient smoothness to be Riemann-integrable, even when it is bounded, so that the integral of the limit may have no meaning.

Example 1.3.1   Let K be a closed subset of the open unit interval $(0,1)$. For $0 < x < 1$, let $f_n(x) = [1 - d(x)]^n$, $n = 1, 2, \ldots$, where $d(x)$ is the distance between x and K; that is,

$$d(x) = \inf_{y \in K} |x - y|.$$

Then $f_1, f_2, \ldots$ is a monotone-decreasing sequence of bounded continuous functions whose limit f is the characteristic function of K; that is, $f(x) = 1$ if x is in K and $f(x) = 0$ otherwise.   Now if K is sufficiently complicated, f will not be Riemann-integrable.   Such a K may be constructed as follows: Let $r_1, r_2, \ldots$ be an enumeration of the rational numbers in $(0,1)$, and suppose $0 < \epsilon < 1$.   In $(0,1)$ choose an open interval E_n about r_n whose length does not exceed $\epsilon/2^n$.   The union E of the intervals E_n is then an open subset of $(0,1)$.   At first glance, the reader might suspect that the closed set $K = (0,1) - E$ is empty, but this is far from being the case.   In fact, it turns out that the Lebesgue measure of K, $m(K)$, satisfies the inequality $m(K) \geq 1 - \epsilon$. In addition, it is almost immediate from the general theory that all upper Riemann sums for f over $(0,1)$ are at least as large as $1 - \epsilon$.   On the other hand, since the points $r_1, r_2, \ldots$ are dense in $(0,1)$, it follows that all lower Riemann sums for f over $(0,1)$ are 0.   This shows that f is not Riemann-integrable.   It is true, however, that f is Lebesgue-integrable over $(0,1)$ and, moreover, that

$$m(K) = \int_0^1 f(x)\, dx = \lim_{n} \int_0^1 f_n(x)\, dx.$$

In the general theory to be presented the difficulty just illustrated does not occur; the limit function is always "measurable" (the term for the type of smoothness relevant to integration theory) whenever the functions f_n in the sequence are such.   It may still not be true that the limit of the integral is equal to the integral of the limit; but this is as it should be, for while this specious statement has verbal simplicity, it is not at all intuitive or correct, as the following elementary example shows.

Example 1.3.2   Let $f_n(x) = ne^{-nx}$, $0 < x < 1$, where $n = 1, 2, 3, \ldots$. Then it is clear that $f_n(x) \to 0$ for every x, and hence that

$$\int_0^1 \lim_{n} f_n(x)\, dx = 0.$$

But $\int_0^1 f_n(x)\, dx = 1 - e^{-n}$, so that $\lim\limits_{n} \int_0^1 f_n(x)\, dx = 1$.

It turns out that the difficulty just encountered is not present when, for example, $f_1, f_2, \ldots$ is a monotone-increasing sequence of nonnegative measurable functions. Then the equality of the limits is true; i.e.,

$$\lim_{n} \int_{a}^{b} f_{n}(x)\, dx = \int_{a}^{b} \lim_{n} f_{n}(x)\, dx.$$

Moreover, this equality is valid in the Lebesgue theory in about the broadest possible interpretation. It is even characteristic of the Lebesgue theory, which may be described as essentially the minimal extension of the Riemann theory in which this kind of commutativity of the integral and passage to the limit is valid. A simple corollary to the commutativity is its extension to the case when the sequence $f_1, f_2, \ldots$ is not necessarily monotone-increasing but does have a limit at every point, and in which there exists a fixed function g whose integral $\int_{a}^{b} g(x)\, dx$ exists, such that $|f_n(x)| \leq g(x)$ for all x and n. This result is adequate for the majority of the concrete cases in analysis where the commutativity is needed.

Another significant feature of the Lebesgue integral is that certain function spaces that arise in the theory of integration are then *complete* as metric spaces, in distinction from the situation for the case of the Riemann integral. In fact, the Lebesgue function spaces can be identified with the completion (in the sense of the theory of metric spaces) of the corresponding "Riemann" function spaces. To illuminate this let us first recall that the limit of a uniformly convergent sequence of continuous functions, say, on the interval $0 \leq x \leq 1$, is again continuous. This may be reformulated as the statement that the collection **C** of all continuous functions on [0,1] is a complete metric space, with respect to the metric

$$d_{\infty}(f,g) = \sup_{0 \leq x \leq 1} |f(x) - g(x)|\,.$$

Now this metric has nothing to do with integration or the notion of measure on the interval [0,1]; it could be adapted to the set of all continuous functions on an arbitrary compact topological space, devoid of any measure. Hence it is not surprising that for questions in analysis in which integration is relevant, certain other metrics which are defined directly in terms of integration are more significant. The simplest such metric is that in which the distance d_1 is defined as a kind of average, or accumulated, absolute difference:

$$d_1(f,g) = \int_{0}^{1} |f(x) - g(x)|\, dx.$$

We leave to the reader the simple verification that this is in fact a metric. Now it is easy to give examples to show that **C** is not complete relative to the metric d_1.

Example 1.3.3 Consider the sequence $f_1, f_2, \ldots$ in which $f_n(x) = 0$, $0 \leq x \leq \frac{1}{2}$; $f_n(x) = n(x - \frac{1}{2})$ for $\frac{1}{2} < x \leq \frac{1}{2} + 1/n$; and $f_n(x) = 1$ for

$\frac{1}{2} + 1/n < x \le 1$. If $n > m$, it is clear by elementary geometry that $d_1(f_n, f_m) < 1/m$. Thus $f_1, f_2, \dots$ is a Cauchy sequence in **C**. Now suppose $f \in$ **C** and $d_1(f_n, f) \to 0$. Then

$$\int_0^{\frac{1}{2}} |f(x)| \, dx + \lim_n \int_{\frac{1}{2}}^1 |f(x) - f_n(x)| \, dx = 0.$$

This implies that $\int_0^{\frac{1}{2}} |f(x)| \, dx = 0$. Since f is continuous, it follows that $f(x) = 0$ for $0 \le x \le \frac{1}{2}$. On the other hand, it is also necessary that

$$\int_a^1 |f(x) - 1| \, dx = 0$$

for a such that $\frac{1}{2} < a < 1$. Hence $f(x) = 1$ for $a \le x \le 1$. Because a can be arbitrarily close to $\frac{1}{2}$, this contradicts the fact that f is continuous and has the value 0 at $\frac{1}{2}$. Thus **C** is not complete with respect to the metric d_1.

. It might be thought that the "natural" set on which to define this metric is not **C**, but the larger class **D**, of all Riemann-integrable functions on $[0,1]$. The first objection to this is that d_1 is not a metric on **D**, but only a pseudo-metric; that is, d_1 satisfies the axioms for a metric except for the fact that the distance between two different elements of **D** may be 0. Nevertheless, **D** may be treated essentially as a metric space and as such is not complete. More precisely, one obtains a metric space from **D** by identifying pairs of functions f and g for which $d_1(f,g) = 0$, and the identification space is not complete. This is not shown by Example 1.3.3, but the fact is that more complicated examples may be given (as in Example 1.3.1), along with a more complicated argument, to show that **D** is not complete.

A great many of the basic techniques of functional analysis ultimately require completeness for their application. Therefore the lack of completeness of the spaces **C** and **D** with regard to the "natural" metric for integration theory, d_1, greatly limits their usefulness in functional analysis. Now an apparent direct way out of this difficulty is to form, in an abstract manner, as in an arbitrary metric space, the completion of **C** or that of **D** and to work with these complete spaces. The difficulty in this approach is that there is no straightforward connection between the new elements of the completion and functions on $[0,1]$. This greatly hampers one's freedom in dealing with these elements. For example, a product analogous to the ordinary product $(fg)(x) = f(x)g(x)$ would not exist. The method used to define the product of irrational numbers when these are defined as the new members in the completion of the rational numbers is inapplicable, and in fact one sees later that the product should not exist as a member of the space; the product of two integrable functions in the ordinary sense will in general not be integrable.

It is therefore a great technical simplification to have a way of representing the new elements of the completion of $\mathbf{C}$ by ordinary functions on [0,1]. The function space one gets in this way is the space of Lebesgue-integrable functions on [0,1]. At the same time, a conceptual clarification is brought about in the theory which is comparable with that introduced in the theory of real numbers by identifying points in the completion of the rationals with infinite decimals.

A specific example from the theory of Fourier series may help to indicate how completeness is valuable in concrete cases. It is well known that if f is a continuous complex-valued function on the interval $0 \le x \le 2\pi$, and if c_n is its nth Fourier coefficient,

$$c_n = \frac{1}{2\pi} \int_0^{2\pi} e^{-inx} f(x) \, dx,$$

then

$$\frac{1}{2\pi} \int_0^{2\pi} |f(x)|^2 \, dx = \sum_{n=-\infty}^{\infty} |c_n|^2,$$

and also the partial sums $f_N(x) = \sum_{n=-N}^{N} c_n e^{inx}$ converge to f in the sense that $\int_0^{2\pi} |f(x) - f_N(x)|^2 \, dx \to 0$. The reader may be more familiar with the development of f into the functions $\cos nx$ and $\sin nx$; the present discussion may equally well be put in terms of these real functions. Now it is natural to ask whether the converse is valid; i.e., given complex numbers c_n, $n = 0$, $\pm 1, \pm 2, \ldots$, such that

$$\sum_{n=-\infty}^{\infty} |c_n|^2 < +\infty,$$

does there exist a continuous or other well-behaved function f such that the foregoing statement holds? The answer is that, in general, there is no continuous function with these properties. There is, however, an essentially unique measurable function f with Fourier coefficients c_n where the integrals defining c_n are taken in the Lebesgue sense. This function f has the property that the square of its absolute value is Lebesgue-integrable over $[0,2\pi]$, and when formulated with proper generality, the result just cited turns out to be equivalent to the statement that the set of all such functions is complete with respect to the metric (strictly speaking, pseudometric, but the distinction is here inessential) d_2, where

$$[d_2(f,g)]^2 = \int_0^{2\pi} |f(x) - g(x)|^2 \, dx.$$

This metric, as well as the analogous metrics d_p, for arbitrary p, such that $1 \le p \le \infty$, will be treated later; in particular, it will be shown that they are metrics on certain spaces. Here we shall only mention that quite generally the use of Lebesgue-measurable functions leads to a more rounded and elegant formulation of the theory of Fourier series, and is in fact essential for modern harmonic analysis on any group. The corollary notion of a Lebesgue-measurable set, i.e., one whose characteristic function is Lebesgue-measurable, has come to play a vital part in the local study of Fourier series on the real line.

The completeness features of the Lebesgue integral serve, like the convergence features, essentially to characterize the integral. That is, the Lebesgue integral is essentially the unique minimal extension of the Riemann integral leading to these completeness properties.

1.4   LIMITATIONS ON THE CONCEPT OF MEASURE SPACE

The Lebesgue theory gives a way to define the volume (or more generally, measure) of a class of subsets of euclidean space that is much more extensive than that measured in the Riemann theory. But the theory does not define a volume for all subsets. The question naturally arises whether it is possible to measure effectively *all* subsets. It would greatly simplify matters if this were the case. It turns out, however, that only partial extensions of Lebesgue measure are effectively possible, and there is no unique or natural extension. More specifically, there exists no countably additive measure on *all* subsets of a set whose cardinal number is in a certain theoretical class which vanishes on the subsets consisting of single elements. Since an individual point in euclidean space, or in most other analytically interesting spaces, has measure zero, this result indicates the inappropriateness of measuring all subsets in the significant cases, even though ambiguities in the foundations of mathematical logic presently leave it uncertain whether the power of the continuum is itself in the class in question.

One might think of turning back to finitely additive spaces, on the theory that the possibility of measuring all subsets might compensate for the lack of countable additivity. It turns out, however, that although it is possible to measure all subsets in euclidean space by a finitely additive measure that vanishes on subsets consisting of one element, the measure may then fail to have certain other highly desirable properties. For example, any reasonable formulation of the idea of euclidean measure will have the property that the measure of a set is unchanged by a euclidean motion, i.e., by a translation, or rotation. It is true that there exists a nonvanishing finitely additive measure m on the collection of all subsets of the plane that is invariant under euclidean motions,

$$m(E + a) = m(E), \quad m[r(E)] = m(E),$$

where a designates an arbitrary vector, and $E + a$ is the set of all vectors of the form $x + a$, with x in E, while r is an arbitrary rotation. But when one turns to euclidean space of dimension 3 or higher, no such measure exists. This is indicated dramatically by the so-called Banach-Tarski paradox, a slight variation of which may be paraphrased as follows: Any two solid spheres in three-dimensional space (think, for example, of the earth or a baseball) may be dissected into finite collections of subsets and then reassembled (by euclidean motions) in such a way as to yield each other. Thus a finitely additive measure defined on all subsets of three-dimensional space and invariant under euclidean motions would have to give the same measure to any two solid spheres.

It will be seen later that Lebesgue measure in euclidean n-space is both countably additive and invariant under all euclidean motions. In fact, these features, together with the elementary requirement that this measure be defined on all parallelepipeds, also make it essentially unique. It measures not all subsets, but in a sense all subsets whose measures are uniquely determined by the intuitive geometrical requirements. These are also all the subsets which have concrete analytical significance. The very existence of other subsets depends on presently ambiguous logical principles.

An important class of Lebesgue-measurable sets consists of the sets of measure zero, which are also known as Lebesgue null sets. Such a set is one whose characteristic function is Lebesgue-integrable with integral zero, or equivalently (as the theoretical development shows), as a set which can be covered by a countable collection of parallelepipeds of arbitrarily small total volume. The Lebesgue null sets commonly occur as the exceptional ones in many questions in analysis. As an example, those numbers whose decimal expansions do not contain all the digits $0, 1, 2, \ldots, 9$ with equal frequency "on the average" form a subset of the reals of Lebesgue measure zero. Similarly, the fractional parts of the powers of a real number $a > 1$, that is, $a - [a], a^2 - [a^2], \ldots, a^n - [a^n], \ldots$ (where $[b]$ denotes, for any real number b, the largest integer m such that $m \leq b$) are distributed "uniformly" in the interval $[0,1]$ for all except a Lebesgue null set; this is the only presently known way to show that there exists even *one* number a for which this is the case. An example within the sphere of classical analysis is provided by the set on which the series $\sum_n c_n e^{inx}$ fails to converge, where $\{c_n\}$ is, as earlier, a given sequence for which the series $\sum_n |c_n|^2$ is convergent; the sum of the series is in fact, except on a null set, the function whose Fourier coefficient sequence is the given one, $\{c_n\}$.

1.5   GENERALIZED SPECTRAL THEORY AND MEASURE SPACES

One of the central parts of linear algebra is concerned with the diagonalization of given linear operators, or suitable sets thereof, by an appropriate orthogonal (or in the case of a complex space, unitary) transformation. A basic result is that any self-adjoint operator on a finite-dimensional euclidean space (or commuting set of such) can be diagonalized (simultaneously) relative to a given basis by transformation by some orthogonal operator. Many problems in functional analysis can be regarded as subsumable under the corresponding problem for the case of an infinite-dimensional space. However, self-adjoint operators in such spaces are not in general diagonalizable in the obvious sense, but only in a generalized sense, which for its full explication requires the use of Lebesgue integration theory; the new phenomenon of the *continuous spectrum*, which Hilbert's theory of integral equations was invented to deal with, intervenes.

To give a very simple example of an operator with continuous spectrum, consider the space **C** of all real-valued continuous functions on the interval [0,1], relative to the inner product

$$(f,g) = \int_0^1 f(x)g(x)\,dx.$$

All the usual axioms for a euclidean space are satisfied except for the finite-dimensionality. Now let A denote the operator

$$f(x) \to xf(x)$$

on **C**. It is easily seen that this is a self-adjoint operator; i.e.,

$$(Af,g) = (f,Ag)$$

for arbitrary f and g in **C**. Additionally, it is continuous relative to the metric determined by the analog of euclidean length: If $(f_n,f_n) \to 0$ for some sequence $\{f_n\}$ in **C**, then $(Af_n,Af_n) \to 0$ (this is continuity at $f = 0$, but continuity at other points follows readily from the linearity of A). However, it is clear that there are no proper vectors for A; that is, for no function $f \neq 0$ and constant c is the equation $Af = cf$ valid. On the other hand, the operator A is an extremely simple one, and is in a certain sense already in diagonal form. If the interval [0,1] is approximated in an intuitive sense by the discrete points $0, 1/n, 2/n, \ldots, (n-1)/n$ for some large integer n, and the function f by its values $f_0 = f(0)$, $f_1 = f(1/n), \ldots, f_k = f(k/n), \ldots$, $f_n = f(1)$, then the transformation A is correspondingly approximated by the operation

$$f_k \to (k/n)f_k.$$

This, in turn, is given by a diagonal matrix with entries $0, 1/n, \ldots, k/n, \ldots, 1$ in the main diagonal for an appropriate choice of basis of the linear space consisting of all functions defined on the $n + 1$ points in question. For such reasons A is regarded as having a continuous range of proper values from 0 to 1, which range is defined as the continuous spectrum of the operator.

This operator A is far too trivial to be of any interest in itself. The point is the extraordinary one that every continuous normal operator in an infinite-dimensional euclidean space may be diagonalized in the same sense as the operator A. That is to say, within an isomorphism of the space (i.e., roughly, a change of basis) the operator is a simple multiplication operator

$$f(x) \rightarrow k(x)f(x),$$

where, however, the fixed function k, multiplication by which defines the operator, need not be continuous, but only smooth in the sense appropriate to integration theory, i.e., "measurable." An additional novelty in the infinite-dimensional case is the possibility of an "unbounded" normal operator A, that is, one which is not everywhere defined; included among these are typically differential operators, for example, d/dx. The same theorem is true for these operators, the only difference being that the function k is then not bounded. [In the case of d/dx, for example, the corresponding k is the function $k(x) = ix$ on the entire real line, with a correspondingly redefined infinite euclidean space; and it is the "Fourier transform" which diagonalizes the operator—in fact, this may be virtually regarded as the defining property of the Fourier transform.]

The Lebesgue integral plays a very essential role in the spectral theory of operators in infinite-dimensional spaces. As a matter of fact, the general theory of commuting normal operators in these spaces, which in the work of von Neumann and others some decades ago appeared as rather reminiscent of integration theory, is now recognized as in essence a specialization of abstract integration theory, following a subsumption which can be carried out rather briefly. The spectral theory of Hilbert and his successors thus attains its most cogent and transparent form through the use of abstract integration theory. This is a theory of basic importance in modern functional analysis, as well as in the quantum theory, which was a major stimulant to the growth of this subject.

EXERCISES

1 In each of the following cases, show that $f_n(x) \rightarrow f(x)$ for every x, and compare $\lim_n \int f_n(x)\,dx$ with $\int f(x)\,dx$.

    a. $f_n(x) = 1$ for $n < x < n + 1$, and $f_n(x) = 0$ otherwise; $f(x) = 0$ for all real x.

    b. $f_n(x) = (\pi n)^{-\frac{1}{2}}e^{-x^2/n}$; $f(x) = 0$ for all real x. $\left(\textit{Hint:}\text{ Use the formula}\right.$
$$\int_{-\infty}^{\infty} e^{-x^2}\,dx = \sqrt{\pi}.\bigg)$$

2 Let $f_n(x)$ denote the function on $[0,1]$ which vanishes at 0, $2/n$, and 1, has the value 1 at $1/n$, and is linear on each of the segments $[0,1/n]$, $[1/n,2/n]$, and $[2/n,1]$. Show that $f_n(x) \rightarrow 0$ for each x, but that the convergence is not uniform on $[0,1]$. Compare the limit of the integral with the integral of the limit.

3 Show that if $x_1, x_2, \ldots$ is any sequence of real numbers and $\epsilon > 0$, then there exists a sequence of open intervals (a_n, b_n) such that $a_n < x_n < b_n$, and

$$\sum_n (b_n - a_n) < \epsilon.$$

(This means that any countable set of real numbers, e.g., the set of rationals, is of Lebesgue measure 0, according to one possible definition.)

4* Show that if f_n is any monotone-decreasing sequence of nonnegative continuous functions on $[0,1]$, converging pointwise to 0 at every point, then $\int f_n(x)\,dx \to 0$. (This fact provides the key classical basis for Daniell's approach to Lebesgue measure.)

5* Show that if E is any linear functional on the space $C[0,1]$ of all real-valued continuous functions on $[0,1]$, with the property that $E(f) \geq 0$ when $f \geq 0$, then, with the same hypothesis as in Exercise 4, $E(f_n) \to 0$.

6 A *Boolean ring* is defined as a commutative ring having the properties that $a + a = 0$ and $a^2 = a$ for all elements a.

  a. Show that the power set of any set is a Boolean ring, relative to the symmetric difference as addition, and intersection as multiplication.

  b. Show that any factor ring of a Boolean ring (i.e., the quotient modulo an ideal) is again one.

  c. Show that, by the definition $a \subset b$ if $ab = a$, any Boolean ring is *partially ordered*, i.e., the following properties hold: $a \subset b$ and $b \subset c$ imply $a \subset c$; $a \subset a$; and $a \subset b$ and $b \subset a$ imply $a = b$.

  d. Define a minimal element in a Boolean ring as an element $a \neq 0$ such that $b \subset a$ only if $b = 0$ or $b = a$; show that the only minimal elements in the power set of a set are the sets consisting of exactly one element.

7* *a.* Show that the quotient ring of the power set of any infinite set modulo the ideal of all finite sets contains no minimal elements. (*Hint:* Use the fact that any infinite set contains infinite complementary subsets.)

  b. Show that there exists a finitely additive measure $\neq 0$ defined on the power set of any uncountable set which vanishes on all countable subsets. (*Hint:* Consider the quotient modulo the ideal of all countable subsets.)

8* Show that any Boolean ring is ring-isomorphic to the subring of the power set of some set (a theorem of M. H. Stone). (*Hint:* Consider the set of all maximal ideals $\mathcal{M}$ of the rings, and to any element x associated the subset of $\mathcal{M}$ consisting of the maximal ideals not containing x.)

9 *a.* Show that if $\mathcal{R}_1$ and $\mathcal{R}_2$ are subrings of the power set of a set, then there exists another subring $\mathcal{R}_3$ which contains $\mathcal{R}_1$ and $\mathcal{R}_2$ and which is contained in every subring having this property. (This minimal ring containing $\mathcal{R}_1$ and $\mathcal{R}_2$ is called the subring *generated* by $\mathcal{R}_1$ and $\mathcal{R}_2$, and sometimes denoted $\mathcal{R}_1 \vee \mathcal{R}_2$.)

  b. Show that if one of $\mathcal{R}_1$ and $\mathcal{R}_2$ is an ideal in the power set, then $\mathcal{R}_3 = \mathcal{R}_1 + \mathcal{R}_2$. (Note that if A and B are any subsets of an Abelian group G, then $A + B$ denotes the set of all elements of G of the form $a + b$, with a in A and b in B.)

c. Show that, for any collection of subrings of the power set of a set, there exists a unique ring containing all elements of the collection and minimal with respect to this property. (*Hint:* Consider the set of all subrings which contain every subring in the given collection and form the intersection. As in (*a*), the minimal ring is said to be that generated by the given ones.)

10   Let m be a finitely additive (nonnegative) measure on a ring of sets $\mathcal{R}$ of a set S.

a. Show that the "null sets" of m, that is, the sets on which m vanishes, form an ideal in $\mathcal{R}$, say, $\mathcal{N}$.

b. Show that the subring of the power set of S which is generated by all subsets of elements of $\mathcal{N}$ is an ideal $\mathcal{N}'$ in the power set.

c. Show that m may be uniquely extended to a finitely additive measure on the ring generated by $\mathcal{R}$ and $\mathcal{N}'$ and that the resulting measure vanishes on $\mathcal{N}'$.

11   A function f on an Abelian group G is called *periodic*, and a nonzero element a of G is called a *period* of f, in case $f(x + a) = f(x)$ for all elements x of G.   A subset of G is similarly designated as periodic in case its characteristic function is such.   Now let G denote the additive group of the rational integers.

a. Show that the periodic subsets of G form a ring $\mathcal{R}$.

b. For any periodic set A, let $m(A)$ denote the limit, as $n \to \infty$, of the number of elements in $A \cap [-n,n]$, divided by $2n$; show that $m(A)$ exists and that m is a finitely additive measure on $\mathcal{R}$.

*c.** Show that m is countably additive.

d. Let $\mathcal{F}$ denote the Boolean ideal of all finite subsets of G, and let $\mathcal{R}'$ denote the ring generated by $\mathcal{R}$ and $\mathcal{F}$; show that there exists a unique measure on $\mathcal{R}'$ which extends m and vanishes on $\mathcal{N}$. (*Hint:* Apply Exercise 10.) Is this measure countably additive?

II

BASIC
INTEGRALS

2.1 BASIC MEASURE SPACES

In developing the theory of integration, our strategy will be to begin with the conceptually simple and concretely often encountered notion of a *basic measure space* and with the integration of the simplest functions on such a space. A full-fledged measure space and the integration of a quite general type of function on such a space will then be derived by a natural species of approximation, in Chap. 3.

Apart from the general theory, which is relatively simple in this chapter, as it develops, the main contents of the chapter are (*1*) the construction of Lebesgue-Stieltjes measures, and (*2*) the construction of product measures. These constructions are important not only in themselves, but as indications to the reader that the general theory under development has a nontrivial sphere of application.

A *basic measure space* is roughly analogous to a neighborhood system for a topological space; a full, rather than basic, measure space may be defined by a variety of different basic spaces, just as a topological space may be defined by a variety of neighborhood systems. In either case it is quite convenient for applications to be able to construct the final space from any of a number of preliminary structures. The explicit determination from a basic space of a full space will be given in Chap. 3.

DEFINITIONS   A *basic measure space* is a system composed of a set S, a ring $\mathcal{R}$ of subsets of S, and a function m on $\mathcal{R}$ to the nonnegative real numbers, with the following property:

If A_1, A_2, ... is a sequence of disjoint sets in $\mathcal{R}$ whose union A is again in $\mathcal{R}$, then $m(A) = \sum_i m(A_i)$. Such a space M will be denoted by the triple $(S,\mathcal{R},m)$. The ring $\mathcal{R}$ will be referred to as the *ring of basic measurable sets on* M, and m will be called the *measure*.

In other terms, a basic measure space is simply a ring of sets together with a countably additive real nonnegative function on the ring. The simplest such spaces are the so-called discrete ones. Intuitively, these are constructed on a given set S by the assignment of various weights to the points of S. Before making a formal definition, it is useful to make the generally useful

DEFINITION   An indexed set of elements of a set K, with *index set* Λ, is a mapping $\lambda \to x_\lambda$ from Λ into K. If $\{x_\lambda ; \lambda \in \Lambda\}$ is an indexed set of nonnegative real numbers, the sum $\sum_\lambda x_\lambda$ (or more specifically, $\sum_{\lambda \in \Lambda} x_\lambda$) is defined as $+\infty$, in case $x_\lambda > 0$ for uncountably many values of λ, and as the usual sum of the nonzero x_λ, in case their number is countable (i.e., as $+\infty$ in case the resulting series is divergent, and as the sum in case it is convergent).

Thus, for a sequence a_1, a_2, ..., $\sum_i a_i$ is the usual limit of the partial sums, or $+\infty$ if the partial sums become unbounded. For an indexed family $\{a_\lambda ; \lambda \in \Lambda\}$, with index set Λ, $\sum_\lambda a_\lambda$ is $+\infty$, unless only countably many of the a_λ are distinct from zero; if all a_λ vanish except for the indices λ_1, λ_2, ..., then $\sum_\lambda a_\lambda$ is as a matter of definition $\sum_i a_{\lambda_i}$.

DEFINITION   A basic measure space $(S,\mathcal{R},m)$ is discrete in case $\mathcal{R}$ includes all one-point subsets of S and m is *completely additive* on $\mathcal{R}$, in the following sense: If $\{A_\lambda : \lambda \in \Lambda\}$ is any indexed family (with index set Λ) of mutually disjoint elements of $\mathcal{R}$, whose union A is again in $\mathcal{R}$, then

$$m(A) = \sum_\lambda m(A_\lambda).$$

This definition suggests a simple mode of construction for a discrete space. Let S be any set, let $\mathcal{R}$ denote the ring of all finite subsets of S, and let w

denote a real nonnegative function on S. For any set A, let

$$m(A) = \sum_{x \in A} w(x).$$

It is easily seen that the sum of the "weights" $w(x)$ in the union of two disjoint finite sets is the sum of the total "weights" in the respective sets; thus m is finitely additive. In the present situation, countable, and indeed complete, additivity are trivial; $(S, \mathscr{R}, m)$ is a basic measure space, and a discrete one according to the definition just made. This remains the case if the ring $\mathscr{R}$ is enlarged to a ring $\mathscr{R}^{\dagger}$ including all subsets A for which the sum $\sum_{x \in A} w(x)$ is finite, with $m(A)$ defined as this sum.

   The proof is a simple deduction from the theory of infinite series of nonnegative terms, and is left as an exercise (a much more general result is deducible from Scholium 2.4).

   Example 2.1.1   Let S denote the real line, and x_1, x_2, ... an enumeration of the rational points in S.   Define $w(x)$ as 2^{-n} if $x = x_n$ for some n, and as 0 otherwise.   The ring $\mathscr{R}^{\dagger}$ is the power set of the reals, the measure of any subset A being the sum of 2^{-n} taken over those indices n such that $x_n \in A$.

   Example 2.1.2   Now let S denote the interval $[0,1]$; let $w(x) = n^{-1}$ if $x = n^{-1}$, and let $w(x)$ otherwise be defined as 0.   Then $\mathscr{R}^{\dagger}$ includes some infinite sets, but not S itself, nor even all countable sets: because of the divergence of the series $\sum_{n} n^{-1}$, the set of all numbers of the form n^{-1} $(n = 1, 2, \ldots)$ is not in $\mathscr{R}^{\dagger}$.

2.2   THE BASIC LEBESGUE–STIELTJES SPACES

The theory of integration over discrete measure spaces will turn out, naturally enough, to be essentially coextensive with the theory of numerical series, with which theory the reader should already be fairly well acquainted.   The first really novel and interesting example of a basic measure space, as well as one which is of great historical and general importance, is that of Lebesgue measure on the real line.   This is essentially simply the usual length function, which assigns to the real interval having the finite endpoints a and b, with $a < b$, the measure $b - a$, with suitable emphasis, use, and development of its countable additivity, as already indicated.   The measure was first defined by Borel; its analytical importance first became apparent in the work of Lebesgue.   Only the *basic* space will be constructed here, in which connection only the countable additivity of the length function on intervals offers any difficulty; the extension of the length function to arbitrary "Lebesgue-measurable" sets is an immediate consequence of this and the general theory to be developed in the next chapter.

   Just before the turn of the century, slightly before the time of Borel and

Lebesgue, the Dutch mathematician Stieltjes had shown the usefulness and scope of an extension of the Riemann integral in which the length function $b - a$ is replaced by the more general measure $\mu(b) - \mu(a)$, where μ is a given function.   When $\mu(x)$ is identically equal to x, the ordinary length is obtained, but this is evidently a quite particular case.   One way to see the naturalness of the Stieltjes generalization is from a probabilistic standpoint; if $\mu(b)$ is the probability that a "random variable" X is less than b, then $\mu(b) - \mu(a)$ is the probability that $a \leq X < b$; it is natural to extend this to a measure defined on more general sets than intervals, the measure of a set representing the probability that X is a member of it.

It is just about as easy to construct the Stieltjes generalization of Lebesgue measure as it is to construct Lebesgue measure; this more general construction of a Lebesgue-Stieltjes basic space is essentially Theorem 2.1, below. The idea of the proof is identical with that for the special case in which $\mu(x) = x$, and the reader may find it helpful to consider only this special case in his initial reading of Theorem 2.1 and its proof.

Throughout this section we shall be concerned with measures on the ring $\mathfrak{R}$ generated by the bounded intervals contained in an interval S on the real line; *interval* here means any convex subset of the reals.   Such a set is characterized by the property that whenever it contains a and b, with $a < b$, then it also contains all points x such that $a < x < b$.   Thus a given interval may be open, closed, a single point, half open (i.e., open on one end and closed on the other), or even empty.   The *length* of the bounded interval with left endpoint a and right endpoint $b(a \leq b)$ is of course $b - a$.   We recall the notation $[a,b]$ for the closed interval from a to b; (a,b) for the open interval; and $(a,b]$ and $[a,b)$ for the half-open intervals, excluding the left and right endpoints, respectively.

THEOREM 2.1   *If μ is a continuous monotone-increasing function on a real interval S, there exists a unique countably additive measure m on the ring $\mathfrak{R}$ generated by the compact intervals in S such that*

$$m(E) = \mu(b) - \mu(a)$$

for every interval E in $\mathfrak{R}$ with left endpoint a and right endpoint b.

Lemma 2.2.1   *The ring $\mathfrak{R}$ is precisely the collection of all finite disjoint unions of bounded intervals whose closures are contained in S.*

Proof   For the moment let the indicated collection be denoted by $\mathcal{C}$. Then, as $\mathcal{C}$ is contained in $\mathfrak{R}$ and contains the generators of $\mathfrak{R}$, it is enough to show that $\mathcal{C}$ is a ring.   The union of any sets A and B may always be expressed as a disjoint union by means of the relation

$$A \cup B = A \cup (B - A).$$

Since $\mathcal{C}$ is obviously closed under disjoint unions, it is therefore enough to prove that $\mathcal{C}$ is closed under differences.   If B is a disjoint union of intervals $F_1, F_2, \ldots, F_n$, the difference $A - B$ may be formed by successive subtraction of these intervals from A.   It therefore suffices to show that the difference of a set in $\mathcal{C}$ with an interval in $\mathcal{C}$ is again in $\mathcal{C}$.   But it is easily seen that if an interval is removed from a disjoint union of intervals, the result is again a disjoint union of intervals.

Lemma 2.2.2   *For any complex-valued function μ defined on S there is a unique finitely additive complex-valued function m defined on $\mathcal{R}$ such that*

$$m(E) = \mu(b) - \mu(a)$$

for all intervals E in $\mathcal{R}$ with left endpoint a and right endpoint b.

Proof   For intervals in $\mathcal{R}$ define m as above.   By Lemma 2.2.1, a general element A of $\mathcal{R}$ is a finite disjoint union of intervals $E_1, E_2, \ldots, E_n$ in $\mathcal{R}$. But unless A reduces to a single point, it has infinitely many such representations.   Thus it is necessary, and also clearly sufficient, to show that the definition

$$m(A) = \sum_{i=1}^{n} m(E_i)$$

is independent of the particular way in which A is represented.   Suppose, then, that A is also represented as the disjoint union of the intervals $F_1, F_2, \ldots, F_r$.   Then E_i is the disjoint union of the intervals $E_i \cap F_1$, $E_i \cap F_2, \ldots, E_i \cap F_r$.   Moreover,

$$m(E_i) = \sum_{j} m(E_i \cap F_j).$$

For after proper arrangement, one gets a telescoping sum on the right which reduces to value on the left.   By symmetry it is immediate that

$$m(F_j) = \sum_{i} m(E_i \cap F_j).$$

It follows that

$$\sum_{i} m(E_i) = \sum_{j} m(F_j).$$

Proof of Theorem 2.1   Let μ be a continuous monotone-increasing function defined on S.   Then the set function m defined in Lemma 2.2.2 is a nonnegative finitely additive measure.   All that remains to be shown is the countable additivity of m, which is in fact the main content of the theorem.

Suppose $A_1, A_2 \ldots$ is a decreasing sequence of sets in $\mathcal{R}$ whose intersection is empty, and let $\epsilon > 0$.   If A_n is an interval, it is obvious from the continuity of μ that A_n contains a closed interval B_n such that

$$m(A_n - B_n) < \frac{\epsilon}{2^n}.$$

But even if A_n is a finite union of disjoint intervals, each of these intervals will contain a closed interval whose measure is arbitrarily close to the measure of the containing interval. Thus, in any event, A_n contains a closed set B_n in $\mathcal{R}$ such that

$$m(A_n - B_n) < \frac{\epsilon}{2^n}.$$

The intersection of the B_n is contained in that of the A_n and is therefore empty. As the B_n are not only closed but bounded as well (i.e., compact), there exists a positive integer r for which

$$\bigcap_{n=1}^{r} B_n = \emptyset.$$

If $s \geq r$, it follows easily that

$$A_s \subset \bigcup_{n=1}^{r} (A_n - B_n).$$

The proof is concluded by showing that

$$m(A_s) \leq \sum_{n=1}^{r} m(A_n - B_n) < \epsilon.$$

First observe that $m(E) \leq m(F)$ when E and F are in $\mathcal{R}$ and $E \subset F$. Hence $m(A_s)$ is bounded by the measure of

$$\bigcup_{n=1}^{r} (A_n - B_n).$$

Next we make use of the *finite subadditivity* of m, that is, the fact that

$$m\left(\bigcup_{n=1}^{r} E_n \right) \leq \sum_{n=1}^{r} m(E_n)$$

for any $E_1, E_2, \ldots, E_r$ in $\mathcal{R}$. To prove this, let $D_n = E_n - \left(\bigcup_{n=1}^{r} E_i \right)$, and note that the D_n are disjoint elements of $\mathcal{R}$ whose union is the same as that of the E_n. Hence

$$m\left(\bigcup_{n} E_n \right) = \sum_{n} m(D_n)$$

by finite additivity. Since $D_n \subset E_n$, it follows that $m(D_n) \leq m(E_n)$, and hence in the equation above the sum on the right is bounded by $\sum_{n} m(E_n)$.

Basic Borel-Lebesgue measure on the reals may now be defined as that given by Theorem 2.1 for the case in which S is the entire real line and $\mu(x) = x$. It may be characterized as the only measure of the type given by Theorem 2.1 which is translation-invariant, that is, $m(E + x) = m(E)$ for any set E in $\mathcal{R}$ and any real number x; here the notation $A + x$ is used to indicate the set of all numbers of the form $a + x$, with a in A.

THE DISCRETE PART OF A STIELTJES MEASURE   Every space $(S,\mathfrak{R},m)$ given by Theorem 2.1 is a *basic Lebesgue-Stieltjes measure space*, where this is defined as one of the form $(S,\mathfrak{R},n)$, such that S and $\mathfrak{R}$ are as in the theorem, and n is a countably additive measure on $\mathfrak{R}$.   But not every basic Lebesgue-Stieltjes space is of this form; e.g., if n assigns the measure 1 to a set in $\mathfrak{R}$ if and only if $\mathfrak{R}$ contains a particular point p_0 of S, it cannot be of the type given by Theorem 2.1, for it is easily seen that any such measure vanishes on the set consisting of a single point.   More generally, for the same reason, any discrete measure on $\mathfrak{R}$ cannot be in the class given by Theorem 2.1.   The property of vanishing on points indeed characterizes the members of this class, and the general Lebesgue-Stieltjes measure is simply the sum, in a unique fashion, of a "purely continuous" measure, defined as one vanishing on one-point sets, and a discrete measure.   The situation may be summarized as follows:

COROLLARY 2.2.1   *If m is a basic Lebesgue-Stieltjes measure on the real interval S, then there exists a monotone-increasing right-continuous function $\mu(x)$ defined on S, unique within an additive constant, such that if $(a,b]$ is any bounded left-open interval in S, then*

$$m((a,b]) = \mu(b) - \mu(a);$$

and every monotone-increasing right-continuous $\mu(x)$ on S is associated with a unique basic Lebesgue-Stieltjes measure in this fashion.

The main point here is that a monotone function can have only a countable number of discontinuities and that these are of a very simple character. More formally, let a (right-continuous) *jump function* on S be defined as a function f of the form $f(x) = d((p,x])$, when $x > p$ and $f(x) = -d((x,p])$ for $x < p$, for some discrete measure d on the ring generated by the compact subsets of S and some point p in S, or the sum of such a function and a constant.   Note that any jump function is monotone-increasing and is actually, by the countable additivity of the discrete measure, right-continuous.

Lemma 2.2.3   *Any monotone-increasing right-continuous function $\lambda(x)$ on an interval S is the sum of a continuous such function and a jump function, each of which is unique, within an additive constant.*

Proof   For any monotone function f, the limits $\lim\limits_{\epsilon \to 0} f(x + \epsilon)$ exist for any point x in the interval of definition, with the convention that $x \pm \epsilon$ is replaced by x in case it is outside the interval of definition (or alternatively, f is suitably extended outside its interval of definition, namely, by making the extension constant on each interval complementary to S and continuous at any boundary point of S).   These limits will be denoted $f(x \pm 0)$, as is customary, when there is little likelihood of confusion with the literal interpretation of $f(x \pm 0)$. That $\lambda(x)$ is right-continuous means that $\lambda(x + 0) = \lambda(x)$ for all x in S. But it is also true that $\lambda(x - 0) = \lambda(x)$ except for at most countably many values of x. For let $C = [a,b]$ denote an arbitrary compact interval contained in S, and let $a = x_0 < x_1 < x_2 < \cdots < x_{n+1} = b$ be any finite set of elements of C beginning with a and terminating with b.

Then $\lambda(b) - \lambda(a)$ may be expressed as the telescoping sum $\sum_{j=0}^{n} [\lambda(x_{j+1}) - \lambda(x_j)]$, each term of which is nonnegative. Setting $\delta(x) = \lambda(x) - \lambda(x - 0)$, it follows that $\sum_{j=1}^{n} \delta(x_j) \leq \lambda(b) - \lambda(a)$. It results that at most a finite number of $\delta(x)$ for x in C can exceed any given positive value; for example, the value m^{-1}, m being an arbitrary positive integer. Now $\delta(x)$ is positive if and only if $\delta(x) > m^{-1}$ for some positive integer m; so there can be at most countably many values of x in C for which $\delta(x) > 0$; since S is a countable union of compact subintervals, there are at most countably many such values x in S. But $\delta(x) > 0$ if and only if $\lambda(x)$ is discontinuous at the point x.

Now let d denote the discrete measure on the ring $\Re$ generated by the compact subintervals of S such that the one-point set consisting of the point x has the measure $\delta(x)$. Let p denote an arbitrary point in S such that $\delta(p) = 0$, and define $\lambda_d(x)$ for $x \in S$ as follows:

$$\lambda_d(x) = d((p,x]) \quad \text{for} \quad x \geq p; \quad \lambda_d(x) = -d((x,p]) \quad \text{for} \quad x < p.$$

From the countable additivity of d (compare the remark immediately preceding Example 2.1.1), it follows easily that λ_d is right-continuous; it is obvious that it is monotone-increasing; and it is constructed so that its "jumps" $\lambda_d(x) - \lambda_d(x - 0)$ are identical with those of $\lambda(x)$ itself. It follows that $\lambda(x) - \lambda_d(x)$ is continuous; and it is easily seen that $\lambda_d(y) - \lambda_d(x) \leq \lambda(y) - \lambda(x)$ whenever $x < y$, from which it follows that $\lambda(x) - \lambda_d(x)$ is monotone-increasing. The decomposition

$$\lambda(x) = [\lambda(x) - \lambda_d(x)] + \lambda_d(x)$$

then has the indicated properties. In case

$$\lambda(x) = \rho(x) + \sigma(x)$$

is another such decomposition, it is easily seen that λ_d and σ have the same jumps, from which it follows that they differ only by an additive constant, which implies that the same is true of $\lambda(x) - \lambda_d(x)$ and $\rho(x)$.

Proof of Corollary 2.2.1   Let p be any point in S such that $m(\{p\}) = 0$, assuming that S does not consist of a single point, in which case the result is trivial, and set $\mu(x) = m((p,x])$ for $x \geq p$ and $\mu(x) = -m((x,p])$ for $x < p$. From the countable additivity of m it follows that μ is right-continuous; from the nonnegativity of m, it follows that μ is monotone-increasing.   Considering separately the cases in which both a and b exceed p, both are exceeded by p, or $p \in (a,b]$, there is no difficulty in verifying that $m((a,b]) = \mu(b) - \mu(a)$.

If, conversely, $\mu(x)$ is a given monotone-increasing right-continuous function on S, let $\rho(x)$ and $\sigma(x)$ be its continuous and jump-function constituents in the sense of Lemma 2.2.1.   Then $\rho(x)$ determines a measure m_ρ by Theorem 2.1 such that $m((a,b]) = \rho(b) - \rho(a)$ whenever $(a,b] \subset S$; $\sigma(x)$ similarly determines a measure, m_σ, by its definition; $m_\rho + m_\sigma$ is then the measure whose existence is asserted by the second part of the corollary.

EXERCISES

1   Let $f(x)$ be a nonnegative continuous function on the reals, and set $F(x) = \int_0^x f(t)\,dt$.   Show that if m is the measure associated by Theorem 2.1, with the monotone function F, then $m((a,b)) = \int_a^b f(x)\,dx$, for an arbitrary interval (a,b) (with $a < b$).

2   Show that any monotone real-valued function f on an interval has at most a countable number of discontinuities.   (*Hint:* Extend the argument given in the text for the case in which f is right-continuous.)

3   Draw the same conclusion as in Exercise 2 by proceeding as follows: First show that the intervals $(f(x - 0), f(x + 0))$ are mutually disjoint.   Next use the countability of a basis for the open subsets of the reals.

4   Show that a monotone real-valued function f may be redefined at its discontinuities so that it remains monotone and $f(x - 0) = f(x)$ for all values of x.   Show that it may similarly be redefined so that $f(x + 0) = f(x)$ for all x; and that, with a third redefinition, $f(x) = \frac{1}{2}[f(x + 0) + f(x - 0)]$ for all x.

5   Suppose f is a monotone-increasing function defined on the closed bounded interval $[a,b]$.   Show that if the range of f is dense in $[f(a), f(b)]$, then f is continuous.

6   Define f on the open unit interval by the equation

$$f(x) = \sum_{1-n^{-1} \leq x} n^{-1}.$$

Show that f is monotone-increasing, unbounded, right-continuous, and continuous except at the points $1 - n^{-1}$, $n = 2, 3, \ldots$.

7   Show that if m is a discrete and bounded measure on the ring generated by the compact subsets of the reals, then the range of values of $m((a,x])$, for any fixed value of a, is a nowhere-dense set, i.e., that its closure has empty interior in the topology of the real line.

8   Let f be a monotone-increasing function on $[a,b]$, and set $g(x) = \sum_{a \leq t \leq x} [f(t + 0) - f(t)]$.   Show that g is monotone-increasing and left-continuous.   On replacing $f(t + 0) - f(t)$ by $f(t) - f(t - 0)$, show that the resulting function is a right-continuous jump function.

9   A measure m on an interval $[a,b]$ such that $m([a,b]) = 1$ is called a *probability measure*. The function $F(x) = m([a,x])$ is called its *cumulative distribution function*. Show that F is the cumulative distribution function of a countably additive probability measure on an interval $[a,b]$ if and only if

     a. F is monotone-increasing.

     b. F is right-continuous.

     c. $F(a) = 0$ and $F(b) = 1$.

10   Show that the ring $\mathcal{R}_0$ generated by the left-open (right-closed) subintervals $(a,b]$ of an interval S consists of all finite disjoint unions of such intervals.

11   *a.* The "Riemann-Stieltjes" integral of a given function g with respect to a given function F as "Stieltjes distribution function," on an interval $[a,b]$, is defined as follows: For any partition $P: a = x_0 < x_1 < \cdots < x_n = b$, and sequence $y_1, y_2, \ldots, y_n$ such that $x_{i-1} \le y_i \le x_i$ $(1 \le i \le n)$ form $s(P,y) = \sum_i f(y_i)[F(x_i) - F(x_{i-1})]$. If $s(P,y)$ tends to a limit s as $\max_i |x_i - x_{i-1}| \to 0$, that is, if for every $\epsilon > 0$ there is a δ such that $|s(P,t) - s| < \epsilon$ whenever $\max_i |x_i - x_{i-1}| < \delta$, then the Riemann-Stieltjes integral $\displaystyle\int_a^b g(x)\,dF(x)$ is said to exist and have the value s.

     Show that if g is continuous and if F is monotone, then $\int g(x)\,dF(x)$ exists.

     b. A function F is said to be of "bounded variation" on the interval $[a,b]$ in case $\sum_i |F(x_{i-1}) - F(x_i)|$ is bounded as the partition P varies; the least upper bound of the indicated expression is called the "total variation" of F over the interval. Show that if g is continuous and F is of bounded variation, then $\int g(x)\,dF(x)$ exists and is bounded by the product of the supremum of $|g|$ and the total variation of F.

12   Show that the Riemann-Stieltjes integral has the usual formal properties of the integral:

     a. If g_1 and g_2 are both integrable with respect to the distribution function Γ, then so also is $g_1 + g_2$, and $\int (g_1 + g_2)\,dF = \int g_1\,dF + \int g_2\,dF$.

     b. If $\int g\,dF$ exists on the intervals $[a,b]$ and $[b,c]$, where $a < b < c$, then
$$\int_a^c g\,dF \text{ exists and equals } \int_a^b g\,dF + \int_b^c g\,dF.$$

     c. If $F(x) = \displaystyle\int_a^x f(t)\,dt$ for some continuous function f, then
$$\int_a^b g(x)\,dF(x) = \int_a^b g(x)f(x)\,dx.$$

13*   Show that if m is a countably additive real-valued function on the ring generated by the subintervals of an interval $[a,b]$, then the function $F(x) = m([a,x])$ is of bounded variation on this interval.

2.3 INTEGRALS OF STEP FUNCTIONS

The theory of integration naturally begins with the integration of the simplest functions. On a basic measure space there is a natural such class, the class of step functions. The theory of integration of these functions is simple, but fundamental for later purposes, and will also be immediately useful for the construction of direct-product measures in the next section.

> DEFINITIONS   If $\mathfrak{R}$ is a ring of subsets of a set S, a *step function* relative to $\mathfrak{R}$ is a complex-valued function defined on (all of) S which assumes only a finite number of distinct values, each of which, with the possible exception of 0, is assumed on a set in $\mathfrak{R}$. It is convenient also to formulate the definition in terms of characteristic functions. The *characteristic function* φ_A of an arbitrary subset A of S has the value 1 at points in A and the value 0 at points outside of A It follows that a function f is a step function relative to $\mathfrak{R}$ if and only if there exist a finite number of disjoint sets $E_1, E_2, \ldots, E_n$ in $\mathfrak{R}$ and distinct complex numbers $a_1, a_2, \ldots, a_n$ such that
>
> $$f = \sum_i a_i \varphi_{E_i}$$
>
> A *step function on the basic measure space* $(S,\mathfrak{R},m)$ is simply a step function relative to $\mathfrak{R}$. Note that the support of a step function, where the *support* of a function is defined as the set on which it assumes its nonzero values, is always in $\mathfrak{R}$.

Any collection of complex functions on a set generates a corresponding algebra of functions, characterized as the intersection of all algebras containing the given collection. In some cases, such as the following, the algebra may be described quite explicitly.

> SCHOLIUM 2.1   *The set of all step functions on a basic measure space is precisely the algebra generated by the characteristic functions of the basic measurable sets. On this algebra there is a unique linear functional, the integral, whose value on the characteristic function of a basic measurable set is its measure.*

Proof   Let f be a step function on the basic measure space $(S,\mathfrak{R},m)$ whose distinct nonzero values $a_1, a_2, \ldots, a_n$ are assumed on the corresponding sets $E_1, E_2, \ldots, E_n$ of $\mathfrak{R}$. If g is another such function with similar values $b_1, b_2, \ldots, b_r$ on the sets $F_1, F_2, \ldots, F_r$ in $\mathfrak{R}$, then all the nonzero values of $f + g$ are included among the values $a_i + b_j$. Now $E_i \cap F_j$ is a set in $\mathfrak{R}$ on which $f + g$ has the value $a_i + b_j$. If $a_i + b_j \neq 0$, the largest set on which $f + g$ assumes this value is a finite union of certain of the $E_p \cap F_q$ and is thus again a set in $\mathfrak{R}$. Hence the collection of all step functions is closed under addition, and by a similar proof the product of two step functions is also a step function.   Since the collection is obviously closed under multiplication by scalars, it forms an algebra.   In particular, it should be noticed that any finite linear combination of characteristic functions of sets in $\mathcal{R}$ is again a step function.

If f is a step function whose distinct nonzero values $a_1, a_2, \ldots, a_n$ are assumed on the sets $E_1, E_2, \ldots, E_n$, let $I(f)$, the integral of f, be defined by the equation

$$I(f) = \sum_i a_i m(E_i).$$

It is then clear that the functional I has the property that $I(\varphi_E) = m(E)$ for all E in $\mathcal{R}$.   Conversely, it is clear that if there exists any linear functional on the step functions whose value on the characteristic function of a basic measurable set is its measure, then it must be the given functional I.   It remains only to show that I is indeed linear, or rather, since its homogeneity under multiplication by scalars is obvious, actually additive.

To this end, let f and g be as above in the first part of the proof.   Let E_0 and F_0 denote the respective subsets of S on which f and g vanish.   Then it is clear that

$$E_i = \bigcup_{j \geq 0} E_i \cap F_j,$$

and in addition $E_i \cap F_0$ belongs to $\mathcal{R}$, for it is the difference of the sets E_i and $\bigcup_{j \geq 1} E_i \cap F_j$, which are in $\mathcal{R}$.   Thus

$$m(E_i) = \sum_{j \geq 0} m(E_i \cap F_j).$$

If $m(F_j)$ is expressed in a similar fashion and $a_0 = b_0 = 0$, then

$$I(f) + I(g) = \sum_{i,j} (a_i + b_j) m(E_i \cap F_j).$$

If the distinct nonzero values among the $a_i + b_j$ are $c_1, c_2, \ldots, c_r$, then the sum on the right may be written

$$\sum_k c_k \left[\sum_{a_i + b_j = c_k} m(E_i \cap F_j) \right],$$

the sum in the brackets being extended over all indices i and j satisfying the indicated restriction.   On the other hand, this sum, in view of the additivity of m, is the measure of the set G_k on which $f + g$ assumes the value c_k.   Thus

$$I(f) + I(g) = \sum_k c_k m(G_k) = I(f + g).$$

The countable additivity of m plays no role in Scholium 2.1, which is equally applicable to finitely additive measures.   Its real importance is its necessity for the development of later chapters, but is exemplified also by the following result, to the effect that, for a countably additive measure, the corresponding integral has a type of continuity property; this property may be described in an elementary way as implying the freedom to "take the limit under the integral sign" in a certain key case where the elementary criterion involving uniform convergence of the integrands is inapplicable.

SCHOLIUM 2.2   *Let $f_1, f_2, \ldots$ be a monotone-increasing sequence of step functions on $(S, \mathcal{R}, m)$ which converges pointwise to a step function f. Then the corresponding sequence of integrals $I(f_1), I(f_2), \ldots$ converges to $I(f)$.*

Proof   It follows from Scholium 2.1 that $g_n = f - f_n$ is also a step function with integral $I(g_n) = I(f) - I(f_n)$.   Moreover, since $f_n(x) \leq f_{n+1}(x)$, $n = 1, 2, \ldots$, and $\lim_n f_n(x) = f(x)$ for all x in S, the sequence $g_1, g_2, \ldots$, is monotone-decreasing and converges pointwise to 0.   Since I is linear and nonnegative on nonnegative functions, it is also true that

$$I(g_n) \geq I(g_{n+1}) \geq 0.$$

Thus it suffices to prove that $\lim_n I(g_n) = 0$.

Let $\epsilon > 0$, and A_n be the set of x for which $g_n(x) > \epsilon$.   Then $A_n \in \mathcal{R}$, $A_n \supset A_{n+1}$ for $n = 1, 2, \ldots$, and

$$\bigcap_n A_n = \emptyset.$$

Let B_n be the intersection of the support of g_n with $S - A_n$.   As noted earlier, the support of a step function is always in $\mathcal{R}$, so that B_n is also in $\mathcal{R}$ and

$$\int g_n(x)\, dm(x) = \int_{A_n} g_n(x)\, dm(x) + \int_{B_n} g_n(x)\, dm(x).$$

If E is the support of g_1 and $b = \sup_x g_1(x)$, then $B_n \subset E$ and $\sup_x g_n(x) \leq b$. Now observe that if a step function f is bounded by a constant k on a set C, then

$$\left| \int_C f(x)\, dm(x) \right| \leq k m(C);$$

this follows at once from the definition of the integral and the properties of the measure.   In particular, as $\lim_n m(A_n) = 0$, it follows that

$$0 \leq \limsup_n \int g_n(x)\, dm(x) \leq \epsilon m(E)$$

for every $\epsilon > 0$; hence $\lim_n I(g_n) = 0$.

EXERCISES

1   Show that for any step function f, $|I(f)| \leq I(|f|)$.

2   Show that, if f and g are real-valued step functions such that $f \leq g$, then $I(f) \leq I(g)$.

3   Show that if f and g are any two step functions, then

   a. $|I(fg)| \leq \sup\limits_{x \in \mathcal{S}} |f(x)| \, I(|g|)$.

   b. $|I(fg)|^2 \leq I(|f|^2)I(|g|^2)$.

[*Hint:* To prove (*b*), refer to and use *Cauchy's inequality*.]

4   Show that the set of all limits of uniformly convergent sequences of step functions is an algebra, to which the integral I may be uniquely extended so as to retain its linearity and the property that $I(f) \geq 0$ in case f is nonnegative-valued.

5   Show that any continuous function on [0,1] is a uniform limit of step functions relative to the ring generated by all closed subintervals.

6   Show that if F is any right-continuous monotone-increasing function on [0,1], and g is any continuous function, then $\int g(x)\, dF(x)$ is the same as the extension to g in the sense of Exercise 4 of the elementary integral for the basic measure space associated with F by Corollary 2.2.1.

7   If $\mathcal{R}$ is a ring of subsets of a set S, a subset A of S is said to be affiliated with $\mathcal{R}$, in case $A \cap E \in \mathcal{R}$ whenever $E \in \mathcal{R}$.   For any set A affiliated with $\mathcal{R}$ and step function f, the *integral of f over A* is defined as $I(f\varphi_A)$, where φ_A denotes the characteristic function of A; $I(f\varphi_A)$ as a function of A is called the *basic indefinite integral of f*.   Show that

   a. The set of all sets affiliated with $\mathcal{R}$ is a *complemented* ring, i.e., includes with every set A also $S - A$.

   b. For any step function f, the basic indefinite integral is countably additive on this ring.

8*   Show that if $\{f_n\}$ is any monotone-decreasing sequence of continuous functions on $[a,b]$ such that $f_n(x) \to 0$ for all x, then $\displaystyle\int_a^b f_n(x)\, dx \to 0$ as $n \to \infty$.

[*Hint:* Show that $\{f_n\}$ converges to zero uniformly.]

9   *a.* Let **A** be an algebra of functions on a set S, and $\mathcal{R}$ the collection of all subsets E of S for which φ_E belongs to **A**.   Show that $\mathcal{R}$ is a ring.

   b. Suppose I is a linear functional on **A** such that $I(f) \geq 0$ when $f \geq 0$. Show that there is a unique finitely additive measure m on $\mathcal{R}$ such that $m(E) = I(\varphi_E)$ for $E \in \mathcal{R}$.

   c. Show that m is countably additive if and only if I has the property that $I(f_n) \to 0$ whenever $f_n(x)$ is monotone-decreasing to 0 for all values of x.

10*   Let **A** denote the algebra of all real polynomials on the interval [0,1], and let I denote a linear functional on **A** which is nonnegative on polynomials that are nonnegative throughout this interval.   Show that there exists a Lebesgue-Stieltjes measure m on $[0,1]$ such that $I(p) = \int_0^1 p(x)\,dF(x)$, where F is the cumulative distribution function for m: $F(x) = m([0,x])$; and that m is unique. [*Hint:* First show that $|I(p)| \leq \text{const} \sup_x |p(x)|$, making use of the nonnegativity property of I. Then extend I to the continuous functions, as in Exercise 5, using the Weierstrass approximation theorem. Then define $m([a,b])$ as the infimum of the $I(f)$ for continuous nonnegative functions f such that $f \geq \varphi_{[a,b]}$. Compare also the first part of chap. 5.]

2.4 PRODUCTS OF BASIC SPACES

An important technique for the construction of basic measure spaces is the formation of the direct product of a finite set of such spaces. The case in which two spaces are involved is the fundamental one, and is described by

THEOREM 2.2 *Let $M_i = (S_i, \mathcal{R}_i, m_i)$ $(i = 1, 2)$ be given basic measure spaces. Then the collection $\mathcal{R}$ of all finite unions of cartesian products $A_1 \times A_2$ of sets in $\mathcal{R}_1$ and $\mathcal{R}_2$ is a ring of subsets of $S_1 \times S_2$ on which there is a unique countably additive measure m such that*

$$m(A_1 \times A_2) = m_1(A_1)m_2(A_2).$$

Proof In some respects the argument is similar to that used in the proof of Theorem 2.1, but of course no topology is involved in this result, and the crucial countable additivity follows in a different fashion. It is convenient to use the term *rectangle* for a set of the form $A_1 \times A_2$, with A_i in $\mathcal{R}_i$.

To show that all finite unions of rectangles form a ring, it suffices to show that all finite disjoint unions of rectangles do so. This reduces, as in the proof of Lemma 2.2.1, to showing that the difference of two disjoint unions is again such, which in turn reduces to showing that the difference of any two rectangles is a disjoint union of rectangles. It is a matter of simple set theory to verify for the rectangles $A_1 \times A_2$ and $B_1 \times B_2$ that

$$A_1 \times A_2 - B_1 \times B_2$$

is the disjoint union of the rectangles $(A_1 \cap B_1) \times (A_2 - B_2)$, $(A_1 - B_1) \times (A_2 \cap B_2)$, and $(A_1 - B_1) \times (A_2 - B_2)$. Geometrically, this is clarified by a plane diagram in which the A_i and B_i are represented by rectangles in the elementary sense.

Evidently, any two additive measures on the ring $\mathcal{R}$ just obtained which agree on rectangles are identical. To show the existence of the measure m satisfying the stated conditions, observe that a function f on $S_1 \times S_2$ is a step function relative to $\mathcal{R}$ if and only if it has an expression of the form

$$f(x,y) = \sum_{i=1}^{n} g_i(x)h_i(y),$$

where the g_i and h_i are step functions on $(S_1,\mathcal{R}_1,m_1)$ and $(S_2,\mathcal{R}_2,m_2)$, respectively. Thus the equation

$$k(x) = \int f(x,y)\, dm_2(y)$$

defines a step function k on $(S_1,\mathcal{R}_1,m_1)$; in fact,

$$k(x) = \sum_{i=1}^{n} I_2(h_i)g_i(x),$$

where I_2 is the integral on $(S_2,\mathcal{R}_2,m_2)$. One can therefore define a functional I (the product integral) on the step functions, relative to $\mathcal{R}$, by setting

$$I(f) = \int \left[\int f(x,y)\, dm_2(y) \right] dm_1(x).$$

If f has the above form $I(f) = \sum_{i \le n} I_1(g_i)I_2(h_i)$, where I_1 is the integral on $(S_1,\mathcal{R}_1,m_1)$; in addition

$$\int f(x,y)\, dm_1(x) = \sum_{i=1}^{n} I_1(g_i)h_i(y).$$

Hence $I(f) = \int [\int f(x,y)\, dm_1(x)]\, dm_2(y)$ as well.

The fact that I is a linear functional follows easily from the linearity of I_1 and I_2. This being the case, we may define a nonnegative finitely additive measure m on $\mathcal{R}$ by setting

$$m(E) = I(\varphi_E)$$

for E in $\mathcal{R}$. With this definition, $m(A_1 \times A_2) = m_1(A_1)m_2(A_2)$.

It remains only to show the countable additivity of m. Let $f_1, f_2, \ldots$ be a monotone-decreasing sequence of step functions relative to $\mathcal{R}$ which converges to 0. Then, for fixed x, $f_1(x,y), f_2(x,y), \ldots$ is a monotone-decreasing sequence of step functions of the variable y which converges to 0; it follows by Scholium 2.2 that

$$\int f_n(x,y)\, dm_2(y) \downarrow 0 \quad \text{as} \quad n \uparrow \infty.$$

On the other hand, the integrals as functions of x are likewise step functions, and it follows similarly that

$$\int \left[\int f_n(x,y)\, dm_2(y) \right] dm_1(x) \downarrow 0 \quad \text{as} \quad n \uparrow \infty.$$

Thus m is countably additive.

The basic measure space $(S_1 \times S_2,\ \mathcal{R},\ m)$ is called the *direct product* of the basic spaces M_1 and M_2, and denoted $M_1 \otimes M_2$. On occasion, $\mathcal{R}$ will be denoted as $\mathcal{R}_1 \otimes \mathcal{R}_2$, and m as $m_1 \otimes m_2$.

It is useful to note that the functional I in the proof of the theorem is the integral defined by m as in Scholium 2.1. It is noteworthy also that these considerations extend without difficulty to the formation of direct products of any finite sequence of basic spaces, and that direct products of basic spaces are associative to precisely the same extent that cartesian products are associative; i.e., with the usual identification of $(S_1 \times S_2) \times S_3$ with $S_1 \times (S_2 \times S_3)$, the basic measure spaces $(M_1 \otimes M_2) \otimes M_3$ and $M_1 \otimes (M_2 \otimes M_3)$ are the same. It follows that if $M_1, M_2, \ldots, M_n$ is any finite sequence of measure spaces, the direct-product measure space $\otimes_i M_i$ is well defined as a basic measure space. However, unlike cartesian products, the situation is altogether different for infinite products of basic measure spaces, which are effectively definable only when all but a finite number of the factors are basic probability spaces; a special case will be considered in the next section.

2.5* COIN-TOSSING SPACE

A quite different type of measure space from any of the preceding is one that occurs in the treatment of sequences of independent random variables in probability. The simplest nontrivial space of this sort, which, however, is fairly typical, is the space used to analyze the behavior of infinite sequences of tosses of a balanced coin. The use of this space permits various laws about what happens "on the average," or "almost certainly," to be formulated with mathematical precision.

To describe the space mathematically, let P denote the space consisting of the two points 1 and 0 (denoting heads and tails) and let P be endowed with the discrete topology. Then, obviously, P is compact, so that by Tychonoff's theorem the direct product S of a countably infinite number of copies of P is again compact. An element x of S is simply an infinite sequence $(x_1, x_2, \ldots)$, where each x_i is 1 or 0. If $E^{(n)}$ is an arbitrary subset of the n-fold direct product P^n of P with itself and

$$A = [x \in S : (x_1, x_2, \ldots, x_n) \in E^{(n)}],$$

define $m(A)$ as 2^{-n} times the number of points in $E^{(n)}$. Since A may be represented in such a form in many ways, it must be shown that $m(A)$ is independent of the representation. Suppose also that

$$A = [x \in S : (x_1, x_2, \ldots, x_r) \in F^{(r)}].$$

Then either $r \leq n$ or $n \leq r$, and it is no loss of generality to assume the latter. It is then easy to see that

$$F^{(r)} = E^{(n)} \times P \times P \cdots \times P,$$

where the factor P occurs $r - n$ times. Hence the number of points in $F^{(r)}$

is 2^{r-n} times the number of points in $E^{(n)}$, which shows that $m(A)$ is well defined.

Let $\mathcal{R}$ be the collection of all subsets of S of the form of the set A above. Then the system $(S, \mathcal{R}, m)$ is a basic measure space. For $\mathcal{R}$ is a ring, and m is countably additive on $\mathcal{R}$.

To see that $\mathcal{R}$ is a ring, suppose A is as before and that

$$B = [x \in S : (x_1, x_2, \ldots, x_r) \in G^{(r)}].$$

Then it may be assumed that $n \leq r$, so that

$$A = [x \in S : (x_1, x_2, \ldots, x_r) \in E^{(r)}],$$

where $E^{(r)} = E^{(n)} \times P \times P \times \cdots \times P$, and P occurs as a factor $r - n$ times. Thus

$$A \cup B = [x \in S : (x_1, x_2, \ldots, x_r) \in E^{(r)} \cup G^{(r)}]$$

and is hence in $\mathcal{R}$. A similar expression shows that $A - B$ is also in $\mathcal{R}$, so that $\mathcal{R}$ is indeed a ring.

To see that m is additive, let A and B be disjoint sets expressed as in the preceding paragraph. Then $m(A \cup B)$ is 2^{-r} times the number of points in $E^{(r)} \cup G^{(r)}$. As A and B are disjoint, so are $E^{(r)}$ and $G^{(r)}$. Thus the number of points in $E^{(r)} \cup G^{(r)}$ is the sum of the numbers in $E^{(r)}$ and $G^{(r)}$. On the other hand, these numbers after multiplication by 2^{-r} are $m(A)$ and $m(B)$, respectively.

It remains to show that m is countably additive. Suppose $A_1, A_2, \ldots$ is a monotone-decreasing sequence of sets in $\mathcal{R}$ with empty intersection. Then, by elementary combinatorial considerations, one can show that A_n is empty from a certain point onward. Thus $\lim_n m(A_n) = 0$ for trivial reasons. A more sophisticated proof that a finite intersection of the A_n is empty may be given along the following lines. One first observes that the sets in $\mathcal{R}$ are both open and closed. This follows easily from the definition of the product topology. Then, since S is compact, an infinite collection of closed subsets of S can have an empty intersection only if some finite subcollection already has empty intersection.

As an example of how the space just defined may be used in connection with non-coin-tossing matters, consider a sequence of real numbers $a_1, a_2, \ldots$ such that $\sum_n a_n^2$ is convergent. Then, in general, the series $\sum_n a_n$ will not be convergent; for example, this is the case for $a_n = n^{-\epsilon}$, with $\frac{1}{2} < \epsilon \leq 1$. However, the series $\sum_n \pm a_n$ may be convergent if the $\pm$ signs are appropriately distributed, e.g., if in the cited example they are alternating. If the $\pm$ signs are chosen "at random," it is plausible that there should virtually always be sufficient cancellation so that the series converges. With the use of the basic measure space just constructed, this can be formulated as a precise mathematical statement, which is in fact true.

EXERCISES

1 Let f be a nonnegative step function on the basic measure space $M = (S, \mathcal{R}, m)$, and let L be the basic Lebesgue measure space on the reals, with ring of sets generated by the compact intervals. Show that the integral of f is the measure in $M \otimes L$ of the "ordinate set" of f, that is, the set of all pairs (x, λ), with $x \in S$ and λ a real number such that $0 \leq \lambda \leq f(x)$.

2 Generalize the construction of the coin-tossing space to the case in which the probability of heads is p, $0 \leq p \leq 1$, and that of tails is $1 - p$, and show that the resulting structure is again a basic measure space. Then generalize again by permitting p to vary with the "trial."

3 *a.* In the space E_n of real n-tuples, let $\mathcal{I}$ denote the collection of all bounded rectangular parallelepipeds, i.e., sets of the form $[x \in E_n : x_i \in I_i; \ i = 1, 2, \ldots, n]$, where I_i is a bounded interval on the real line. Show that the set of all finite disjoint unions of elements of $\mathcal{I}$ is a ring $\mathcal{R}$.

    b. Define the volume of the indicated element of $\mathcal{I}$ as the product of the lengths of the intervals I_i and of any finite disjoint union of elements of $\mathcal{I}$ as the sum of the volumes of the respective elements. Show that the latter volume is unique, i.e., the same for all representations of the element as a disjoint union of elements of $\mathcal{I}$, and that the resulting volume function is an additive measure on $\mathcal{R}$.

    c. Show that m is countably additive on $\mathcal{R}$.

[*Remark:* $(E_n, \mathcal{R}, m)$ is called the n-dimensional basic Lebesgue measure space; it is also definable as the n-fold direct product of one-dimensional basic Lebesgue spaces, but it is instructive to give direct proofs of (a), (b), and (c).]

4 Show that if $\mathcal{R}_n$ is a monotone-increasing sequence of subrings of a set S, if m_n is a countably additive measure on $\mathcal{R}_n$ such that m_{n+1} agrees on $\mathcal{R}_n$ with m_n, then the union of the $\mathcal{R}_n$ is a ring $\mathcal{R}$ on which the function m which extends each of the m_n is a finitely additive, but in general not countably additive, measure.

2.6 INFINITY IN INTEGRATION THEORY

Integrals which exist and are finite, with which we are largely concerned in this book, are determined by measures which are likewise finite. Nevertheless, infinite measures arise naturally in other connections; for example, geometrical ones. In addition, as already noted, the usual definition of (full) measure space admits possible infinite values for the measure.

The "infinity" which arises in integration theory is a somewhat distinctive one, related to, but more structured than, the elementary or geometrical concepts of infinity. It originates in the circumstance that, while the real line is simple from an algebraic viewpoint, being in fact a "field," it is incomplete from the viewpoint of its order properties; the extension of the real line by the adjunction of $\pm \infty$ somewhat complicates its algebraic structure; in fact, algebraic operations are not always defined, but the system becomes order-complete, as defined below, as well as topologically compact.

Although only relatively specific ordered systems occur in this book, and this section is largely concerned with one particular case, that of the extended real number system, it will be helpful to orient our considerations relative to the general theory of ordered systems. A *partially ordered set* is the system composed of a set S and a relation $\leq$, defined for certain ordered pairs of elements of S and satisfying certain conditions; the relation $a \leq b$ is expressed as "a is less than or equal to b." The following properties are required of this relation for it to be designated as a "partial ordering on S": (*1*) if $a \leq b$ and $b \leq c$, then $a \leq c$; (*2*) if $a \leq b$ and $b \leq a$, then $a = b$; (*3*) $a \leq a$. A partially ordered set is said to be *simply ordered* in case any two elements are *comparable*, by which is meant that either $a \leq b$ or $b \leq a$. The real numbers are simply ordered by their usual ordering; the set of all subsets of a given set containing more than one element is partially ordered by set inclusion as the order relation, but not simply ordered. A subset T of a partially ordered set S is said to have the *upper bound x* in case $a \leq x$ for every element $a \in T$; similarly, it has the *lower bound y* in case $y \leq a$ for all $a \in T$. An upper bound x for a set T is called the *least upper bound*, or *supremum*, of T in case it is less than or equal to all other upper bounds; the concept of *greatest lower bound*, or *infimum*, of a set is similarly defined.

In general, of course, a bounded subset of a partially ordered set will fail to have an infimum or supremum; the property of a partially ordered set, that every bounded set should have an infimum and supremum in the given partially ordered set, is called *conditional completeness*. (The reader may recall that the construction of the real number system from the rationals by the method of Dedekind cuts can be looked on as a means of attaining conditional completeness by a minimal enlargement of the rational number system; similar completions are available for general partially ordered sets.) Still more rare is the property of (unconditional) *completeness*, by which is meant the existence of an infimum and a supremum, for any subset of the given ordered set. An example of a complete partially ordered set is the power set of any given set, relative to set inclusion as the ordering; the supremum, for example, of any given collection of subsets is simply their set-theoretic union. The real number system is of course not complete, although it will become so after the adjunction of $\pm \infty$.

An *open interval* in a simply ordered set S may be defined as a subset consisting of all elements x satisfying any one of the following types of inequalities: $x < a$; $x > b$; $a < x < b$, for arbitrary but fixed elements a and b of S; the relation $x < y$ is, of course, defined by the conjunction of the relations $x \leq y$ and $x \neq y$. The open intervals form a basis for the open sets in a topology on S, called the (order-) interval topology; in this topology, any open interval containing a given point is a neighborhood of the point.

It is a simple matter to apply these concepts to the extension of the real number system described in the

DEFINITION The *extended-real number system* is a set N with partially defined algebraic operations and a simple ordering as follows:

(*1*) As a set, N consists precisely of the set R of all real numbers and two distinguished elements, called plus/minus infinity and denoted ∞ and $-\infty$, respectively.

(*2*) N is simply ordered in extension of the usual ordering on the real numbers by the relations $-\infty \leq x \leq \infty$ for all elements $x \in N$.

(*3*) Algebraic operations on R are extended by the definitions

$$\infty + \infty = \infty.$$

$$(-\infty) - \infty = -\infty = -\infty + (-\infty).$$

$$a + \infty = \infty + a = \infty.$$

$$a - \infty = a + (-\infty) = (-\infty) + a = -\infty \qquad \text{if } a \in R.$$

$$a \cdot \infty = \infty \cdot a = \infty \qquad \text{if } a \in N \text{ and } a > 0.$$

$$a \cdot \infty = \infty \cdot a = -\infty \qquad \text{if } a \in N \text{ and } a < 0.$$

$$\infty \cdot 0 = 0 \cdot \infty = 0 \cdot (-\infty) = (-\infty) \cdot 0 = 0.$$

(*4*) N is topologized by the interval topology.

SCHOLIUM 2.3 *The extended-real number system N is a complete simply ordered set, order-isomorphic to the interval $[-1,1]$; where fully defined, multiplication and addition are associative and commutative, and multiplication is distributive with respect to addition; these operations are jointly continuous in the two variables in question where defined, with the following exception: multiplication at the points $(0,\pm\infty)$ and $(\pm\infty,0)$ of $N \times N$.*

The much-used facts summarized in this scholium are quite elementary in their proof, which will for the most part be left to the reader. As already indicated, the order-completeness is the main reason for the extension, and is valid virtually by definition. The mapping $x \rightarrow \tan^{-1} x$ maps R into the interval $(-\pi/2,\pi/2)$, in an order-preserving fashion, and is easily seen to extend to an order-preserving mapping of N onto the closed interval $[-\pi/2, \pi/2]$, which is in turn obviously order-isomorphic with the interval $[-1,1]$.

While the mapping $(a,b) \rightarrow ab$ is not jointly continuous in a and b at the points $(0,\pm\infty)$, it is separately continuous as a function of b for fixed values of a. This limited continuity is in practice quite applicable in integration theory, where the typical product $0 \cdot \infty$ involves 0 as a measure (usually fixed) and ∞ as the value of a function (subject to variation).

The definition given earlier for arbitrary sums of nonnegative real numbers is extended by the following definition, applicable to extended-real numbers.

DEFINITION   If $\{x_\lambda; \lambda \in \Lambda\}$ is an indexed set of extended-real numbers, the sum $\sum_\lambda x_\lambda$ (where it is understood that the summation index λ ranges over all of Λ unless otherwise indicated) is said to *exist* (unconditionally) in case for every finite subset Φ of Λ, the sum $\sum_{\lambda \in \Phi} x_\lambda = s_\Phi$, say, is defined, and if the limit of the net, $\lim_\Phi s_\Phi$, exists. The indicated sum is then defined as this limit.

We leave to the reader the simple proof that this definition agrees with the one in Sec. 2.1 for the case in which the x_λ are all real and nonnegative: It may also be easily shown that the sum always exists when the x_λ are all nonnegative.

Among other simple properties of the generalized sum here defined are its complete additivity (see the exercises below).

LOCAL MEASURABILITY AND INFINITE MEASURES   With this development of the extended-real numbers, it is straightforward to extend the earlier considerations to measures whose values are possibly infinite, i.e., contained in the set of all nonnegative extended-real numbers. It turns out that all sets of importance in the theory of integration, even when of infinite measure, are approximable in a certain sense by sets of finite measure. The nature of this approximation, and the general means of referring questions concerning "large" measurable sets to those of finite measure, are indicated by the following developments.

DEFINITION   If $\mathfrak{R}$ is any ring of subsets of a set S, a subset E of S such that $E \cap A \in \mathfrak{R}$ for all A in $\mathfrak{R}$ is said to be affiliated with $\mathfrak{R}$; a set affiliated with the ring $\mathfrak{R}$ of a basic measure space is called a *basic locally measurable set*; the set of all such sets is denoted $\mathfrak{R}^+$.

Example 2.7.1   The entire "space" S in a basic measure space $(S,\mathfrak{R},m)$ is always locally measurable, although in general it will not be contained in $\mathfrak{R}$. Every set in $\mathfrak{R}$ is evidently also in $\mathfrak{R}^+$; or in other terms, basic measurability implies local basic measurability.

SCHOLIUM 2.4   *If $(S,\mathfrak{R},m)$ is a basic measure space, the basic locally measurable sets form a complemented ring, to which m may be extended with preservation of its countable additivity by the definition*

$$m(E) = \sup_{A \in R} m(E \cap A).$$

To amplify this statement slightly, it must be shown that if the function m' on $\mathfrak{R}^+$, with values in the set of all nonnegative extended-real numbers, is defined by the equation

$$m'(E) = \sup_{A \in R} m(E \cap A),$$

then m' extends m and is countably additive; additionally, it must be shown that $\mathfrak{R}^+$ is a ring, and that if E is any element of $\mathfrak{R}^+$, then so also is $S - E$. These Boolean-algebraic properties of $\mathfrak{R}^+$ are entirely straightforward to derive; for example, if $E \in \mathfrak{R}^+$, then for any $A \in \mathfrak{R}, (S - E) \cap A = (S \cap A) - (E \cap A) = A - (E \cap A)$, which is in $\mathfrak{R}$ since it is the difference of two elements of $\mathfrak{R}$. That m' extends the original measure m follows from the observation that if $E \in \mathfrak{R}$, then for all $A \in \mathfrak{R}, E \cap A \subset E$, so that $m(E \cap A) \le m(E)$, showing that the indicated supremum cannot exceed $m(E)$; on the other hand, it attains this value when $A = E$.

To prove the countable additivity of m' (the main content of the scholium), let $E_1, E_2, \ldots$ be a sequence of disjoint sets in $\mathfrak{R}^+$ whose union E is again in $\mathfrak{R}^+$; it must be shown that $m'(E) = \sum_i m'(E_i)$. To this end, let A be an arbitrary element of $\mathfrak{R}$. Then, since

$$E \cap A = \bigcup_i E_i \cap A,$$

and since m is countably additive, it follows that

$$m(E \cap A) = \sum_i m(E_i \cap A) \le \sum_i m'(E_i).$$

Since A is arbitrary in $\mathfrak{R}$, it results that

$$m'(E) \le \sum_i m'(E_i).$$

To obtain the reverse inequality (and thereby conclude the proof), let $A_1, A_2, \ldots$ be an arbitrary sequence of sets in $\mathfrak{R}$; then the set $B_n = \bigcup_{i \le n} E_i \cap A_i$ belongs to $\mathfrak{R}$, and $E \cap B_n = B_n$. This implies the inequality

$$m'(E) \ge m(B_n) = \sum_{1 \le i \le n} m(E_i \cap A_i).$$

Now, using the arbitrariness of the A_i in $\mathfrak{R}$, it follows that

$$m'(E) \ge \sum_{1 \le i \le n} m'(E_i).$$

This inequality is indeed valid for every positive integer n; consequently, it may be concluded that

$$m'(E) \ge \sum_i m'(E_i).$$

At this point there is no significant loss of clarity in denoting m' simply as m, as in the statement of the scholium.

EXERCISES

1 Show that $\mathcal{R}^+ = \mathcal{R}$ if and only if $S \in \mathcal{R}$.

2 Suppose that the basic measure space $(S,\mathcal{R},m)$ has the property that if $E_1, E_2, \ldots$ is any sequence of mutually disjoint elements of $\mathcal{R}$ for which $\sum_i m(E_i)$ is convergent, then $\bigcup_i E_i \in \mathcal{R}$. Show that $\mathcal{R}^+$ is then closed under countable unions (i.e., is then a σ-ring).

3 Show that if m'' is any countably additive measure extending the measure m in Scholium 2.4, then $m'(E) \le m''(E)$ for all $E \in \mathcal{R}^+$.

4 Define a basic nonnegative locally measurable function on a basic measure space as a function having only a finite number of values, each of which is a nonnegative extended-real number and is assumed on a basic locally measurable set. Let **M** denote the set of all such functions; if the element f of **M** has the value x_i on the locally measurable set E_i, where i ranges over a finite set and $\bigcup_i E_i = S$, define the integral $I(f)$ as $\sum_i x_i m(E_i)$. Show that $I(f)$ is uniquely defined (i.e., the same for all representations of f in the indicated form) and that $I(f + g) = I(f) + I(g)$ for arbitrary f and g in **M**.

5 Show that if $\{x_\lambda; \lambda \in \Lambda\}$ is an indexed set of nonnegative extended-real numbers, and if Λ is the disjoint union of subsets Λ_μ, then $\sum_\lambda x_\lambda = \sum_\mu \left(\sum_{\lambda \in \Lambda_\mu} x_\lambda \right)$.

6 The concept of infinity in integration theory may be refined by taking it as an infinite cardinal number. Let *nonnegative quantity* be defined as either a nonnegative real number or an infinite cardinal number; let these be simply ordered so that the latter are ordered in the usual fashion and exceed all the former; let the sum of a nonnegative real number and an infinite cardinal be defined as the cardinal, and the product defined as the cardinal or 0, according as the real number is positive or zero. Show that an associative, commutative, distributive system is obtained, with the properties that if $a \le b$, then $a + c \le b + c$ and $ac \le bc$.

7* For any indexed set x_λ of nonnegative quantities (that is, $\lambda \to x_\lambda$ is a mapping from a given set Λ into the nonnegative quantities), let $\sum_\lambda x_\lambda$ be defined as follows: Let Λ_1 be the set of all indices λ such that x_λ is real and positive; let Λ_2 be the set of indices for which x_λ is an infinite cardinal; for $\lambda \in \Lambda_2$, let A_λ denote any set of cardinal number x_λ, and let B_λ denote the set of all pairs of the form (λ, a), with $a \in A_\lambda$ (or let the B_λ be any mutually disjoint sets such that B_λ has the power x_λ); let y_1 denote the power of Λ_1, and y_2 the power of $\bigcup_{\lambda \in \Lambda_2} B_\lambda$; then $\sum_\lambda x_\lambda$ is defined as the sum earlier defined in case all the x_λ are real and their sum is finite as previously defined; and otherwise as $y_1 + y_2$.

a. Show that if Λ is the disjoint union of subsets Λ_μ, then

$$\sum_{\lambda \in \Lambda} x_\lambda = \sum_\mu \left(\sum_{\lambda \in \Lambda_\mu} x_\lambda \right).$$

b. Let f denote the mapping which carries a nonnegative quantity into itself or ∞, according as it is real or an infinite cardinal. Show that

$$f\left(\sum_{\lambda} x_{\lambda}\right) = \sum_{\lambda} f(x_{\lambda}).$$

8* First extend the result of Exercise 4 to the case where the values of the function are nonnegative quantities. Then extend it further to the case where the values of the measure are also nonnegative quantities.

III

MEASURABLE FUNCTIONS AND THEIR INTEGRALS

3.1 THE EXTENSION PROBLEM

For a given basic measure space, the basic measurable sets and the corresponding step functions form quite limited classes. Many of the sets, and actually most of the functions, which arise in analytical theory and practice will not be in these classes. It is apparent, for example, in the case of the Lebesgue basic measure space on the real line, that the integration of bounded continuous functions over finite intervals is not covered by the theory of integration for step functions. This is, however, as it should be. The general program in the development of the theory of integration should commence with a limited, yet transparent and readily constructed notion of integration such as that given in Chap. 2. The next step is to extend the integral to a wide class of functions, including, hopefully, all those of analytical interest, in such a way as to maintain all such useful properties as those already derived. That this should be possible is not at all obvious, nor does it have even an appearance of intuitive inevitability. That it is possible in fact, in such a relatively unique fashion, is in its way a quite striking phenomenon.

The problem of extending the integral to a sufficiently wide class of functions breaks naturally into two parts.   One part, which is independent of the measure and depends only on the ring of basic measurable sets, is concerned with the development of an appropriate notion of regularity for functions and with the approximation of regular functions by step functions. The other part, in which the measure plays a crucial role, then deals with the problem of extending the integral to those regular functions which are, roughly speaking, not too large.   The first problem is treated in the next section.

3.2   MEASURABILITY RELATIVE TO A BASIC RING

Perhaps the simplest analytically nontrivial desideratum for the class of functions which should be considered admissible from the standpoint of the theory of integration is that it should include the step functions and should, additionally, be closed under the operation of formation of pointwise limits of sequences.   This is actually sufficient, and in fact essentially just right. The corresponding functions are called "measurable," and play, roughly, the same role in measure theory as the continuous functions do in connection with general topology, a parallel which also exists between basic measurable sets in measure theory and basic open sets in topology.

Although the case of numerical functions is the primary one, it is useful also to consider measurable functions with values in more general topological spaces.   All topological spaces considered in this book will be assumed to be *Hausdorff*, and the convention is used throughout that topological space means *Hausdorff space*.

> DEFINITIONS   Let S be a set, $\Re$ a ring of subsets of S, and Y a topological space with a distinguished point 0.   Suppose f is a function defined on S with values in Y.   Then f is a *step function based on* $\Re$, or a *basic measurable function with respect to* $\Re$, if it assumes only a finite number of distinct values, each of which, with the possible exception of 0, is assumed on a set in $\Re$.   On the other hand, f is *measurable with respect to* $\Re$ if it is in the least class $\mathbf{M} = \mathbf{M}(S,\Re,Y)$ of functions from S to Y which contains all step functions based on $\Re$ and is closed under pointwise convergence of sequences.   A *measurable function on a basic measure space* $(S,\Re,m)$ is one that is measurable with respect to $\Re$.

To amplify on the meaning of the condition concerning pointwise convergence, let $\mathcal{F}$ be an arbitrary collection of functions on S to Y.   Then $\mathcal{F}$ is closed under pointwise convergence of sequences if and only if the following condition is satisfied.   Whenever $f_1, f_2, \ldots$ is a sequence of functions in $\mathcal{F}$ and $f_n(x) \to f(x)$ at all points x of S, then f belongs to $\mathcal{F}$.   That a smallest class $\mathcal{M}$ with the two indicated properties actually exists may be seen in the following way. First, the collection of all functions from S to Y has these properties. Next, the intersection of all collections having these properties again has these properties, and is least among them.

One of the fundamental properties of a measurable function is that it necessarily assumes the value 0 outside a countable union of sets in the ring $\mathfrak{R}$. For the subclass of **M** for which this is true contains the step functions, by definition, and is easily seen to be closed under pointwise convergence of sequences. Of course, if the entire space S is itself a countable union of sets in $\mathfrak{R}$, a measurable function need not assume the value 0 at all. In most applications S has this property relative to $\mathfrak{R}$. When this is true, it turns out (and is easy to see) that the notion of measurability does not depend on the distinguished point 0; at the same time, the measurable functions form an entirely adequate and satisfactory class for the general theory. However, if S is not a countable union of sets in $\mathfrak{R}$, the notion of measurability depends on the choice of the distinguished point, and many respectable functions will fail to be measurable, e.g., the "nonzero" constant functions. These particular difficulties can be avoided if one considers a larger class of functions.

DEFINITIONS If S is a set and $\mathfrak{R}$ a ring of subsets of S, a function f from S to a topological space Y is a *basic locally measurable function* with respect to $\mathfrak{R}$ if it assumes only a finite number of distinct values, each of which is assumed on an element of the ring $\mathfrak{R}^+$ of basic locally measurable sets. A function from S to Y is *locally measurable with respect to $\mathfrak{R}$* if it is in the least class $\mathbf{M}^{\mathrm{loc}} = \mathbf{M}^{\mathrm{loc}}(S,\mathfrak{R},Y)$ of functions from S to Y which contains all basic locally measurable functions and is closed under pointwise convergence of sequences.

If a distinguished point is chosen in Y, it is easy to see that a function on S to Y is locally measurable with respect to $\mathfrak{R}$ if and only if it is measurable with respect to the ring $\mathfrak{R}^+$ of basic locally measurable sets. Thus the notion of measurability with respect to $\mathfrak{R}^+$ is independent of the distinguished point.

The class $\mathbf{M}^{\mathrm{loc}}$ always contains the constant functions, and hence is in general properly larger than **M**. In fact, it is elementary to verify that $\mathbf{M}^{\mathrm{loc}} = \mathbf{M}$ if and only if S is a countable union of sets in $\mathfrak{R}$.

Example 3.2.1 Let S be an uncountable set, and $\mathfrak{R}$ the ring of all finite subsets of S. For Y take the space $[-\infty,\infty]$ of *extended-real numbers*, with the real number 0 as distinguished point. Then Y is a Hausdorff space, and an extended-real-valued function on S is measurable with respect to $\mathfrak{R}$ if and only if it assumes the value 0 at all but a countable number of points of S. On the other hand, every extended-real-valued function on S is locally measurable with respect to $\mathfrak{R}$. The proof of this involves an important method of approximation, which is used and described in detail in the next example.

Example 3.2.2   Let S be an arbitrary set, and f an extended-real-valued function on S.   Then the sets $[x:f(x) > a]$ and $[x:f(x) < -a]$, $0 < a < +\infty$, generate a certain ring $\mathcal{R}$ of subsets of S.   We shall show that f is measurable with respect to $\mathcal{R}$.

Let n be a positive integer and i an integer such that $0 < |i| \leq n2^n - 1$. If $i < 0$, let f_{ni} be the characteristic function of

$$[x:(i - 1)2^{-n} \leq f(x) < i2^{-n}],$$

and for $i > 0$, let f_{ni} be the characteristic function of

$$[x:i2^{-n} < f(x) \leq (i + 1)2^{-n}].$$

Set $f_n = \sum_{i<0} i2^{-n}f_{ni} + \sum_{i>0} i2^{-n}f_{ni}$.   Then f_n is a step function based on $\mathcal{R}$ whose support is contained in the set E_n, where $|f(x)| \leq n$.   Now let $g_n(x) = -n$ when $f(x) < -n$, $g_n(x) = f_n(x)$ for x in E_n, and $g_n(x) = n$ if $f(x) > n$.   Then g_n is also a step function based on $\mathcal{R}$, and $|f(x) - g_n(x)| \leq 2^{-n}$ for every x in E_n.   It is now quite easy to see that $g_n(x) \to f(x)$ for every x in S.   Thus f is indeed measurable with respect to $\mathcal{R}$.   Moreover, the approximating sequence has the following two important properties:

(*1*) The convergence of $\{g_n\}$ to f is *dominated* in the sense that $|f(x)| \geq |g_n(x)|$ for all x in S.

(*2*) $g_n \to f$ uniformly on every set where f is bounded by a positive real constant.   For any such set is contained in one of the sets E_n, and hence in all the sets E_k for $k > n$.

The set of all measurable functions can be regarded as the closure in a certain topology of the set of all step functions; similarly, the collection of all locally measurable functions is the closure in the same topology of the class of all basic locally measurable functions.   The topology occurs quite naturally in a variety of applications, and will be used not only here, but later as well.   In this context, it provides a generally effective and convenient approach to a number of problems that arise in the development of the integral.

DEFINITION   Let S be a set, Y a topological space, and $\mathbf{F} = \mathbf{F}(S,Y)$ the collection of all functions from S to Y.   Then the family of closed sets for the topology of *sequential pointwise convergence on* $\mathbf{F}$ consists precisely of those subsets of $\mathbf{F}$ which are closed under pointwise convergence of sequences.

The collection of all closed subsets of $\mathbf{F}$ satisfies the usual axioms involved in the definition of a topology by closed sets.   Specifically, (*1*) $\mathbf{F}$ is itself closed; (*2*) any intersection of closed sets is closed; (*3*) the union of two closed sets is closed.   Only (*3*) fails to follow immediately, and it may be proved as follows.   Let $\mathbf{A}$ and $\mathbf{B}$ be closed subsets of $\mathbf{F}$, and suppose that $f_1, f_2, f_3, \ldots$ is a sequence of elements in $\mathbf{A} \cup \mathbf{B}$ which converges pointwise to a function f.

Then infinitely many of the f_n must be in one of **A** or **B**, say, for definiteness, **A**. Thus **A** contains a subsequence $f_{n_1}, f_{n_2}, \ldots$ which converges to f, so that f is in **A**.

If $f_1, f_2, \ldots$ is a sequence in **F** which converges pointwise to f, then $f_n \to f$ in the topology of **F**. For if not, there is an open neighborhood of f whose complement contains a subsequence $f_{n_1}, f_{n_2}, \ldots$ converging pointwise to f. Conversely, if $f_n \to f$ in the topology of **F**, then $f_n(x) \to f(x)$ for all x in S. For if x is a point in S and U is an open set in Y,

$$[f \in \mathbf{F} : f(x) \in U]$$

is readily seen to be an open set in the topology of sequential pointwise convergence. Since two distinct elements of **F** have disjoint neighborhoods of the above type, it follows also that **F** satisfies the Hausdorff separation axiom.

It should be recognized that the closure $\mathbf{A}^c$ of a given set **A** in **F** does not, in general, consist of the class $\mathbf{A}'$ of all limits of sequences of functions in **A**. For $\mathbf{A}'$ will itself have sequential limits which it need not contain. This distinction, at first glance rather easily overlooked, is in fact quite material, and to a considerable extent underlies the major difficulty in the extension of the integral from the step functions to the measurable ones.

Example 3.2.3 Let S be the closed unit interval $[0,1]$, and set

$$f_{mn}(x) = (\cos m! \ \pi x)^{2n}, \qquad 0 \le x \le 1,$$

where m and n vary over the nonnegative integers. If $\mathbf{A} = \{f_{m,n}\}$, then $\mathbf{A}'$ contains the sequence $f_1, f_2, \ldots$, defined by

$$f_m(x) = \lim_n f_{mn}(x).$$

Now $f_m(x) = 1$ when $x = k/m!$, with $k = 0, 1, \ldots, m!$, and for all other $x, f_m(x) = 0$. From this it is easy to see that $f_1, f_2, \ldots$ converges pointwise to Dirichlet's function f, which assigns the value 1 to a rational point in $[0,1]$ and the value 0 to an irrational. Since **A** consists of continuous functions and f is discontinuous at all points, it is intuitively plausible that f is not in $\mathbf{A}'$. In fact, it can be shown that the functions in $\mathbf{A}'$ are necessarily continuous on a dense subset of $[0,1]$ for any collection **A** of continuous functions on $[0,1]$.

Next we shall prove a simple but useful result about continuity of mappings for the topology of sequential pointwise convergence.

SCHOLIUM 3.1 *For a mapping T from the space $\mathbf{F} = \mathbf{F}(S,Y)$ into a topological space H, the following statements are equivalent:*

(1) T is continuous.

(2) $T(\mathbf{L}^c) \subseteq [T(\mathbf{L})]^c$ for every subset $\mathbf{L}$ of $\mathbf{F}$.

(3) $T(f_n) \to T(f)$ whenever $f_n \to f$ in $\mathbf{F}$.

Proof   That (*1*) and (*2*) are equivalent and imply (*3*) is an elementary fact from general topology.   On the other hand, suppose (*3*) is true.   Then, for any closed subset of H, the inverse image under T is also closed under pointwise convergence of sequences.   Hence the inverse image is a closed set, and T is continuous.

With the aid of the preceding result, we shall now derive a number of elementary properties of measurable functions.

> SCHOLIUM 3.2   *Let f be a function from S to Y which is locally measurable with respect to a ring $\Re$ of subsets of S.   If E is a basic locally measurable set, let $f_E(x)$ be defined as $f(x)$ when x is in E, and as 0 when x is outside E.   Then f_E is locally measurable with respect to $\Re$, and measurable when f is measurable or $E \in \Re$.*

Proof   Let E be a fixed basic locally measurable set, and define f_E as above for all functions f from S to Y.   Then, by Scholium 3.1, the map T on $\mathbf{F}(S,Y)$, taking f into f_E, is continuous in the topology of pointwise convergence.   For it is obvious that $Tf_n \rightarrow Tf$ when $f_n \rightarrow f$.   Let $\mathbf{M}^{\mathrm{loc}}$ be the class of locally measurable functions from S to Y.   If $\mathbf{L}$ is the class of basic locally measurable functions, it is immediate that $T(\mathbf{L})$ consists of basic locally measurable functions as well, and that these are basic measurable functions in the case that $E \in \Re$.   Since the continuity of T implies $T(\mathbf{L}^c) \subset [T(\mathbf{L})]^c$, it follows that $T(\mathbf{M}^{\mathrm{loc}})$ consists of locally measurable functions which are in fact measurable when $E \in \Re$.   On the other hand, if $\mathbf{L}$ is the class of basic measurable functions from S to Y, then $T(\mathbf{L})$ also consists of such functions, so that f_E is measurable when f is measurable, and E is a basic locally measurable set.

> DEFINITION   If Y and Z are topological spaces, a *Baire function from Y to Z* is a function in the least class including all continuous mappings from Y to Z which is closed under pointwise convergence of sequences. Alternatively, the class of Baire functions may be defined as the closure in $\mathbf{F}(Y,Z)$ of the class of continuous functions.

> SCHOLIUM 3.3   *Let f be a function from S to Y which is locally measurable with respect to a ring $\Re$ of subsets of S.   Suppose λ is a Baire function mapping Y into Z.   Then $\lambda \circ f$ is a function from S to Z and is locally measurable with respect to $\Re$.   If f is measurable with respect to $\Re$ and $\lambda(0) = 0$, then $\lambda \circ f$ is also measurable with respect to $\Re$.*

Proof   Suppose first that λ is continuous, and let T be the map on $\mathbf{F}(S,Y)$ defined by $T(f) = \lambda \circ f$.   If $f_n \rightarrow f$, then $\lambda[f_n(x)] \rightarrow \lambda[f(x)]$ for all x in S; hence $T(f_n) \rightarrow T(f)$, and T is continuous by Scholium 3.1.   Let $\mathbf{L}$ be the class of basic locally measurable functions from S to Y.   Then $T(\mathbf{L})$

consists of basic locally measurable functions from S to Z.   Since $T(L^c) \subset [T(L)]^c$, it follows at once that T maps locally measurable functions to locally measurable functions.

Now for a fixed locally measurable f, let **B** denote the class of all Baire functions λ from Y to Z such that $\lambda \circ f$ is locally measurable.   Then **B** contains all continuous maps, and is easily seen to be closed under pointwise convergence of sequences.   Hence **B** is precisely the class of all Baire functions from Y to Z.

The case when f is measurable and $\lambda(0) = 0$ is treated in a similar fashion. That the latter condition cannot be dispensed with is clear from the fact that a constant function need not be measurable, but only locally measurable on certain spaces.

Suppose Z is a closed subset of Y which contains the distinguished point 0. If $\Re$ is a ring of subsets of S, and Z is given the induced topology, the inclusion map of Z into Y is continuous and carries 0 into 0.   Under these conditions it follows easily from the preceding result that $\mathbf{M}(S,\Re,Z)$ may be identified with a closed subset of $\mathbf{M}(S,\Re,Y)$.   It may be true, however, that the collection of f in $\mathbf{M}(S,\Re,Y)$ which map S into Z is properly larger than $\mathbf{M}(S,\Re,Z)$. This somewhat awkward situation is likely to occur if, for example, the complement of Z is dense in Y but Z is discrete in the induced topology. A simple but useful case in which such pathology does not arise is given in the next result.

SCHOLIUM 3.4   *Let $\Re$ be a ring of subsets of S, and E a basic locally measurable set with respect to $\Re$.   If $[a,b]$ is an interval in the extended reals which contains 0, then the set of all step functions with support in E and values in $[a,b]$ is dense in the corresponding subset of measurable functions.*

Proof   If f is any function from S to $[-\infty,+\infty]$, define $f_E(x)$ as $f(x)$ when $x \in E$, and as 0 when x is outside E.   Let $\lambda(t) = t$ for $a \leq t \leq b$, $\lambda(t) = a$ for $t < a$, and $\lambda(t) = b$ for $t > b$.   Then λ is a continuous map of $[-\infty, +\infty]$ into $[a,b]$ and $\lambda(0) = 0$.   If T is defined on the space $\mathbf{F}$ of extended-real-valued functions on S by the relation $T(f) = \lambda \circ f_E$, it follows from Scholium 3.1 that T is continuous in the topology of sequential pointwise convergence.   For it is easily verified that $T(f_n) \to T(f)$ when $f_n \to f$.   If $\mathbf{L}$ is the class of basic measurable functions from S to the extended reals, it also follows that $T(\mathbf{L}^c) \subset [T(\mathbf{L})]^c$.   Moreover, $T(\mathbf{L})$ consists of step functions based on $\Re$ which vanish outside E and have values in $[a,b]$.   Since $T(f) = f$ when f has support in E and values in $[a,b]$, this concludes the proof.

SCHOLIUM 3.5   *If f and g are measurable functions from S to Y and Z, respectively, relative to a ring $\Re$ of subsets of S, then the map $x \to [f(x),g(x)]$ from S into $Y \times Z$ is measurable with respect to $\Re$.*

Proof First suppose g is a fixed step function based on $\mathfrak{R}$. Then the set of all f such that the indicated map is measurable includes all step functions. Here it is understood that the distinguished point of $Y \times Z$ is the pair $(0,0)$ consisting of the distinguished points of Y and Z, respectively. In addition, the set of all such f is closed under pointwise convergence of sequences, so that it must include all measurable f. Next, suppose f is a fixed measurable function. Then the set of all g such that the indicated map is measurable includes all step functions, as just proved, and is closed under pointwise convergence of sequences; hence it includes all measurable g.

SCHOLIUM 3.6 *Let f and g be functions from S to the extended reals $[-\infty,\infty]$ which are measurable with respect to a ring $\mathfrak{R}$ of subsets of S, and c a nonnegative extended-real number. Then the functions $\max(f,g)$, $\min(f,g)$, $\max(f,-c)$, and $\min(f,c)$ are all measurable with respect to $\mathfrak{R}$.*

Proof The function $\max(f,g)$ is the composition of $x \to [f(x),g(x)]$ with the continuous map λ which assigns the value $\max(y,z)$ to the pair of extended-real numbers (y,z). Thus $\max(f,g)$ is measurable by Scholiums 3.3 and 3.5. The measurability of $\min(f,g)$ is proved similarly. The function $\min(f,c)$ assumes the value $\min[f(x),c]$ at the point x. Thus $\min(f,c)$ is the composition of f with the continuous map λ which assigns the value $\min(y,c)$ to the point y. Since $c \geq 0$, it follows that $\lambda(0) = 0$, and therefore $\min(f,c)$ is measurable by Scholium 3.3. For similar reasons $\max(f,-c)$ is also measurable.

We turn next to the question of the measurability of sums and products of extended-real-valued functions.

If f and g are extended-real-valued functions on a set S, their sum and product are defined as usual, in terms of pointwise sums and products, except for the convention that $f + g$ is partially undefined when there exists an x in S for which $f(x)$ and $g(x)$ are infinite with opposite signs.

SCHOLIUM 3.7 *Let f and g be extended-real-valued functions on a set S which are measurable with respect to a ring $\mathfrak{R}$ of subsets of S. If c is an extended-real number, then $cf, f \cdot g$, and $f + g$ (if everywhere defined) are also measurable with respect to $\mathfrak{R}$.*

Proof Let $\lambda_n(y) = \min[\max(y,-n),n]$ for $n = 1, 2, \ldots$. Then λ_n is a continuous map from $[-\infty,\infty]$ onto $[-n,n]$, and by Scholium 3.3, the functions $f_n = \lambda_n \circ f$ and $g_n = \lambda_n \circ g$ are measurable with respect to $\mathfrak{R}$. The maps $(y,z) \to y + z$ and $(y,z) \to yz$ are both continuous from $[-n,n] \otimes [-n,n]$ into $[-\infty,\infty]$. Thus it follows from Scholiums 3.3 and 3.5 that $f_n + g_n$ and $f_n g_n$ are measurable as functions from S to $[-\infty,\infty]$. Since $f_n g_n \to fg$, it follows that fg is measurable. If $f + g$ is defined, it is also true that $f_n \dotplus g_n \to f \dotplus g$, so that $f \dotplus g$ is measurable as well. A similar argument shows that cf is measurable.

Because of our extensive use of the following concepts, already indicated in Chap. 1, their definitions are repeated, in more formal fashion, at this point.

> DEFINITION   A *σ-ring* is a ring of subsets of a set S which is closed under countable unions. Any collection of sets is contained in a least σ-ring, namely, the intersection of all σ-rings containing the given collection; this least σ-ring is called the *σ-ring generated by the given collection*.

The σ-ring generated by the ring $\Re$ of subsets of S may also be described as the closure $\Re^c$ of $\Re$ in the topology of *sequential convergence*. This topology may be defined as follows. Let Y be the discrete space consisting of the two points 0 and 1. There is then a one-to-one correspondence between the space $\mathbf{F}(S, Y)$ of functions from S to Y and subsets of S in which any given subset A of S corresponds to its characteristic function φ_A. Hence there is a unique topology on the collection of all subsets of S such that the map $\varphi_A \to A$ is a homeomorphism. If $A_1, A_2, \ldots$ is an arbitrary sequence of subsets of S, it is easy to see that the set

$$\liminf_n A_n = \bigcup_{i \geq 1} \bigcap_{n \geq i} A_n$$

corresponds to the function $\liminf_n \varphi_{A_n}$; similarly,

$$\limsup_n A_n = \bigcap_{i \geq 1} \bigcup_{n \geq i} A_n$$

corresponds to $\limsup_n \varphi_{A_n}$. Thus a sequence of subsets $A_1, A_2, \ldots$ converges to a set A if and only if

$$A = \liminf_n A_n = \limsup_n A_n.$$

> SCHOLIUM 3.8   *The closure $\Re^c$ of a given ring $\Re$ of subsets of S is the σ-ring generated by $\Re$.*

Proof   For the moment, let $\Re'$ denote the σ-ring generated by $\Re$. If $A_1, A_2, \ldots$ is any sequence of sets in $\Re'$, then

$$\bigcap_n A_n = A_1 - \bigcup_{n > 1} (A_1 - A_n),$$

so that $\Re'$ is closed under countable intersections as well as countable unions. This implies the fact that $\liminf_n A_n$ and $\limsup_n A_n$ also belong to $\Re'$. Thus $\Re'$ is closed in the relevant topology, and $\Re^c \subset \Re'$. To show that $\Re^c = \Re'$, it is enough to show that $\Re^c$ is itself a σ-ring. Because $\Re^c$ is closed, this reduces to showing that $\mathfrak{R}^c$ is a ring.   For it is easily verified that

$$\bigcup_n A_n = \lim_n \left(\bigcup_{i \leq n} A_i \right)$$

for any sequence $A_1, A_2, \ldots$ of subsets of S.

If A is an arbitrary element of $\mathfrak{R}^c$, let $\mathcal{C}(A)$ denote the collection of all sets B in $\mathfrak{R}^c$ such that $A \cup B$, $A - B$, and $B - A$ all belong to $\mathfrak{R}^c$.   It should be noted that $B \in \mathcal{C}(A)$ if and only if $A \in \mathcal{C}(B)$.   Moreover, $\mathcal{C}(A)$ is always closed.   For if $B_1, B_2, \ldots$ is a sequence of sets in $\mathcal{C}(A)$ and $B_n \to B$, it is a matter of simple verification to check that $A \cup B_n \to A \cup B$, $A - B_n \to A - B$, and $B_n - A \to B - A$.   If $A \in \mathfrak{R}$, then clearly, $\mathcal{C}(A)$ contains $\mathfrak{R}$, and since $\mathcal{C}(A)$ is closed, it follows that $\mathcal{C}(A) = \mathfrak{R}^c$.   Hence, as noted above, $A \in \mathcal{C}(B)$ for every A in $\mathfrak{R}$ and all B in $\mathfrak{R}^c$.   Thus $\mathcal{C}(B) \supset \mathfrak{R}^c$ for every B in $\mathfrak{R}^c$, and consequently $\mathfrak{R}^c$ is a ring.

> DEFINITIONS   A subset of a given set S is said to be *measurable with respect to a ring* $\mathfrak{R}$ of basic measurable subsets of S if it is contained in the σ-ring $\mathfrak{R}^c$ generated by $\mathfrak{R}$.   A set is *locally measurable with respect to* $\mathfrak{R}$ if it is measurable with respect to the ring $\mathfrak{R}^+$ of basic locally measurable sets.

> **THEOREM 3.1**   *Let S be a set, and $\mathfrak{R}$ a ring of subsets of S.   Then an extended-real-valued function f on S is measurable with respect to $\mathfrak{R}$ if and only if $[x : f(x) > a]$ and $[x : f(x) < -a]$ are measurable with respect to $\mathfrak{R}$ for every real $a > 0$.*

Proof   Let $\mathbf{M}'$ be the class of measurable functions f for which $[x : f(x) > a]$ belongs to $\mathfrak{R}^c$ for all real $a > 0$.   Then $\mathbf{M}'$ obviously contains the step functions.   Next we shall show that $\mathbf{M}'$ is closed under pointwise convergence of sequences.

Suppose $f_1, f_2, \ldots$ is a sequence of functions in $\mathbf{M}'$ which converges pointwise to a function f and that a is a positive real number.   If x is a point for which $f(x) > a$, then there is a positive integer k such that $f(x) > a + 1/k$. Since $f_n(x) \to f(x)$, there are at the same time infinitely many n such that $f_n(x) > a + 1/k$.   Conversely, if x is a point for which $f_n(x) > a + 1/k$ for infinitely many n, with k a positive integer, then certainly $f(x) \geq a + 1/k > a$. Thus

$$[x : f(x) > a] = \bigcup_{k \,:\, 1} \bigcap_{j \,>\, 1} \bigcup_{n \,\geq\, j} [x : f_n(x) > a + 1/k].$$

By assumption, $[x : f_n(x) > a + 1/k]$ is an element of $\mathfrak{R}^c$.   Since $\mathfrak{R}^c$ is closed under countable unions and intersections, therefore it follows that $\mathfrak{R}^c$ contains $[x : f(x) > a]$.   Thus $\mathbf{M}'$ is closed under pointwise convergence of sequences. Thus $\mathbf{M}'$ is the class $\mathbf{M}$ of all extended-real-valued functions which are measurable with respect to $\mathfrak{R}$.   Since the measurability of f implies that of $-f$, it follows immediately that $[x : f(x) < -a]$ also belongs to $\mathfrak{R}^c$ for all f in $\mathbf{M}$.

One simple consequence of the preceding deduction is that $\Re^c$ contains all subsets of S whose characteristic functions are measurable with respect to $\Re$. The converse, that is, the fact that the characteristic function of every set in $\Re^c$ is measurable as an extended-real-valued function with respect to $\Re$ is an immediate corollary of Scholium 3.8 and the remarks following Scholium 3.3.

Now suppose f is an extended-real-valued function on S such that $[x:f(x) > a]$ and $[x:f(x) < -a]$ belong to $\Re^c$ for all real $a > 0$. Then, by Example 3.2.2, f is the pointwise limit of a sequence g_1, g_2, $\ldots$ of step functions based on the ring generated by these sets. From Scholium 3.7 and the preceding paragraph, it follows that each g_n is measurable with respect to $\Re$. Hence f is also measurable with respect to $\Re$. This completes the proof.

The proof shows slightly more than was stated. In fact, the following corollary is an immediate consequence of the theorem, its proof, and the results given in Example 3.2.2.

COROLLARY 3.2.1 *An extended-real-valued function f on S is measurable with respect to a σ-ring $\Re$ of subsets of S if and only if it is the pointwise limit of a sequence g_1, g_2, $\ldots$ of step functions based on $\Re$ such that*

$$(1) \qquad\qquad |g_n(x)| \leq |f(x)|, \; n = 1, 2, \ldots$$

for all x in S, and

(2) g_n converges to f uniformly on every set where f is bounded by a positive real constant.

COROLLARY 3.2.2 *For any collection of extended-real-valued functions on a set, there is a unique smallest σ-ring of subsets with respect to which all functions in the collection are measurable.*

Proof Let $\mathbf{C}$ be a given collection of extended-real-valued functions on a set S and $\mathfrak{U}$ the σ-ring generated by the sets $[x:f(x) > a]$ and $[x:f(x) < -a]$, where f is arbitrary in $\mathbf{C}$ and a is an arbitrary positive real number. Then the functions in $\mathbf{C}$ are all measurable with respect to $\mathfrak{U}$. On the other hand, if $\Re$ is any σ-ring with respect to which the functions in $\mathbf{C}$ are measurable, then $\Re$ contains the generators of $\mathfrak{U}$, and hence $\mathfrak{U}$ itself.

DEFINITION The ring $\mathfrak{U}$ described in the proof of the corollary will be called the *σ-ring determined by* $\mathbf{C}$.

There is one other point which is immediate but important enough to be mentioned explicitly.

COROLLARY 3.2.3 *If $\Re_1$ and $\Re_2$ are rings of subsets of S, the class of all extended-real-valued functions on S measurable with respect to $\Re_1$ coincides with the class measurable with respect to $\Re_2$ if and only if $\Re_1$ and $\Re_2$ generate the same σ-ring.*

EXERCISES

1   A *Borel set* in a topological space S is an element in the σ-ring $\mathcal{B}$ generated by the open subsets of S.   If f is a continuous function from S to $[-\infty, \infty]$, show that f is Borel-measurable, i.e., that f is measurable with respect to $\mathcal{B}$.   Next show that a Baire function from S to the extended reals is also Borel-measurable.

2   *a.* If $(S, \mathcal{R}, m)$ is the n-fold direct product of one-dimensional basic Lebesgue-Stieltjes measure spaces, show that a continuous function from S to the extended reals is measurable with respect to $\mathcal{R}$.

   b. Show that $\mathcal{R}$ generates the σ-ring of all Borel subsets of S.

3   *a.* Let f be a function from a set S to a metrizable space Y with distinguished point 0.   If f is measurable with respect to a ring $\mathcal{R}$ of basic measurable subsets of S, show that $f^{-1}(N)$ is measurable for every open set N which does not contain 0.   (*Hint:* Show that an open set N is the union of an increasing sequence N_1, N_2, ... of open sets whose closures are contained in N, and generalize part of the proof of Theorem 3.1.)

   b. Formulate and prove the analog of (*a*) for locally measurable functions.

   c. If f is any function from S to Y, show that the set of subsets B of Y such that $f^{-1}(B)$ is measurable is a σ-ring (which may of course be trivial).   Show that the set of B such that $f^{-1}(B)$ is locally measurable is a σ-ring with unit, i.e., contains Y.

   d. Deduce that for any locally measurable function f and for any Borel subset B, $f^{-1}(B)$ is locally measurable.   Formulate and prove the corresponding result for measurable functions.

4   Let $\mathcal{R}$ be a ring of subsets of a set S, and E a set in $\mathcal{R}$, and $\mathcal{R}'$ the ring $E \cap \mathcal{R}$ consisting of all sets of the form $E \cap A$ with A in $\mathcal{R}$.   If f is a function defined on E which is measurable with respect to $\mathcal{R}'$, show that the function g defined as f on E and as 0 on the complement of E is measurable with respect to $\mathcal{R}$.

5   If S is a locally compact Hausdorff space with a countable base for open sets, show that the following are identical:

   a. The σ-ring determined by the continuous real-valued functions on S with compact supports;

   b. The σ-ring generated by the compact subsets of S;

   c. The σ-ring of Borel sets.

6   Let $\mathcal{R}$ be a given ring of subsets of a set S.   Show that the class of complex-valued functions on S measurable with respect to $\mathcal{R}$ forms an ideal in the algebra of all locally measurable such functions.

3.3   THE INTEGRAL

If we are given a basic measure space and a numerical function f on the space, we want to define an integral $\int f$ of the function over the space in a maximally natural and analytically viable fashion.   The integral should be linear as a function of f, and nonnegative on nonnegative functions; and should of course relate naturally to the given measure m, which is to say that if f is the characteristic function of a set in the ring of the basic space, then $\int f = m(E)$. In addition, one wants some sort of continuity condition, and the properties of the integral on step functions suggest specifically the desideratum

$$\int \lim_n f_n = \lim_n \int f_n$$

in case $\{f_n\}$ is a monotone-increasing sequence of nonnegative functions which is pointwise-convergent to f.

It is in a way extraordinary that these few desiderata suffice to determine the integral uniquely for a natural class of functions embracing all that are analytically reasonable, and that the resulting integral has in addition a great many useful properties. This integral may be called the *general (abstract) Lebesgue integral.*

The original work of Lebesgue (1902) involved primarily the length function on the reals, which provides the most important single example, and the germ from which the general theory grew; moreover, it is to a large extent typical, in the sense that a method due originally to Lebesgue and F. Riesz for setting up an equivalence between fairly general measure spaces and the one indicated on the reals permits the basic general theorems to be carried over directly. However, certain features which may be quite important (e.g., group invariance) may be lost in this way, and the theory lacks a certain logical directness, as well as inner completeness. The extension of Lebesgue's work to more general measures in euclidean spaces by Radon, Young, F. Riesz, and Lebesgue himself (ca. 1910) led to the consideration of integration in abstract spaces by E. H. Moore (1912) and by Frechet (1915). In 1918, Daniell published an elegant and general construction of virtually definitive simplicity for an abstract integral satisfying the desiderata indicated above. In our presentation of the integral we shall essentially follow the method of Daniell as have many authors in recent decades; however, certain modifications are desirable, inasmuch as Daniell's work is slightly ambiguous in its treatment of infinite-valued functions, and depends for its full justification on the negligibility of null sets.

It should be mentioned that the concrete needs of other parts of mathematics have more recently led to somewhat different formulations of integration theory, involving such notions as regular (Radon) measures in locally compact spaces, or that of an integration algebra. The development of the theory that arises from these concepts, which will be treated later, is different from the classical one, but in spirit is quite similar to that of Daniell.

Although these formulations provide in principle methods for bypassing the more classical approach, it has not yet been found generally convenient to do so, and however developments may take place in the future, the present theory will remain of basic importance for the classical theory of real functions.

The fundamental existence theorem for the integral may be stated in the following fashion.

THEOREM 3.2   *For any basic measure space there exists a unique additive functional I, the integral, from the nonnegative measurable functions to the nonnegative extended reals, which assigns to the characteristic function of any measurable set its measure, and has the property that if $f_1, f_2, \ldots$ is any monotone-increasing sequence of such functions, then $\lim_n I(f_n) = I(\lim_n f_n)$.*

The method of Daniell actually applies to a more general situation than that covered by the theorem.   It is therefore appropriate to prove a stronger result, which includes the above as a special case.   For this we shall need to make a number of definitions.

DEFINITIONS   A *function lattice* is a collection L of extended-real-valued functions defined on a set S which is a lattice with respect to the usual partial ordering on functions.   This means that if f and g are in L, so also are $\max(f,g)$ and $\min(f,g)$.   A given function lattice need not contain any constant functions; however, the lattices of interest in the theory of integration satisfy a kind of substitute condition.   We shall call these lattices *admissible*, and the condition is the following: For every f in L and every real number $c > 0$, the functions $\min(f,c)$ and $\max(f,-c)$ are also in L.   By a *linear function lattice* we shall mean a lattice L of finite-valued functions, which, in addition, is a vector space over the reals with respect to the usual operations.   The final concept is that of an *integration lattice*, defined as a system (S,L,J) composed of a set S, an admissible linear function lattice L on S, and a real-valued linear functional J on L, which is *monotone* and *continuous with respect to increasing sequences*.   This means that $J(f) = \lim_n J(f_n)$ if $f_1, f_2, \ldots$ is a monotone-increasing sequence of functions in L whose limit f is also in L.

Example 3.3.1   Let M be the class of measurable extended-real-valued functions on a basic measure space $(S,\mathcal{R},m)$.   Then M is an admissible function lattice by Scholium 3.6.   The class of finite-valued functions in M is an admissible linear function lattice, by Scholium 3.7.   Finally, if L is the collection of real-valued step functions on $(S,\mathcal{R},m)$ and I the integral, the results of Chap. 2, notably, Scholium 2.2, show that (S,L,I) is an integration lattice.

Example 3.3.2   Let S be a locally compact Hausdorff space, and L the collection of continuous real-valued functions on S with compact supports.   Then it is elementary to verify that L is an admissible linear function lattice.

Suppose J is a real-valued linear functional on **L** such that $J(f) \geq 0$ when $f \geq 0$. Then, evidently, $J(f) \leq J(g)$ when f and g belong to **L** and $f \leq g$. In addition, for each compact subset K of S, there exists a constant C_K such that

$$|J(f)| \leq C_K \|f\|_\infty$$

for all f in **L** with supports in K, where

$$\|f\|_\infty = \sup_x |f(x)|.$$

To see that this is true, choose an element g in **L** that is nonnegative and assumes the value 1 on K. That such a g exists is a fundamental property of locally compact spaces. Then $\pm f \leq \|f\|_\infty g$ for all f with supports in K, so that $\pm J(f) \leq \|f\|_\infty J(g)$. Thus we may take $C_K = J(g)$.

Next suppose $f_1, f_2, \ldots$ is a monotone-increasing sequence of elements of **L** whose limit f is also in **L**. We may assume that $f_n \geq 0$. Let K be a compact set containing the support of f. Then since $f_1, f_2, \ldots$ is monotone-increasing, K also contains the supports of all the f_n. It now follows by a classical result, namely, *Dini's theorem*, that *the sequence converges uniformly to f*. To prove this, let $\epsilon > 0$ and set

$$K_{n,\epsilon} = [x \in K : f(x) - f_n(x) \geq \epsilon].$$

Then $K_{n,\epsilon}$ is compact, $K_{n,\epsilon} \supset K_{n+1,\epsilon}$, and

$$\bigcap_n K_{n,\epsilon} = \emptyset.$$

Thus a finite intersection of the $K_{n,\epsilon}$ is empty, that is, $|f(x) - f_n(x)| < \epsilon$ for all x provided n is sufficiently large. Hence $\|f - f_n\|_\infty \to 0$, and since

$$|J(f) - J(f_n)| \leq C_K \|f - f_n\|_\infty,$$

it follows that $J(f_n) \to J(f)$, which shows that $(S,\mathbf{L},J)$ is an integration lattice.

The existence theorem for the integral may now be formulated in terms of integration lattices as follows: The terminology "continuous with respect to increasing sequences," applied to a function F from an ordered topological space to another topological space, means that $F(x_n) \to F(x)$ whenever $\{x_n\}$ is a monotone-increasing sequence which is convergent to x.

THEOREM 3.3 *For any integration lattice $(S,\mathbf{L},J)$ there exists a unique additive functional I, the integral, from the nonnegative functions measurable with respect to the σ-ring determined by **L** to the nonnegative extended reals, which agrees with J on the nonnegative elements of **L** and is continuous with respect to increasing sequences.*

In the lemmas that follow we shall be dealing throughout with a given integration lattice $(S,\mathbf{L},J)$ which is fixed once and for all.

Lemma 3.3.1   *The limit of a monotone-increasing sequence of functions in* **L** *always exists as an extended-real-valued function.   Let* **L**′ *be the set of all such limits.   Then* **L**′ *is an admissible function lattice containing* **L**.   *If* f *and* g *are in* **L**′ *and* $c \geq 0$, *then* cf *and* $f + g$ *are also in* **L**′.   *The limit of an increasing sequence of functions in* **L**′ *is again in* **L**′.   *On* **L**′ *there is a unique functional* $J′$, *assuming finite real values or the value* $+\infty$, *that extends* J *and is continuous with respect to increasing sequences.   In addition,* $J′$ *is monotone and* $J′(cf + g) = cJ′(f) + J′(g)$ *when* $c \geq 0$.

Proof   The fact that **L**′ is an admissible function lattice containing **L** and has the indicated algebraic properties is for the most part a matter of simple verification.   Perhaps the only point worth mentioning in this connection is that the functions in **L**′ never assume the value $-\infty$, so that the sum of two functions in **L**′ is always defined.

It is more difficult to show the existence of the functional $J′$.   For this, suppose f is an element of **L**′.   Then there is an increasing sequence $f_1, f_2, \ldots$ of functions in **L** which converges to f.   The corresponding sequence $J(f_1)$, $J(f_2), \ldots$ of real numbers is monotone-increasing and converges to a finite real number or to the value $+\infty$.   If the functional $J′$ exists, it is obviously unique.   For it is necessary that $J′(f) = \lim_n J(f_n)$.   We shall show next that $\lim_n J(f_n)$ depends only on f, and not on the particular sequence $f_1, f_2, \ldots$.   Let g be an element of **L**′ such that $f \leq g$, and $g_1, g_2, \ldots$ an increasing sequence of functions in **L** which converges to g.   Set $h_{ni} = \min(f_n, g_i)$.   Then $h_{ni} \in$ **L**, $h_{ni} \leq g_i$, and $h_{n1}, h_{n2}, \ldots$ is an increasing sequence with limit f_n.   Since J is monotone and continuous with respect to increasing sequences, it follows that

$$J(f_n) = \lim_i J(h_{ni}) \leq \lim_i J(g_i)$$

for all n.   Thus it is also true that $\lim_n J(f_n) \leq \lim_i J(g_i)$.   If $f = g$, it follows by symmetry that the reverse inequality is also valid, and hence that

$$\lim_n J(f_n) = \lim_i J(g_i).$$

This shows that on **L**′ there exists a single-valued functional $J′$ such that

$$J′(f) = \lim_n J(f_n)$$

whenever $f_1, f_2, \ldots$ is a monotone-increasing sequence of functions in **L** which converges pointwise to f.   It is clear from the last inequality above that $J′(f) \leq J′(g)$ when $f \leq g$ and straightforward to verify that $J′(cf + g) = cJ′(f) + J′(g)$ for $c \geq 0$.

Next suppose f is the limit of an increasing sequence $f_1, f_2, \ldots$ of functions in **L**′.   Then for each n, there exists an increasing sequence $f_{n1}, f_{n2}, \ldots$ of elements of **L** such that $f_n = \lim_i f_{ni}$.   Let $g_i = \max(f_{1i}, f_{2i}, \ldots, f_{ii})$.

Then $g_1, g_2, \ldots$ is an increasing sequence of functions in **L**, and

$$f_{ni} \leq g_i \leq f_i, \qquad 1 \leq n \leq i.$$

It follows that $f_n \leq \lim_i g_i \leq f$.   Since this is true for all n, it follows in turn that $f = \lim_i g_i$.   Hence $f \in \mathbf{L}'$ and $J'(f) = \lim_i J(g_i)$.   On the other hand, J and J' are monotone, and J' agrees with J on **L** because J is continuous with respect to increasing sequences.   Thus, by the inequality above,

$$J(f_{ni}) \leq J(g_i) \leq J'(f_i), \qquad 1 \leq n \leq i.$$

Hence $J'(f_n) \leq \lim_i J(g_i) \leq \lim_i J'(f_i)$, so that $\lim_n J'(f_n) \leq J'(f) \leq \lim_i J'(f_i)$. This shows that J' is continuous with respect to increasing sequences, and concludes the proof.

In general, **L**$'$ is considerably larger than **L**, but in most cases it contains only a small fraction, roughly speaking, of the class **M**$^+$ of nonnegative measurable functions, and hence also needs to be extended. We shall construct a new class **Q** which contains both **L**$'$ and **M**$^+$, and a functional I on **Q** which extends J'.   It turns out that the functional I, when restricted to **M**$^+$, will then be the integral, with the properties stated in the theorem. In fact, I will be additive and continuous with respect to increasing sequences on the class **Q**$^+$ of nonnegative elements of **Q**.   But in general, **Q**$^+$ is much larger than **M**$^+$, and the extension of J with the indicated properties is only unique on **M**$^+$.

DEFINITIONS   If f is any extended-real-valued function on S, let

$$I^+(f) = \inf\,[J'(g) : f \leq g \text{ and } g \in \mathbf{L}']$$

when such g exist, and otherwise set $I^+(f) = +\infty$.

Define $I^-(f)$ as $-I^+(-f)$, and let **Q** be the set of all f for which $I^+(f) = I^-(f)$.   For f in **Q**, let $I(f)$ denote the common value of $I^+(f)$ and $I^-(f)$.

Lemma 3.3.2   *If f is an extended-real-valued function on S, then $I^-(f) \leq I^+(f)$.*

Proof   Let f be an extended-real-valued function on S.   If $I^-(f) = -\infty$ or $I^+(f) = +\infty$, the inequality is certainly true.   Thus we may suppose $I^-(f) > -\infty$ and that $I^+(f) < +\infty$.   Then there exists elements g and h in **L**$'$ such that $-f \leq g, f \leq h, J'(g) < +\infty$, and $J'(h) < +\infty$.   Let g and h be such functions.   Then $I^+(-f) \leq J'(g)$ and $I^+(f) \leq J'(h)$.   Moreover, the relations $-f \leq g$ and $f \leq h$ imply the fact that f assumes finite values at all points where g and h are both finite.   Since g and h never assume the value $-\infty$, it follows that $g + h \geq 0$.   By Lemma 3.3.1, $g + h \in \mathbf{L}'$, and

$$J'(g) + J'(h) = J'(g + h) \geq J'(0) = 0.$$

Thus $-J'(g) \le J'(h)$ for all g and h with the above properties. Next note that, in general,

$$I^-(f) = \sup\,[-J'(g)\colon -f \le g \quad \text{and} \quad g \in \mathbf{L}'].$$

Thus, if h is fixed, it follows that

$$I^-(f) \le J'(h).$$

Since this is true for every h, it follows in turn that $I^-(f) \le I^+(f)$.

Lemma 3.3.3 $\mathbf{Q}$ *contains* $\mathbf{L}'$, *and* I *agrees with* J' *on* $\mathbf{L}'$.

Proof   Suppose f is a function in $\mathbf{L}'$. Then $f \le f$, and as J' is monotone, it is immediate that $I^+(f) = J'(f)$. In addition, there is an increasing sequence $g_1, g_2, \ldots$ of functions in $\mathbf{L}$ such that $f = \lim_n g_n$. Thus $-f \le -g_n$, and $I^+(-f) \le J'(-g_n) = -J(g_n)$. Hence $J(g_n) \le I^-(f)$ for every n. Taking the limit on n, we see that

$$J'(f) = I^+(f) \le I^-(f).$$

On the other hand, $I^-(f) \le I^+(f)$, by Lemma 3.3.2. Thus $f \in \mathbf{Q}$ and $I(f) = J'(f)$.

Lemma 3.3.4   *If* $f \in \mathbf{Q}$ *and* c *is a real number, then* $cf \in \mathbf{Q}$ *and* $I(cf) = cI(f)$.

Proof   First note that $0f = 0$. Thus, for $f \in \mathbf{Q}$, $0f \in \mathbf{Q}$, and $I(0f) = J'(0) = 0 = 0I(f)$.

Next suppose $c > 0$. Then it is easily verified that $I^+(cf) = cI^+(f)$ for any f. Hence $I^-(cf) = -I^+[c(-f)] = c[-I^+(-f)] = cI^-(f)$. If $I^+(f) = I^-(f)$, it follows at once that $cf \in \mathbf{Q}$ and that $I(cf) = cI(f)$. Moreover, $I(cf) = -I^+(-cf)$, so that $-I(cf) = I^+(-cf) = -I^+[-(-cf)] = I^-(-cf)$. Thus $-cf \in \mathbf{Q}$ and $I(-cf) = -I(cf) = -cI(f)$, which proves the lemma.

Lemma 3.3.5   *Suppose that* f *and* g *are in* $\mathbf{Q}$ *and that both* $f + g$ *and* $I(f) + I(g)$ *are defined.   Then* $f + g \in \mathbf{Q}$ *and* $I(f + g) = I(f) + I(g)$.   *In addition, if* $I(f)$ *and* $I(g)$ *are finite, then* $\max\,(f,g)$ *and* $\min\,(f,g)$ *are in* $\mathbf{Q}$ *and* $I[\max\,(f,g)] + I[\min\,(f,g)] = I(f) + I(g)$.

Proof   We shall show first that $I^+(f + g) \le I(f) + I(g)$. This is obvious if $I(f) = +\infty$ or $I(g) = +\infty$. Thus we may assume $I(f) < +\infty$ and $I(g) < +\infty$. Then there exist f_1 and g_1 in $\mathbf{L}'$ such that $f \le f_1$, $g \le g_1$, $J'(f_1) < +\infty$, and $J'(g_1) < +\infty$. Let f_1 and g_1 be any such functions. Then

$$I(f) + I(g) \le J'(f_1) + J'(g_1).$$

On the other hand, $f + g \le f_1 + g_1$ and $J'(f_1 + g_1) = J'(f_1) + J'(g_1)$. Thus

$$I^+(f + g) \le J'(f_1) + J'(g_1).$$

This is true for all f_1 and g_1 as above; hence

$$I^+(f + g) \le I(f) + I(g).$$

Now by Lemma 3.3.4, $-f$ and $-g$ also satisfy the conditions of the present lemma. Therefore, by what has just been shown,

$$I^+(-f - g) \le I(-f) + I(-g).$$

This may be written in the equivalent form

$$I(f) + I(g) \le I^-(f + g).$$

When these inequalities are combined with the result of Lemma 3.3.2, it follows that $I^-(f + g) = I^+(f + g)$, that $f + g \in \mathbf{Q}$, and also that $I(f + g) = I(f) + I(g)$.

Next suppose that $I(f)$ and $I(g)$ are finite. Then there exist f_1 and g_1 in $\mathbf{L}'$ such that $f \le f_1, g \le g_1, J'(f_1) < +\infty$, and $J'(g_1) < +\infty$. For any such f_1 and g_1, $\max(f_1,g_1) + \min(f_1,g_1) = f_1 + g_1$, and

$$J'[\max(f_1,g_1)] + J'[\min(f_1,g_1)] = J'(f_1) + J'(g_1).$$

This implies the inequality

$$I^+[\max(f,g)] + I^+[\min(f,g)] \le J'(f_1) + J'(g_1).$$

Since this is valid for all f_1 and g_1 as above, it follows that

$$I^+[\max(f,g)] + I^+[\min(f,g)] \le I(f) + I(g).$$

The inequality just obtained also applies to $-f$ and $-g$, and since $\max(-f,-g) = -\min(f,g)$ and $\min(-f,-g) = -\max(f,g)$, it follows that

$$I(f) + I(g) \le I^-[\max(f,g)] + I^-[\min(f,g)].$$

The final statement of the lemma is a consequence of these inequalities and Lemma 3.3.2.

Lemma 3.3.6 *Suppose* $f \in \mathbf{Q}$ *and* $I(f)$ *is finite. Then* $\min(f,a) \in \mathbf{Q}$ *and* $-\infty < I[\min(f,a)] \le I(f)$ *for all real* $a \ge 0$.

Proof Let g and h be functions in $\mathbf{L}'$ such that $-f \le g, f \le h, J'(g) < +\infty$, and $J'(h) < +\infty$. Then it follows, as in the proof of Lemma 3.3.2, that $g + h \ge 0$. Thus, for real a,

$$\max(g,-a) + \min(h,a) \le g + h.$$

Suppose a is real and $a \geq 0$. Then max $(g,-a)$ and min (h,a) are in $\mathbf{L}'$, and

$$J'[\max (g,-a)] + J'[\min (h,a)] \leq J'(g) + J'(h)$$

by Lemma 3.3.1. This shows in particular that the terms on the left side of the inequality are finite. Since max $(-f,-a) \leq$ max $(g,-a)$, it follows that $I^+[\max (-f,-a)] < +\infty$. On the other hand, $-\min (f,a) = \max (-f,-a)$. Thus $-\infty < I^-[\min (f,a)] \leq I^+[\min (f,a)]$ and at the same time, $I^+[\min (f,a)] \leq J'[\min (h,a)] < +\infty$. Hence we have the inequality

$$I^+[\min (f,a)] - I^-[\min (f,a)] \leq J'(g) + J'(h)$$

valid for all g and h subject to the above conditions. This implies the fact that

$$I^+[\min (f,a)] - I^-[\min (f,a)] \leq 0,$$

and completes the proof.

Lemma 3.3.7 *Let f be the limit of an increasing sequence $f_1, f_2, \ldots$ of everywhere-finite elements of $\mathbf{Q}$ with $I(f_1) > -\infty$. Then $f \in \mathbf{Q}$ and $I(f_n) \uparrow I(f)$.*

Proof It is clear from the definitions that I^+, I^-, and I are monotone functionals. Thus $I(f_n) = I^-(f_n) \leq I^-(f)$, and the sequence $I(f_1), I(f_2), \ldots$ is monotone-increasing. Thus $\lim\limits_n I(f_n) \leq I^-(f) \leq I^+(f)$. If $\lim\limits_n I(f_n) = +\infty$, there is nothing more to prove. Suppose that $\lim\limits_n I(f_n) < +\infty$. Then $-\infty < I(f_1) \leq I(f_n) < +\infty$ for every n. Let $f_0(x) = 0$ for all x. Then $f_n - f_{n-1} \in \mathbf{Q}$ and $I(f_n - f_{n-1}) = I(f_n) - I(f_{n-1})$, by Lemmas 3.3.4 and 3.3.5. Let ϵ be an arbitrary positive number. Then, since $I(f_n - f_{n-1})$ is finite, there exists an element g_n of $\mathbf{L}'$ such that $f_n - f_{n-1} \leq g_n$ and $J'(g_n) \leq I(f_n - f_{n-1}) + \epsilon/2^n$. Let $h_n = g_1 + g_2 + \cdots + g_n$. Then $h_n \in \mathbf{L}'$, and as each g_n is nonnegative, $h_1, h_2, \ldots$ is an increasing sequence. If $h = \lim\limits_n h_n$, it follows from Lemma 3.3.1 that $h \in \mathbf{L}'$. On the other hand,

$$f = \sum_n (f_n - f_{n-1}) \leq \sum_n g_n = h.$$

Thus $I^+(f) \leq J'(h) = \lim\limits_n J'(h_n)$. Now

$$J'(h_n) = \sum_{i \leq n} J'(g_i) \leq \sum_{i \leq n} I(f_i) - I(f_{i-1}) + \epsilon/2^i.$$

Thus $J'(h_n) \leq I(f_n) + \epsilon$ for all n, and

$$I^+(f) \leq \lim\limits_n I(f_n) + \epsilon$$

for every $\epsilon > 0$; hence $I^+(f) \leq \lim\limits_n I(f_n)$. This inequality, together with the one obtained at the beginning of the proof, establishes the result.

Next we shall show that the conclusion of Lemma 3.3.7 is valid under weaker assumptions.

Lemma 3.3.8 *Let f be the limit of an increasing sequence $f_1, f_2, \ldots$ of elements in $\mathbf{Q}$, with f_1 everywhere finite and $I(f_1) > -\infty$. Then $f \in \mathbf{Q}$ and $I(f_n) \uparrow I(f)$.*

Proof We may assume, as in the preceding lemma, that $\lim_n I(f_n) < +\infty$, and hence that $I(f_n)$ is finite for all n. Then, by Lemma 3.3.6, $\min(f_n, a) \in \mathbf{Q}$ and $I[\min(f_n, a)]$ is finite for real $a \geq 0$. In addition, $\min(f_n, a)$ is finite-valued and $\min(f_n, a) \uparrow \min(f, a)$. Thus $\min(f, a) \in \mathbf{Q}$ and $I[\min(f, a)] = \sup_n I[\min(f_n, a)]$ by Lemma 3.3.7. Now let $a = 1, 2, \ldots$. Then $f = \lim_a \min(f, a)$, and by a second application of Lemma 3.3.7, $f \in \mathbf{Q}$ and $I(f) = \sup_a I[\min(f, a)]$. For the same reasons, $I(f_n) = \sup_a I[\min(f_n, a)]$. Thus

$$
\begin{aligned}
I(f) &= \sup_a \sup_n I[\min(f_n, a)] \\
&= \sup_{a,n} I[\min(f_n, a)] \\
&= \sup_n \sup_a I[\min(f_n, a)] \\
&= \sup_n I(f_n).
\end{aligned}
$$

Since $I(f_1), I(f_2), \ldots$ is a monotone-increasing sequence, this concludes the proof.

Lemma 3.3.9 *Suppose f is a nonnegative finite-valued element of $\mathbf{Q}$ for which $I(f)$ is finite. Let $\mathbf{Q}(f) = [g \in \mathbf{Q} : |g| \leq f]$. Then $\mathbf{Q}(f)$ is closed and I is continuous on $\mathbf{Q}(f)$ in the topology of sequential pointwise convergence.*

Proof First observe that, by Lemma 3.3.7, $\mathbf{Q}(f)$ contains the limit of any monotone-increasing sequence of its elements.

Now suppose g is the limit of an arbitrary convergent sequence $g_1, g_2, \ldots$ of functions in $\mathbf{Q}(f)$, and let $h_{ni} = \max(g_n, g_{n+1}, \ldots, g_{n+i})$. Then, obviously, $-f \leq g_n \leq h_{ni} \leq f$, and by Lemma 3.3.5, $h_{ni} \in \mathbf{Q}$. Thus $h_{ni} \in \mathbf{Q}(f)$, and as $h_{n1}, h_{n2}, \ldots$ is monotone-increasing, its limit h_n is also in $\mathbf{Q}(f)$. Since $h_n(x) = \sup_{i \geq n} g_i(x)$, it follows that $h_1, h_2, \ldots$ is a monotone-decreasing sequence which converges to g. Now $-h_n$ is also in $\mathbf{Q}(f)$, and $-h_1, -h_2, \ldots$ is an increasing sequence with limit $-g$. Thus $-g$, and hence g itself, belong to $\mathbf{Q}(f)$. In addition, it follows easily from Lemma 3.3.7 that $I(h_n) \downarrow I(g)$. This implies the inequality

$$
\limsup_n I(g_n) \leq I(g),
$$

for $g_n \le h_n$ and $I(g_n) \le I(h_n)$.   This inequality is also true for $-g_n$ and $-g$, and since $-\lim\sup_n I(-g_n) = \lim\inf_n I(g_n)$, it follows that

$$I(g) \le \lim\inf_n I(g_n) \le \lim\sup_n I(g_n) \le I(g).$$

This shows that $I(g) = \lim_n I(g_n)$, and concludes the proof.

Proof of Theorem 3.3   Let $\mathfrak{R}$ be the smallest ring containing all sets of the form $[x:f(x) > a]$ and $[x:f(x) < -a]$, where $0 < a < +\infty$ and f ranges over **L**.   It is clear from the definition following Corollary 3.2.2 that **L** and $\mathfrak{R}$ determine the same σ-ring $\mathcal{S}$.   By Corollary 3.2.3, the class of extended-real functions measurable with respect to $\mathfrak{R}$ is precisely the class measurable with respect to $\mathcal{S}$.

Since **L** is a linear function lattice, the generators of $\mathfrak{R}$ may be described more simply as the sets of the form $[x:f(x) > 1]$, where f ranges over the nonnegative elements of **L**.   If $0 \le f \in \mathbf{L}$, then $g = f - \min(f,1)$ also belongs to **L**, $g \ge 0$, and if $A = [x:g(x) > 0]$, then $A = [x:f(x) > 1]$.   Moreover, the characteristic function φ_A of A satisfies the relation $0 \le \varphi_A \le \min(f,1)$.   For $\min(f,1) > 0$ and $\min[f(x),1] = 1$ for x in A.   Thus

$$0 \le I^+(\varphi_A) \le J[\min(f,1)] < +\infty.$$

If g is an arbitrary nonnegative function in **L** and $A = [x:g(x) > 0]$, $I^+(\varphi_A)$ need not be finite, but in any event $\varphi_A \in \mathbf{Q}$; in fact, $\varphi_A \in \mathbf{L}'$.   For $\min(ng,1) \in \mathbf{L}$, where $n = 1, 2, \ldots$, and $\min(ng,1) \uparrow \varphi_A$.

Next we shall show that $\varphi_A \in \mathbf{Q}$ and that $I(\varphi_A)$ is finite for every A in $\mathfrak{R}$; however, the proof will actually show more.   Let $\mathbf{L}^*$ be the set of all finite-valued f in $\mathbf{Q}$ such that $I(f)$ is finite.   Then $(S,\mathbf{L}^*,I)$ is an integration lattice by Lemmas 3.3.4 to 3.3.7.   If A and B are sets whose characteristic functions belong to $\mathbf{L}^*$, then $\varphi_{A \cup B}$ and φ_{A-B} also belong to $\mathbf{L}^*$.   For $\varphi_{A \cup B} = \max(\varphi_A,\varphi_B)$ and $\varphi_{A-B} = \varphi_A - \min(\varphi_A,\varphi_B)$.   Thus the collection $\mathfrak{R}^*$ of all sets A such that $\varphi_A \in \mathbf{L}^*$ is a ring.   By the preceding paragraph, $\mathfrak{R}^*$ contains the generators of $\mathfrak{R}$, and hence $\mathfrak{R}$ itself.

Since $\mathbf{L}^*$ is a linear space, $\mathbf{L}^*$ contains all real-valued step functions based on $\mathfrak{R}^*$.   If $E \in \mathfrak{R}^*$ and $0 < c < +\infty$, let $\mathbf{M}^*(E,c)$ be the class of all functions measurable with respect to $\mathfrak{R}^*$, with supports in E and values in $[-c,c]$.   Then $\mathbf{M}^*(E,c)$ is contained in $\mathbf{L}^*$ by Scholium 3.4 and Lemma 3.3.9.

Now suppose f is an arbitrary nonnegative extended-real-valued function which is measurable with respect to $\mathfrak{R}^*$.   Then f vanishes outside a countable union of sets $A_1, A_2, \ldots$ in $\mathfrak{R}^*$.   If

$$E_n = \bigcup_{i \le n} A_i,$$

then $E_n \in \mathcal{R}^*$, and by Scholiums 3.2 and 3.6 the functions

$$f_n = \min(f_{E_n}, n), \qquad n = 1, 2, \ldots,$$

are also measurable with respect to $\mathcal{R}^*$. By the preceding paragraph, $f_n \in \mathbf{L}^*$. In addition, $f_1, f_2, \ldots$ is a monotone-increasing sequence converging pointwise to f; hence $f \in \mathbf{Q}$, by Lemma 3.3.7.

Next observe, by Lemma 3.3.8, that I is continuous with respect to arbitrary increasing sequences of nonnegative functions measurable with respect to $\mathcal{R}^*$. The additivity of I on such functions is a consequence of Lemma 3.3.5.

Since the class $\mathbf{M}^+$ of nonnegative functions measurable with respect to $\mathcal{R}$ is included in the corresponding class measurable with respect to $\mathcal{R}^*$, all that remains to be shown is the uniqueness of I on $\mathbf{M}^+$. Let I' be any functional satisfying the conclusion of the theorem. Then, as every nonnegative g in $\mathbf{L}'$ is the limit of an increasing sequence of nonnegative functions in $\mathbf{L}$, it follows by continuity that $I(g) = I'(g)$ for such g.

Next, if $E \in \mathcal{R}$ and $0 < c < +\infty$, let $\mathbf{M}^+(E,c)$ be the class of all functions measurable with respect to $\mathcal{R}$ with values in $[0,c]$ and supports in E. Suppose $f \in \mathbf{M}^+(E,c)$ and $\epsilon > 0$. Then there exists an element g in $\mathbf{L}'$ such that $f \leq g$ and

$$I(f) \leq I(g) \leq I(f) + \epsilon.$$

Since I' is additive and nonnegative on nonnegative measurable functions, I' is monotone as well. Thus $I'(f) \leq I'(g) = I(g)$. Since ϵ is arbitrary, it follows that $I'(f) \leq I(f)$. Let $h = c\varphi_E - f$. Then $I'(h) \leq I(h)$ for the same reasons. Thus $I'(f) + I'(h) \leq I(f) + I(h)$. But actually, here we have equality. For $I'(f) + I'(h) = I(c\varphi_E) = I(f) + I(h)$. Thus $I'(f) = I(f)$ and I and I' agree on $\mathbf{M}^+(E,c)$. Since every f in $\mathbf{M}^+$ is the limit of a monotone-increasing sequence of such functions, it follows that $I'(f) = I(f)$ for all f in $\mathbf{M}^+$. This finishes the proof.

EXERCISES

1   Let $(S,\mathbf{L},J)$ be an integration lattice in which the functions in $\mathbf{L}$ are bounded and such that

$$(*) \qquad\qquad |J(f)| \leq c \sup_x |f(x)|$$

for all f in $\mathbf{L}$ with c a fixed constant. Let $\mathbf{U}$ be the class of all limits of uniformly convergent sequences of functions in $\mathbf{L}$.

    a. If $f \in \mathbf{U}$ and $f_1, f_2, \ldots$ is any sequence of functions in $\mathbf{L}$ which converges uniformly to f, show that the numerical sequence $J(f_1), J(f_2), \ldots$ is convergent and that its limit depends only on f. Hence the equation $I(f) = \lim_n J(f_n)$ defines a functional I on $\mathbf{U}$.

    b. Show that $\mathbf{U}$ is an admissible linear lattice of bounded functions and that I is a linear functional on $\mathbf{U}$ which satisfies equation $(*)$.

   c. If $f_1, f_2, \ldots$ is a sequence of functions in **U** which converges uniformly to a function f, show that $f \in$ **U** and that $I(f) = \lim_n I(f_n)$.

   d. Show that **U** is contained in the lattice **L*** constructed in the proof of Theorem 3.3 and that the integral I on **L*** is an extension of the functional I just defined on **U**.

   e. Use (*d*) to show that $(S, \mathbf{U}, I)$ is an integration lattice, and try to prove this directly.

2   Construct an integration lattice $(S, \mathbf{L}, J)$ in which the functions in **L** are not all bounded.

3   Give an example of an integration lattice $(S, \mathbf{L}, J)$ in which the functions in **L** are all bounded but condition (*) of Exercise 1 is not satisfied.

4   Let $(S, \mathcal{R}, m)$ be a basic Lebesgue-Stieltjes measure space on the line in which S is a closed bounded interval. Let **L** be the lattice of real-valued step functions based on $\mathcal{R}$, and J the integral on **L**. Show that **L*** contains all continuous real-valued functions on S and that, for such a function f, $I(f)$ is given as a limit of Riemann sums, as in Sec. 2.3.

3.4   DEVELOPMENT OF THE INTEGRAL

A number of significant properties and extensions of the integral follow easily from Theorem 3.3 and the lemmas of the preceding section. This section develops certain of these properties and extensions that will be particularly useful.

First, it is appropriate to make several definitions in connection with results obtained in the proof of the theorem.

   DEFINITIONS   We shall call the integration lattice $(S, \mathbf{L}^*, I)$ the *completion* of the original integration lattice $(S, \mathbf{L}, J)$. Next, note that if $m(A)$ is defined as $I(\varphi_A)$ for A in $\mathcal{R}^*$, then m is a finite-valued measure on $\mathcal{R}^*$. Moreover, m is countably additive by the sequential continuity of I for increasing sequences. Thus the systems $M = (S, \mathcal{R}, m)$ and $M^* = (S, \mathcal{R}^*, m)$ are basic measure spaces. We shall say that M is the *basic space associated with* $(S, \mathbf{L}, J)$ and that M^* is its *completion*. If the lattice $(S, \mathbf{L}, J)$ arises from a basic measure space, as in Example 3.3.1, it should be observed that M is, then, just the original basic space. We shall say that a function f is *essentially measurable on* M if f is measurable on M^*. An *essentially measurable subset* of M is defined similarly as a subset of S which is measurable with respect to $\mathcal{R}^*$. A *null set on* M is an essentially measurable subset N such that $m(N) = I(\varphi_N) = 0$, while a *null function* is one which vanishes outside a null set. The term *almost everywhere* (abbreviation a.e.) means except on a null set. An extended-real-valued function f defined on S will be called *integrable on* M if $f \in \mathbf{Q}$ and $I(f)$ is finite. Finally, the *integral* of such an f is simply the value $I(f)$ of I on f.

COROLLARY 3.4.1   *An extended-real-valued function f on the basic measure space M is integrable on M if and only if it is essentially measurable and $I(|f|)$ is finite. The set on which such an integrable function assumes infinite values is a null set.   Any extended-real-valued function which agrees a.e. with an integrable such function is likewise integrable and has the same integral.   In particular, an extended-real-valued null function is integrable with integral 0.   Moreover, a nonnegative essentially measurable function with $I(f) = 0$ is a null function.*

Proof   Suppose f is integrable on M.   Let $f^+ = \max(f,0)$ and $f^- = (-f)^+$. Then f^+ and f^- are also integrable, and $I(f) = I(f^+) - I(f^-)$ by Lemmas 3.3.4 and 3.3.5.   Since $|f| = f^+ + f^-$, Lemmas 3.3.4 and 3.3.5 also show that $|f|$ is integrable with integral $I(|f|) = I(f^+) + I(f^-)$.   To prove that f is essentially measurable, it is enough to prove the essential measurability of f^+ and f^-.   In other words, we may assume $f \geq 0$.   Then, by Lemma 3.3.6, each of the functions

$$f_n = \min(f,n), \qquad n = 1, 2, \ldots,$$

belongs to $\mathbf{L}^*$.   Since f is the pointwise limit of the sequence $f_1, f_2, \ldots$, it suffices to show that nonnegative functions in $\mathbf{L}^*$ are measurable with respect to $\mathcal{R}^*$.   If f is such a function and $c > 0$, let $A = [x:f(x) > c]$.   Set $g = f - \min(f,c)$ and $g_n = \min(ng,1)$.   Then $g_n \in \mathbf{L}^*$ and $g_n \uparrow \varphi_A$.   Since $g_n \geq 0$, it follows from Lemma 3.3.7 that $\varphi_A \in \mathbf{Q}$.   Moreover, $0 \leq c\varphi_A \leq f$ and $0 \leq cI(\varphi_A) \leq I(f)$.   Thus $I(\varphi_A)$ is finite and $A \in \mathcal{R}^*$.   Hence $\mathcal{R}^*$ contains $[x:f(x) > c]$ for all real $c > 0$, and since $f \geq 0$, Theorem 3.1 implies that f is measurable with respect to $\mathcal{R}^*$.   This argument shows, incidentally, that $(S,\mathcal{R}^*,m)$ is the basic measure space associated with integration lattice $(S,\mathbf{L}^*,I)$.

Now suppose, conversely, that f is essentially measurable on M.   Then, by the lattice and algebraic properties of measurable functions, f^+, f^-, and $|f| = f^+ + f^-$ are also essentially measurable.   Since it was shown in the proof of Theorem 3.3 that $\mathbf{Q}$ contains all nonnegative essentially measurable functions, the quantities $I(f^+)$, $I(f^-)$, and $I(|f|)$ are all defined.   Suppose now, in addition, that $I(|f|)$ is finite.   Then, as $f^+ \leq |f|$ and $f^- \leq |f|$, it follows that $I(f^+)$ and $I(f^-)$ are finite as well.   Thus by Lemma 3.3.5, $f = f^+ - f^- \in \mathbf{Q}$ and $I(f) = I(f^+) - I(f^-)$, so that f is indeed integrable.

Suppose, next, that f is an extended-real-valued function which is integrable on M.   Let $A = [x:f(x) = +\infty]$ and $B = [x:f(x) = -\infty]$.   Then

$$A = \bigcap_{n \geq 1} [x:f(x) > n]$$

and
$$B = \bigcap_{n \geq 1} [x:f(x) < -n].$$

Since f is essentially measurable, it follows that A and B are essentially measurable subsets of M.   Now $0 \leq n\varphi_A \leq f^+$ and $0 \leq nI(\varphi_A) \leq I(f^+) < +\infty$ for $n = 1, 2, \ldots$.   Thus $I(\varphi_A) = 0$ and A is null set.   For similar reasons, B is also a null set.   If A and B are any null sets, then so is their union.   For $\varphi_{A \cup B} \leq \varphi_A + \varphi_B$, so that $m(A \cup B) \leq m(A) + m(B)$.   More generally, the countable subadditivity of the measure implies the fact that the union of any countable collection of null sets is again a null set.

Next, suppose that f is a nonnegative essentially measurable function on M for which $I(f) = 0$.   The characteristic function f_n of the set

$$A_n = [x : f(x) > n^{-1}], \qquad n = 1, 2, \ldots,$$

is then essentially measurable, and since $f \geq n^{-1}\varphi_{A_n}$, it follows that A_n is a null set.   As noted above, a countable union of null sets is again a null set.   Thus, from the fact that

$$[x : f(x) > 0] = \bigcup_n A_n$$

it results that f is a null function.

Now suppose f is an extended-real-valued function which vanishes outside a null set N.   We shall show that f is integrable with integral 0.   Since f^+ and f^- also vanish outside N, it is enough, in view of Lemma 3.3.5, to consider the case in which $f \geq 0$.   Let $f_n = \min(f, n)$ for $n = 1, 2, \ldots$.   Then

$$0 \leq I^-(f_n) \leq I^+(f_n) \leq nI(\varphi_N) = 0,$$

so that $f_n \in \mathbf{Q}$ and $I(f_n) = 0$.   Since $f_n \uparrow f$, it follows from Lemma 3.2.7 that $f \in \mathbf{Q}$ and $I(f) = 0$.   In particular, note that every null function is essentially measurable and that every subset of a null set is a null set.

Next, suppose $f = f'$ a.e. on $M = (S, \mathscr{R}, m)$, where f and f' are extended-real-valued functions and f' is integrable.   Then there is a null set N such that $f(x) = f'(x)$ for x outside N, and $S - N$ is automatically a locally measurable subset of M^*.   Since f' is measurable on M^*, it follows from Scholium 3.2 that

$$g = f'_{S-N}$$

is essentially measurable on M.   Now $|g| \leq |f'|$, $I(|g|) \leq I(|f'|) < +\infty$, and g is integrable by the first statement of the corollary.   Since $f = g + h$, where h is a null function supported by N, it follows from Lemma 3.3.5 and the preceding paragraph that f is integrable and that $I(f) = I(g)$.   At the same time, $f' = g + h'$, where h' is also a null function supported by N.   Thus $I(f') = I(g) = I(f)$, and this concludes the proof.

Next we shall prove a variant of the sequential continuity condition, which is similar to Lemma 3.3.9, and often useful.

COROLLARY 3.4.2 (FATOU'S LEMMA) *For any sequence* $f_1, f_2, \ldots$ *of nonnegative essentially measurable functions on a basic measure space,*

$$\liminf_n I(f_n) \geq I(\liminf_n f_n).$$

Proof If $g_n(x) = \inf_{i \geq n} f_i(x)$ for $n = 1, 2, \ldots$, then g_n is essentially measurable and nonnegative and $g_n \uparrow \liminf_n f_n$. Thus $I(g_n) \uparrow I(\liminf_n f_n)$. But at the same time $f_n \geq g_n$ and $I(f_n) \geq I(g_n)$, so that $\liminf_n I(f_n) \geq \liminf_n I(g_n)$.

At this point it is appropriate to consider the integration of complex-valued functions and, somewhat more generally, the integration of functions with values in a finite-dimensional euclidean space.

DEFINITION If $M = (S,\Re,m)$ is a basic measure space, we shall say that a function f from S to a finite-dimensional euclidean vector space V is *integrable on* M if f is essentially measurable on M and $I(|f|)$ is finite.

To see that this makes sense, note that the mapping $v \to |v|$ from a vector to its length is continuous from V to the nonnegative reals. Thus the essential measurability of f implies that of $|f|$. Here it is understood that the distinguished point of V is the vector 0.

COROLLARY 3.4.3 *Let V be a finite-dimensional euclidean space and M a basic measure space. Then the set of all integrable functions on M with values in V is a real vector space. On this space there exists a unique linear functional I such that*

$$I(fv) = I(f)v$$

for all v in V and all real-valued integrable functions f.

Proof Let f and g be measurable functions on M^* with values in V. Then the same proof as in the case of real-valued functions shows that $f + g$ is likewise measurable. Actually, the proof is simpler than that of Scholium 3.7, inasmuch as f and g do not assume infinite values. Since $|f + g| \leq |f| + |g|$, so that $I(|f + g|) \leq I(|f|) + I(|g|)$, it follows that $f + g$ is integrable provided both f and g are. If c is a real number, cf is also measurable, $|cf| \leq |c||f|$, and $I(|cf|) \leq |c| I(|f|)$. Thus the integrability of f implies that of cf, and the collection of integrable vector-valued functions is indeed a linear space over the reals.

Next, suppose that f is an arbitrary integrable function on M with values in V. Let $v_1, v_2, \ldots, v_n$ be a basis for V, and let f_i denote the component of f relative to v_i.   The components $f_1, f_2, \ldots, f_n$ are defined by the equation

$$f(x) = \sum_i f_i(x)v_i, \qquad x \in S.$$

Since the map taking a vector into any one of its coordinates, in the basis $v_1, v_2, \ldots, v_n$, is continuous, the measurability of f on M^* implies that of the f_i, and since $|f_i| \leq |f|$, each f_i is integrable.   Now let $I(f)$ be defined as $\sum_i I(f_i)v_i$.   If g is another integrable function with components g_i, then $cf + g$ has components $cf_i + g_i$, so that

$$I(cf + g) = \sum_i I(cf_i + g_i)v_i.$$

Since $I(cf_i + g_i) = cI(f_i) + I(g_i)$, it follows that $I(cf + g) = cI(f) + I(g)$. Thus I is a linear functional on the vector space of integrable maps from M to V.

If f is a real-valued integrable function and $v \in V$, then fv is the function whose value at a point x in S is the vector $f(x)v$.   Suppose that, in terms of the basis $v_1, v_2, \ldots, v_n$,

$$v = \sum_i c_i v_i.$$

Then $f(x)v = \sum_i c_i f(x)v_i$, and hence $I(fv) = \sum_i I(c_i f)v_i = I(f)v$.   This condition, together with linearity, determines I, and I is independent of the particular basis used in its construction.   For if f is integrable and has components $f'_1, f'_2, \ldots, f'_n$ in some other basis $v'_1, v'_2, \ldots, v'_n$, then $f = \sum_i f'_i v'_i$, and

$$I(f) = \sum_i I(f'_i v'_i) = \sum_i I(f'_i)v'_i.$$

DEFINITIONS   If f is an integrable vector-valued function on the basic measure space M, the vector $I(f)$ is called the *integral of f over M*, and is variously denoted as $I(f)$, $I_M(f)$, $\int_S f(x)\, dm(x)$, $\int f(x)\, dm(x)$, or even $\int f(x)\, dx$, depending on the needs of the context.   The *integral* of an integrable function *over a locally measurable subset E of M^** is defined by the equation

$$\int_E f(x)\, dm(x) = I(f_E).$$

This is meaningful, for f_E is measurable on M^*, by Scholium 3.2, and $|f_E| \leq |f|$.   The set function $E \to I(f_E)$ is called the *indefinite integral of f*.

In the special case in which one is dealing with complex-valued functions, it is easy to improve upon the result of Corollary 3.4.3.

COROLLARY 3.4.4   *The set of complex-valued integrable functions on a basic measure space is a vector space over the complex field.   On this space the integral I is a complex linear functional and has the additional property that*

$$|I(f)| \leq I(|f|)$$

for all integrable f.

Proof   Suppose f is a complex-valued integrable function on a basic measure space M.   Let $g = \mathrm{Re}\, f$ and $h = \mathrm{Im}\, f$.   Then g and h are real-valued and $f = g + ih$.   The proof of the preceding corollary shows that g and h are integrable and that $I(f) = I(g) + iI(h)$.   Now $if = -h + ig$ and since g and h are essentially measurable, it is readily proved that if is also essentially measurable.   Since $|if| = |f|$ and $I(|f|)$ is finite, it follows that if is an integrable complex-valued function on M.   Moreover, $I(if) = I(-h) + iI(g) = iI(f)$.   From this it is routine to verify that cf is integrable and that $I(cf) = cI(f)$ for any complex number c.   This proves the first part of the result.

To prove the inequality, choose a complex number c of absolute value 1 such that $|I(f)| = cI(f)$.   Let $cf = g + ih$, where g and h are real-valued.   Then, since

$$0 \leq cI(f) = I(cf) = I(g) + iI(h),$$

it follows that $I(h) = 0$.   Thus $|I(f)| = I(g)$, and since $|g| \leq |cf| = |f|$, so that $I(|g|) \leq I(|f|)$, it suffices to show that $I(g) \leq I(|g|)$.   But this is clear from the fact that $I(g) = I(g^+) - I(g^-)$, while $|g| = g^+ + g^-$ and $I(|g|) = I(g^+) + I(g^-)$.

With some additional effort, one could extend the result above to the case of integrable vector-valued functions in which the euclidean space is obtained from a complex inner-product space by restriction of the scalar field to the reals.   However, we shall leave this point to a more appropriate occasion.

The following extension of the integral is frequently useful.

DEFINITIONS   Suppose that $M = (S, \mathcal{R}, m)$ is a basic measure space and that f is a function defined on a subset of S.   We shall then say that f is an *essentially integrable function on M* if f differs only on a null set from an integrable vector-valued function g, whether undefined on part or all of this null set, or defined there in any fashion, e.g., having values outside the vector space in question.   The *integral $I(f)$* of such an f is then defined as $I(g)$.

To see that this integral is single-valued, first observe that different g's will themselves differ only on a null set.   Thus it is sufficient by linearity to prove that $I(g) = 0$ when g is integrable, vector-valued, and 0 almost everywhere. But this is evident from the construction given for $I(g)$.

The sequential continuity properties of the integral for nonnegative functions carry over to vector-valued functions in the following way.

COROLLARY 3.4.5   (DOMINATED CONVERGENCE THEOREM)   *If $f_1, f_2, \ldots$ is a sequence of essentially integrable vector-valued functions on a basic measure space which converges almost everywhere to a function f, and if there exists a fixed nonnegative integrable function h such that $|f_n(x)| \leq h(x)$ almost everywhere for all n, then f is essentially integrable, and*

$$I(f) = \lim_n I(f_n).$$

Proof   By the assumptions above, there exists a null set A on the basic space $M = (S, \Re, m)$ such that $f_n \to f$ at all points of $S - A$. In addition, there exist null sets A_n such that $f_n \leq h$ and f_n agrees with an integrable function on $S - A_n$. Then, in particular, it follows from Scholium 3.2 that

$$f'_n = (f_n)_{S-A_n}$$

is essentially measurable on M. Let N be the union of A and the A_n and $E = S - N$. If $g = f_E$ and $g_n = (f_n)_E$, then $g_n \to g$ on the whole space. For $A \subset N$ and $S - A \supset E$. Moreover, $g_n = (f'_n)_E$, and E is locally measurable on M^*, so that g_n is essentially measurable on M. Thus the same is true of g. Since $A_n \subset N$ and $S - A_n \supset E$, it follows that $|g_n| \leq h$ on the whole space. Hence

$$0 \leq I(|g_n|) \leq I(h) < +\infty.$$

This shows that g_n is integrable. Furthermore, since N is a null set, $g_n = f_n$ a.e. and $I(f_n) = I(g_n)$. By Fatou's lemma and the inequality above,

$$0 \leq I(|g|) \leq \lim_n \inf I(|g_n|) < +\infty,$$

so that g is also integrable. Thus f is essentially integrable, since it differs from g only on N, and $I(f) = I(g)$.

To show that $I(f_n) \to I(f)$, it therefore suffices to show that each coordinate of $I(g_n)$ converges to the corresponding coordinate of $I(g)$, relative to a fixed basis. Thus the result follows from the special case in which g_n and g are real-valued. Then $g_n + h \geq 0$, and Fatou's lemma implies

$$0 \leq I(g + h) \leq \lim_n \inf I(g_n + h).$$

Since $I(g + h) = I(g) + I(h)$ and $I(g_n + h) = I(g_n) + I(h)$, this implies

$$-\infty < I(g) \leq \lim_n \inf I(g_n).$$

On the other hand, the sequence $-g_1, -g_2, \ldots$ satisfies the same conditions as the original one, so that

$$- \infty < I(-g) \leq \liminf_n I(-g_n).$$

Now this is equivalent to the inequality

$$\infty > I(g) \geq \limsup_n I(g_n),$$

and since the lim inf never exceeds the lim sup, it follows that $\lim_n I(g_n)$ exists and equals $I(g)$.

EXERCISES

1   Let f be a nonnegative Borel measurable function on $[a,b]$, where $-\infty \leq a < b \leq \infty$. If $b < \infty$, show that

$$\int_a^b f(x)\, dx = \lim_{\epsilon \to 0+} \int_a^{b-\epsilon} f(x)\, dx,$$

where dx denotes the element of Lebesgue measure.   On the other hand, if $b = \infty$, show that

$$\int_a^b f(x)\, dx = \lim_{c \to +\infty} \int_a^c f(x)\, dx$$

2   Show the following:

a. $\displaystyle \int_0^1 \frac{dx}{x} = +\infty.$

b. $\displaystyle \int_0^1 x^{a-1}\, dx = \frac{1}{a}, \qquad 0 < a.$

c. $\displaystyle \int_1^\infty x^{a-1}\, dx = -\frac{1}{a}, \qquad a < 0.$

d. $\displaystyle \int_0^\infty e^{-ax}\, dx = \frac{1}{a}, \qquad a > 0.$

e. $\displaystyle \int_0^\infty \frac{x^3}{(x^9 + 1)^{\frac{1}{2}}}\, dx < \infty.$

f. $\displaystyle \int_2^\infty \frac{dx}{x \log x} = \infty.$

g. $\displaystyle \int_1^\infty \frac{\log x}{x}\, dx = \infty.$

h. $\displaystyle \int_1^\infty \frac{\log x}{x^{1+a}}\, dx < \infty, \qquad a > 0.$

i. $\displaystyle \int_0^1 \frac{dx}{-x \log x} = \infty.$

3 Show that $\lim\limits_{b\to\infty}\int_0^b \dfrac{\sin x}{x}\,dx$ exists and is finite, but that $\int_0^\infty \dfrac{|\sin x|}{x}\,dx = \infty$.

4 Show that $\lim\limits_{a\to 0+}\int_a^1 \dfrac{\sin\,(1/x)}{x^{\frac{3}{2}}}\,dx$ exists but that the integrand is not integrable over $(0,1)$.

5 If f is a Lebesgue integrable function on $(-\infty,\,\infty)$, find

$$\lim_{n\to\infty}\int_{-\infty}^\infty \left(\frac{\sin nx}{nx}\right)^2 f(x)\,dx.$$

6 Suppose f is defined and integrable relative to Lebesgue measure on the half-line $t > 0$. If

$$g(x) = \int_0^\infty \frac{f(t)}{x-t}\,dt, \qquad 0 < x < \infty,$$

is g a continuous function? Does g have a limit as $x \to \infty$, and if so, what is its value?

7 Suppose f is a Lebesgue-integrable function on $(-\infty,\infty)$ such that $f(0) = 0$ and f has a derivative at 0. Prove that $\int_{-\infty}^\infty \dfrac{f(x)}{x}\,dx$ exists.

8 Show that

$$\lim_{a\to 0+}\int_a^1 \left(2x \sin\frac{1}{x^2} - \frac{2}{x}\cos\frac{1}{x^2}\right) dx$$

exists, but that the integrand is not Lebesgue-integrable over $(0,1)$.

9 The Fourier transform of a complex-valued Lebesgue-integrable function f on the reals is the function $\hat{f}$, given by the equation

$$\hat{f}(t) = \int_{-\infty}^\infty e^{itx} f(x)\,dx.$$

Show that the integral exists and defines a continuous function.

10 *a.* Suppose f is an integrable complex-valued function on the reals such that $xf(x)$ is also integrable. Show that

$$\int_{-\infty}^\infty xf(x)\,dx = \lim_{t\to 0+}\int_{-\infty}^\infty \left(\frac{e^{itx}-1}{it}\right) f(x)\,dx.$$

b. Suppose as a partial converse that f is integrable and that

$$\sup_{t>0}\int_{-\infty}^\infty \frac{|e^{itx}-1|}{|t|}\,|f(x)|\,dx < \infty.$$

Show that $xf(x)$ is then integrable.

11 *a.* Let μ be a bounded right-continuous monotone-increasing function on the reals. The Fourier-Stieltjes transform of μ is the function F, given by

$$F(t) = \int_{-\infty}^\infty e^{itx}\,d\mu(x)$$

where $d\mu(x)$ is the element of measure determined by μ, as in Chap. 2. Show that the integral exists and defines a continuous function.

   b. Suppose $(S,\mathcal{R},m)$ is a basic measure space such that $S \in \mathcal{R}$.   If h is any real-valued measurable function on the space, show that

$$\int_S e^{ith(x)}\, dm(x)$$

exists for real t and is continuous as a function of t.   Show, moreover, that this function is differentiable provided h is integrable.

12   Let f be an integrable vector-valued function on a basic measure space M.

   a. If T is a linear transformation on the vector space in question, show that

$$T[I(f)] = I(T \circ f)$$

   b. Show that $|I(f)| \leq I(|f|)$.

[*Hint:* Use (a) where T is an appropriately chosen linear functional.]

13   Show that the indefinite integral of an integrable vector-valued function on a basic measure space M is a countably additive (vector-valued) set function on the ring of locally measurable sets on M^*.

14   Let f be an integrable vector-valued function on a basic measure space $(S,\mathcal{R},m)$, with the property that

$$\int_E f(x)\, dm(x) = 0$$

for every measurable set E.   Show that $f = 0$ a.e.

15   Let f and g be nonnegative integrable functions on a basic measure space $(S,\mathcal{R},m)$ such that

$$\int_E f(x)\, dm(x) = \int_E g(x)\, dm(x)$$

for all basic measurable sets E in $\mathcal{R}$.

   a. Show that $\displaystyle\int_E f(x)\, dm(x) = \int_E g(x)\, dm(x)$ for all measurable sets E.

   b. Prove that $f = g$ a.e.

16   Let f be an integrable vector-valued function on a basic measure space $(S,\mathcal{R},m)$ such that

$$\int_E f(x)\, dm(x) = 0$$

for every basic measurable set E in $\mathcal{R}$.   Show that $f = 0$ a.e.

17   Suppose f is a complex-valued Lebesgue-integrable function on the unit interval such that $\displaystyle\int_0^x f(y)\, dy = 0$ for all x in $[0,1]$.   Show that $f = 0$ a.e.   Is this still true if it is assumed only that the integral vanishes for rational values of x?

3.5   EXTENSIONS AND COMPLETIONS OF MEASURE SPACES

In this section also we shall be concerned with results of a corollary nature to Theorem 3.3; however, the material to be treated here is somewhat more technical than that of the previous section.

COROLLARY 3.5.1   (HAHN-KOLMOGOROFF)   *The measure on a basic measure space $(S,\Re,m)$ extends uniquely to a countably additive measure on the σ-ring S of measurable sets.*

Proof   First note that since the domain of the integral includes all non-negative essentially measurable functions, the equation

$$m(A) = I(\varphi_A)$$

may be used to extend the measure from $\Re$ to the σ-ring S^* of essentially measurable sets. The extended measure is countably additive, by the sequential continuity of I, but need not be finite-valued. Since S^* contains S, all that needs to be shown is the uniqueness of the extended m on S. Suppose m' is an arbitrary countably additive measure on the σ-ring S which extends the original measure on $\Re$. Let $\Re'$ be the collection of all sets in S on which m' assumes finite values. Then the system $(S,\Re',m')$ is a basic measure space. The class of nonnegative measurable functions on $(S,\Re',m')$ is the same as that on $(S,\Re,m)$, inasmuch as S is also the σ-ring generated by $\Re'$. Let I' be the integral on $(S,\Re',m')$. Since m' extends the original m, it follows from the uniqueness of the integral on the nonnegative measurable functions that $I' = I$ on these functions. Thus m and m' agree on $\Re'$. Now it is easy to verify that every set in S is a countable disjoint union of sets in $\Re'$, so that m and m' agree on S.

The full measure space determined by a basic measure space may now be described.

DEFINITIONS   An *improper measure space* is a system $(S,\Re,m)$ consisting of a set S, a σ-ring $\Re$ of subsets of S, and a countably additive non-negative function m on $\Re$. An element of $\Re$ is called *σ-finite* in case it is a countable union of elements on which m is finite; in case every element of $\Re$ is σ-finite, the system $(S,\Re,m)$ is called a *(proper) measure space*, or simply a *measure space*.

For comparison with the literature, it may be noted that there is some variability in the precise usage of the term "measure space," which, for example, is on occasion used to signify what is here called an improper measure space. However, for integration-theoretical purposes, only the subring $\Re'$ of all σ-finite elements of $\Re$ really matters; an improper space becomes proper when $\Re$ is replaced by $\Re'$, and the integral, the various L_p-spaces, etc., are thereby unaffected. Only (proper) measure spaces are considered in the remainder of this book.

We shall say that a measure space $(S,\mathcal{S},m)$ is *σ-finite* if $S \in \mathcal{S}$. A *finite measure space* is defined as a σ-finite space $(S,\mathcal{S},m)$ in which $m(S) < +\infty$.

The measure space determined by a given basic measure space $(S,\mathcal{R},m)$ is the system $(S,\mathcal{S},m)$, where $\mathcal{S}$ is the set of all measurable sets as defined above, and m denotes also the extension of the original measure on $\mathcal{R}$ to $\mathcal{S}$. If $(S,\mathcal{R},m)$ is the basic space associated with an integration lattice (S,L,J), it is also convenient to call $(S,\mathcal{S},m)$ the *measure space determined* by (S,L,J). Two basic spaces or integration lattices are equivalent if they give rise in the indicated fashion to the same measure space.

Measure spaces determined by basic spaces may be characterized as the proper ones. For by Corollary 3.5.1, any proper space $(S,\mathcal{S},s)$ is determined by the basic space $(S,\mathcal{S}_0,s_0)$, with $\mathcal{S}_0$ taken as the set of elements of $\mathcal{S}$ on which s is finite, and $s_0 = s \mid \mathcal{S}_0$; the converse that the full space is determined by a basic space is obvious.

Next we shall prove a result concerning the extent to which integrable functions may differ from measurable ones.

COROLLARY 3.5.2 *Let M be the basic measure space associated with an integration lattice (S,L,J), and $\mathbf{L}''$ the class of all limits of monotone-decreasing sequences of functions in $\mathbf{L}'$. Then an extended-real-valued function f is integrable on M if and only if there exist integrable functions g and h in $\mathbf{L}''$ such that $-g \le f \le h$, and*

$$I(-g) = I(h).$$

In this event, $I(-g) = I(f) = I(h)$.

Proof Suppose $-g \le f \le h$, where g and h are integrable elements of $\mathbf{L}''$ such that $I(-g) = I(h)$. Then, since

$$I(-g) \le I^{-}(f) \le I^{+}(f) \le I(h),$$

it follows that f is integrable and $I(-g) = I(f) = I(h)$.

Conversely, suppose f is integrable on M. Then there is a sequence k_1, $k_2, \ldots$ of functions in $\mathbf{L}'$ such that $f \le k_n$ and $I(k_n) \to I(f)$. Let $h_1 = k_1$, and define h_n inductively by the equation

$$h_n = \min(h_{n-1}, k_n), \qquad n > 1.$$

Then $h_n \in \mathbf{L}'$, $f \le h_n$, and $I(h_n) \downarrow I(f)$. If $h = \lim_n h_n$, then $f \le h \le h_n$; $h \in \mathbf{L}''$, and

$$I(f) \le I^{-}(h) \le I^{+}(h) \le I(h_n).$$

Thus h is integrable, and $I(f) = I(h)$. Since $-f$ is also integrable, there is another integrable element g of $\mathbf{L}''$ such that $-f \le g$ and $I(-f) = I(g)$ Then, of course, $-g \le f$ and $I(-g) = I(f)$.

REMARK   The functions g and h above are obviously measurable on M, and once this observation is made, it is an elementary matter to prove the less precise fact that an extended-real-valued function f is integrable on a basic measure space M if and only if there exist measurable and integrable functions g and h such that $g \leq f \leq h$ and $I(g) = I(h)$.

With the aid of the above result, the σ-ring of essentially measurable sets on a basic measure space may be characterized as the ring generated by the measurable sets and the collection of null sets.   Actually, we shall prove the following more precise statement.

COROLLARY 3.5.3   *A subset A of a basic measure space M is essentially measurable if and only if there exists a measurable set B and a subset N of a measurable null set such that*

$$A = B \cup N.$$

Proof   Suppose that $M = (S, \mathfrak{R}, m)$ and $A = B \cup N$, where B is measurable and N is a subset of a measurable null set D.   Since

$$0 \leq I^-(\varphi_N) \leq I^+(\varphi_N) \leq I(\varphi_D) = 0,$$

it follows at once that N is a null set; hence A is essentially measurable.

Conversely, suppose that A is essentially measurable.   Then A may be expressed as a countable union of sets in $\mathfrak{R}^*$.   For in any case A is contained in a countable union of sets $A_1, A_2, \ldots$ in $\mathfrak{R}^*$.   Thus

$$A = \bigcup_i A \cap A_i,$$

and since $m(A \cap A_i) = I(\varphi_{A \cap A_i})$ is finite, it follows that $A \cap A_i \in \mathfrak{R}^*$. Since a countable union of measurable null sets is again a measurable null set, it suffices by an elementary argument to prove the result for A in $\mathfrak{R}^*$. If $A \in \mathfrak{R}^*$, there exist, by Corollary 3.5.2, measurable and integrable functions g and h such that $g \leq \varphi_A \leq h$ and also $I(g) = m(A) = I(h)$.   Let $B = [x : 0 < g(x) \leq 1]$ and $D = [x : h(x) \geq 1]$.   Then B and D are measurable sets such that

$$g \leq \varphi_B \leq \varphi_A \leq \varphi_D \leq h.$$

Thus $B \subset A \subset D$ and $m(B) = m(A) = m(D)$.   To complete the proof, observe that $A = B \cup N$, where $N = A - B$, and that N is a subset of the measurable null set $D - B$.

The corollary shows that equivalent basic spaces have the same completion, and that it is legitimate to speak of an essentially measurable function or set on a measure space.   Moreover, since the integral is unique on the nonnegative measurable functions, it is also unique on the measurable and integrable functions.   Thus equivalent integration lattices have the same completion by the remark following Corollary 3.5.2.

DEFINITION   The *completion of a measure space* $(S,\mathcal{S},m)$ is the system $(S,\mathcal{S}^*,m)$ in which $\mathcal{S}^*$ is the σ-ring of essentially measurable sets, and m is the measure on $\mathcal{S}^*$ defined by the integral, as in the proof of Corollary 3.5.1.

If the measure space $(S,\mathcal{S},m)$ is defined by a given basic space M, then $(S,\mathcal{S}^*,m)$ may also be characterized as the measure space determined by the completion M^* of M.

The completion M^* of a (basic or full) measure space M is again a (basic or full) measure space, and hence also has a completion $(M^*)^*$. But it is readily verified by means of Corollary 3.5.3 that $(M^*)^* = M^*$; that is, M^* is already complete.   A moment's reflection shows that the same is also true of the completion of an integration lattice.

It should be mentioned that our notation, $\mathcal{R}^*$ and $\mathcal{S}^*$, for the rings of the completions of the basic space $(S,\mathcal{R},m)$ and corresponding measure space $(S,\mathcal{S},m)$ is somewhat inadequate.   The rings $\mathcal{R}^*$ and $\mathcal{S}^*$ depend not only on $\mathcal{R}$ and $\mathcal{S}$, but on the basic measure as well.   This point is brought out rather clearly in the following example.

Example 3.5.1   Let $(S,\mathcal{R},m)$ be a one-dimensional basic Lebesgue-Stieltjes measure space.   Since S is an interval and can be expressed as a countable union of (bounded) intervals in $\mathcal{R}$, there is no distinction between measurability and local measurability.   The measurable sets are those in the σ-ring $\mathcal{S}$ generated by the intervals in $\mathcal{R}$, which is clearly also the σ-ring generated by the open intervals in $\mathcal{R}$.   These sets are the *Borel* subsets of S, as are the sets in the σ-ring generated by the open sets in any topological space.   It can be shown that the ring of Borel subsets of S has cardinal number equal to that of the continuum, but we shall not prove this here. If E is any uncountable subset of S, it follows that the ring of all subsets of E contains sets which are not Borel sets.   Thus it is not too surprising that $\mathcal{S}^*$ is considerably larger than the ring $\mathcal{S}$ of Borel sets.   This is most easily seen when m is a discrete measure.   For then $\mathcal{S}^*$ contains all subsets of S.   To prove this, suppose that A is an arbitrary subset of S and that m is discrete. Let B be the countable Borel set consisting of the points $x_1, x_2, \ldots$ which have positive measure.   Then $A = A \cap B \cup (A - B)$, $A \cap B$ is a Borel set, and $A - B$ is a subset of the Borel null set $S - B$.   In the general case, it is true, but not so obvious, that there exist null sets which are not Borel sets.   To prove this it suffices, in view of the cardinality of $\mathcal{S}$, to show that there exist uncountable Borel sets of measure 0.

For example, if m is purely continuous, there always exist perfect nowhere-dense sets, similar to the Cantor "middle third" set, which are of measure 0 (see the exercises for more details on this point).   On the other hand, a nowhere-dense set may very well have positive measure.   The set K constructed in Example 1.3.1 has this property.   In the case of Lebesgue measure, the sets in $\mathcal{S}^*$ are called the *Lebesgue-measurable sets*.   With sufficiently strong logical assumptions, including transfinite induction, one can construct sets of real numbers which are not Lebesgue-measurable. However, the question of the existence of such sets is closely connected with currently unresolved questions in the foundations of mathematical logic.

Next we shall prove a useful result concerning the measure of an essentially measurable set.

COROLLARY 3.5.4 *Let M be the basic measure space associated with an integration lattice (S,L,J), and $\mathcal{C}$ the collection of all limits of increasing sequences of sets of the form $[x:f(x) > 1]$, where f ranges over $\mathbf{L}$. Suppose A is an essentially measurable subset of M. Then there exist sets E containing A which belong to $\mathcal{C}$ and $m(A) = \inf m(E)$, where the infimum is taken over all such E.*

Proof If $M = (S,\mathcal{R},m)$, then A may be expressed as a countable union of sets A_1, A_2, $\ldots$ in $\mathcal{R}^*$. Moreover, since $m(A_n)$ is finite, there exists a function g_n in $\mathbf{L}'$ such that $\varphi_{A_n} \leq g_n$. This implies the existence of an element g in $\mathbf{L}'$ such that $\varphi_A \leq g$. We may, for example, take $g = \sum_n g_n$.

For any such g, let

$$B_a = [x:g(x) > a],$$

where $0 < a < 1$. Then $A \subseteq B_a$ and $B_a \in \mathcal{C}$. For there exists a monotone-increasing sequence h_1, h_2, $\ldots$ of functions in $\mathbf{L}$ such that $g = \lim h_n$, and hence

$$B_a = \bigcup_n [x:h_n(x) > a].$$

If $m(A)$ is finite and $\epsilon > 0$, the function g may be chosen so that

$$m(A) \leq I(g) \leq m(A) + \epsilon/2.$$

Then, setting

$$a = \frac{m(A) + \epsilon/2}{m(A) + \epsilon}$$

and $B = B_a$, and using the fact that $a\varphi_B \leq g$, we find that

$$m(A) \leq m(B) \leq m(A) + \epsilon.$$

Since ϵ is arbitrary, this completes the proof for the case in which $m(A)$ is finite. On the other hand, if $m(A) = +\infty$, then $m(B) = +\infty$ for any measurable B containing A.

EXERCISES

1   *a.* Let T be an arbitrary set, U the interval $[0,1]$, and S the set of all pairs (t,x), with $t \in T$ and $x \in U$. Let $\mathcal{R}$ be the ring of all subsets of S which intersect only a finite number of the sets $\{t\} \times U$ in a nonempty set and for which this set is of the form $\{t\} \times B$, with B a Borel set; define the measure

of a set in $\mathcal{R}$, say, $\bigcup\limits_{i} \{t_i\} \times B_i$, as the sum of the Lebesgue measures of the Borel sets B_i.   Show that $(S,\mathcal{R},m)$ is then a basic measure space.

b. Show that the measurable sets are those which meet at most countably many of the $\{t\} \times U$ and for which each such intersection is of the form $\{t\} \times B$, with B a Borel set.

c. Show that the sets locally measurable with respect to $\mathcal{R}$ are those which meet each set of the form $\{t\} \times U$ in a set of the form $\{t\} \times B$, with B a Borel set, but are otherwise arbitrary.

d. Show that $(S,\mathcal{R},m)$ can be described as the direct product of two basic spaces.

2   Suppose a measure space has the property that any collection of mutually disjoint measurable sets is at most countable.   Is the entire space then measurable?

3   Let $M = (S,\mathcal{S},m)$ be a measure space, and $\mathcal{R}$ the subring of $\mathcal{S}$ of sets of finite measure.   A *simple function* on M is defined as a step function based on $\mathcal{R}$.

a. Show that an extended-real-valued function f measurable with respect to $\mathcal{S}$ is the pointwise limit of a sequence $f_1, f_2, \ldots$ of simple functions such that $|f_n(x)| \leq |f(x)|$ for all n and x.

b. If the function f in (a) is nonnegative, show that it is the pointwise limit of a monotone-increasing sequence $f_1, f_2, \ldots$ of nonnegative simple functions.

c. If the measure space is finite, show that the sequence constructed in (a) can be chosen to have the additional property that it converges uniformly on every set where f is bounded.

d. Construct an example to show that the condition in (c) that the measure space be finite is necessary for the conclusion.

e. Extend the results of (a) and (c) to complex-valued functions.

f. If f is an integrable complex-valued function, show that there exists a sequence $f_1, f_2, \ldots$ of simple functions converging pointwise to f such that $I(f_n) \to I(f)$ and $I(|f - f_n|) \to 0$.

4   If f is a nonnegative measurable function on a measure space M, show that $I(f) = \sup\limits_{g} I(g)$, where g ranges over the class of all nonnegative simple functions bounded from above by f.

5   a. Let $(S,\mathcal{S},m)$ be a measure space, and T a mapping of S into a set R. Suppose $\mathcal{R}$ is a σ-ring of subsets of R such that $T^{-1}(E) \in \mathcal{S}$ for all E in $\mathcal{R}$; define $n(E) = m[T^{-1}(E)]$ for all E in $\mathcal{R}$.   Show that n is a nonnegative countably additive measure on $\mathcal{R}$.

b. Let $\mathcal{R}'$ denote the collection of all sets in $\mathcal{R}$ which are countable unions of sets of finite measure in $\mathcal{R}$.   If a function g is integrable relative to n, that is, if g is integrable on the space $(R,\mathcal{R}',n)$, show that $g \circ T$ is integrable on $(S,\mathcal{S},m)$ and that

$$\int_R g(y)\,dn(y) = \int_S g[T(x)]\,dm(x)$$

(*Hint:* Look first at the case in which g is a simple function.)

6 Let f be a real-valued measurable function on a σ-finite measure space $(S,\mathcal{S},m)$.   Denote by $\mathcal{R}$ the collection of all sets of the form $f^{-1}(B)$, where B is a Borel subset of the reals.

    a. Show that $\mathcal{R}$ is a sub-σ-ring of $\mathcal{S}$ and that $(S,\mathcal{R},m)$ is a measure space.

    b. Prove that f is integrable on $(S,\mathcal{R},m)$.

    c. In what sense is $(S,\mathcal{R},m)$ the smallest measure space on which f is integrable?

7 *a.* The *distribution* of a given real-valued measurable function f on a σ-finite measure space M is the nonnegative countably additive measure F on the Borel subsets of the reals, defined by

$$F(B) = \text{meas } [f^{-1}(B)].$$

    Show that f is integrable on M if and only if the identity function is integrable on the reals relative to F and that in this case

$$I(f) = \int_{-\infty}^{\infty} t\, dF(t).$$

    b. If f and F are as in (*a*) and λ is a complex-valued Baire function on the reals, show that

$$I(\lambda \circ f) = \int_{-\infty}^{\infty} \lambda(t)\, dF(t)$$

    in the sense that if either side exists, both sides are defined and equal.

8 If f is an essentially measurable complex-valued function on a measure space, show that there exists a measurable null set E such that the function defined as 0 on E and as $f(x)$ when x is not in E is measurable.

9 Let $(S,\mathcal{R},m)$ be the basic Lebesgue measure space in which S is the closed unit interval $[0,1]$.

    a. If $E \in \mathcal{R}$, $0 < m(E) < 1$, and $0 < c < 1 - m(E)$, show that there exists an open set F in $\mathcal{R}$ whose closure is disjoint from $\bar{E}$ such that $m(F) = c$.

    b. If $\sum_n c_n = 1$, where $c_n > 0$ for $n = 1, 2, \ldots$, show that there exists a sequence $E_1, E_2, \ldots$ of open sets in $\mathcal{R}$ with disjoint closures such that $m(E_n) = c_n$.

    c. If $E = \bigcup_m E_n$, show that E is dense in S.

    d. Prove that $S - E$ is a closed set of measure 0, which is nowhere dense, perfect (i.e., every point of $S - E$ is a limit point of $S - E$), and hence uncountable.

    e. Verify that the Cantor set may be obtained in this fashion, i.e., is an $S - E$ for suitable E.

10 Suppose $(S,\mathcal{S},m)$ is a Lebesgue-Stieltjes measure space on the line and A is a Borel subset of S.   Show that

$$m(A) = \inf m(E),$$

where E ranges over the open subsets of S containing A.   (*Hint:* Use Corollary 3.5.4.)

11 If $(S,\mathcal{S},m)$ is a Lebesgue-Stieltjes measure space on the line, show that for every Borel subset A of S,

$$m(A) = \sup m(K),$$

where K varies over the compact subsets of A. (*Hint:* First consider the case in which S is a closed bounded interval.)

3.6 MULTIPLE INTEGRATION

If M and M' are basic measure spaces $(S,\mathcal{R},m)$ and $(S',\mathcal{R}',m')$, the integral of a step function f over the direct product $M \otimes M'$ is equal, virtually by definition, to either one of the iterated integrals. In other words,

$$I_{M \otimes M'}(f) = I_{M'}(I_M f) = I_M(I_{M'} f).$$

It is natural to consider the question of the extent to which this holds for more general functions. The answer is particularly satisfying and useful for nonnegative measurable functions.

> **THEOREM 3.4** (FUBINI) *For any nonnegative measurable function f on the direct product $M \otimes M'$ of the basic measure spaces $M = (S,\mathcal{R},m)$ and $M' = (S',\mathcal{R}',m')$, $I_M(f)$ and $I_{M'}(f)$ are measurable functions on M' and M, respectively, and*
>
> $$I_{M \otimes M'}(f) = I_{M'}(I_M f) = I_M(I_{M'} f).$$
>
> *More specifically,*
>
> > *(1) $f(x,y)$ is a measurable function of x for each fixed y in S' and a measurable function of y for each fixed x in S.*
> > *(2) The functions $I_M f$ and $I_{M'} f$, given by*
> >
> > $$I_M f(y) = \int_M f(x,y)\, dm(x), \qquad y \in S'$$
> >
> > $$I_{M'} f(x) = \int_{M'} f(x,y)\, dm'(y), \qquad x \in S$$
>
> *are measurable on M' and M, respectively, and*
>
> $$(3)\ \ I_{M \otimes M'}(f) = \int_{M'} \left[\int_M f(x,y)\, dm(x) \right] dm'(y)$$
>
> $$= \int_M \left[\int_{M'} f(x,y)\, dm'(y) \right] dm(x).$$

Proof Let **C** denote the collection of all nonnegative measurable functions f on $M \otimes M'$ which satisfy (*1*), (*2*), and (*3*). Obviously, **C** contains all nonnegative step functions. Moreover, **C** is closed under sequential convergence of increasing sequences. For if $f_n \in \mathbf{C}$ and $f_n \uparrow f$, then (*1*) follows directly, (2) holds by the sequential continuity of the integral for increasing sequences, and (3) holds for the same reason.   In addition, **C** is closed under sequential convergence of sequences bounded by finite-valued integrable elements of **C** by a quite similar argument based on the dominated convergence criterion.

Now let A be an arbitrary set in the ring $\mathcal{R} \otimes \mathcal{R}'$ of all finite unions of rectangles, and $\mathbf{C}(A,n)$ the set of all elements of **C** which vanish outside A and have values in the interval $[0,n]$.   It is evident that $\mathbf{C}(A,n)$ contains the step functions supported by A, with values in $[0,n]$.   Since the functions in $\mathbf{C}(A,n)$ are dominated by $n\varphi_A$, $\mathbf{C}(A,n)$ is closed in the topology of pointwise convergence of sequences.   By the density of the indicated class of step functions in the measurable functions supported by A with values in $[0,n]$, these functions are likewise contained in $\mathbf{C}(A,n)$.

Suppose f is an arbitrary nonnegative measurable function on $M \otimes M'$. Then there exists an increasing sequence A_1, A_2, ... of sets in $\mathcal{R} \otimes \mathcal{R}'$ such that f vanishes outside the union of the A_n.

If
$$f_n = \min (f_{A_n}, n),$$

then $f_n \uparrow f$, and since $f_n \in \mathbf{C}(A_n, n)$, it follows that $f \in \mathbf{C}$.   This concludes the proof.

COROLLARY 3.6.1   *If f is a measurable extended-real- or vector-valued function on $M \otimes M'$, then $f(x,y)$ is a measurable function of x for each fixed y and a measurable function of y for fixed x.*

Proof   In the vector-valued case, it suffices to consider each vector component separately, and it may therefore be assumed that f is an extended-real-valued function.   The result then follows from the theorem and the decomposition $f = f^+ - f^-$.

COROLLARY 3.6.2   *If f is an integrable extended-real- or vector-valued function on $M \otimes M'$, then $I_M f$ and $I_{M'} f$ exist a.e. and are essentially integrable functions on M' and M, respectively, such that*

$$I_{M \otimes M'} f = I_{M'}(I_M f) = I_M(I_{M'} f).$$

When, in addition, f is measurable on $M \otimes M'$, then $I_M f$ and $I_{M'} f$ agree where defined with measurable and integrable functions.

Proof   For the proof in the vector-valued case, it suffices by linearity to consider each vector component separately, and it may therefore be assumed that f is an extended-real-valued function.

We shall suppose at first that f is measurable in addition to being integrable.   Then $f = f^+ - f^-$, where f^+ and f^- are nonnegative measurable functions on $M \otimes M'$ for which $I_{M \otimes M'}(f^+)$ and $I_{M \otimes M'}(f^-)$ are finite.   Thus, by the theorem, $I_M(f^+)$ and $I_M(f^-)$ are both nonnegative measurable functions on

M', with finite integrals, and so each is finite-valued except on a null set. Combining these null sets, we obtain a null set N such that $x \to f(x,y)$ is integrable on M, and

$$\int_M f(x,y)\, dm(x) = \int_M f^+(x,y)\, dm(x) - \int_M f^-(x,y)\, dm(x)$$

for all fixed y outside N. Define g and h to be 0 on N and to agree with $I_M(f^+)$ and $I_M(f^-)$, respectively, outside N. Then g and h are measurable, finite-valued, and integrable on M'. Moreover, $(I_M f)(y)$ exists, and

$$(I_M f)(y) = g(y) - h(y)$$

for y outside N. Thus $I_M f$ is essentially integrable and agrees with the measurable and integrable function $g - h$ outside the measurable null set N. Hence

$$I_{M'}(I_M f) = I_{M'}(g - h)$$
$$= I_{M'}(g) - I_{M'}(h)$$
$$= I_{M'}(I_M f^+) - I_{M'}(I_M f^-)$$
$$= I_{M \otimes M'}(f^+) - I_{M \otimes M'}(f^-)$$
$$= I_{M \otimes M'}(f).$$

The proof for the case under consideration is now completed by an entirely similar argument.

Next suppose f is an arbitrary integrable extended-real-valued function on $M \otimes M'$. Then there exist, by Corollary 3.5.2, measurable and integrable functions g and h such that $g \leq f \leq h$ and

$$I_{M \otimes M'}(g) = I_{M \otimes M'}(f) = I_{M \otimes M'}(h).$$

It follows from the properties of the functionals $I_M{}^+$ and $I_M{}^-$ that

$$(I_M{}^- g)(y) \leq (I_M{}^- f)(y) \leq (I_M{}^+ f)(y) \leq (I_M{}^+ h)(y)$$

for all y.

Now, by what was proved above, $I_M g$ and $I_M h$ exist a.e., and combining the null sets where $I_M g$ and $I_M h$ are undefined, we obtain a measurable null set N such that

$$(I_M g)(y) = (I_M{}^- g)(y) \leq (I_M{}^+ h)(y) = (I_M h)(y)$$

for y outside N. Moreover, since $I_{M'}(I_M g) = I_{M'}(I_M h)$, we may assume, after adding another null set to N, that $(I_M g)(y) = (I_M h)(y)$ for y outside N. For such y it follows that

$$(I_M g)(y) = (I_M{}^- f)(y) = (I_M{}^+ f)(y) = (I_M h)(y).$$

Hence $I_M f$ exists a.e. and is an essentially integrable function on M', for which

$$I_{M \otimes M'}(f) = I_{M'}(I_M f).$$

The proof is completed by a similar argument.

One of the difficulties in applying the Fubini theorem is due to the fact that there is no general easily applied criterion for determining when a given function f is measurable on the direct product of two basic spaces. For example, it is not enough that $f(x,y)$ be measurable as a function of x for each fixed y and measurable in y for fixed x.

Example 3.6.1   Let M be the basic Lebesgue measure space on the reals, and M' the discrete basic space on the reals, in which the ring consists of all finite subsets and the measure of such a set is the number of points it contains. Let f be the characteristic function of the "diagonal" subset of the product of the reals with itself, i.e., the set in the plane where $x = y$. Then f is separately measurable in each variable and

$$\int_M f(x,y) \, dm(x) = 0$$

for all y, while

$$\int_{M'} f(x,y) \, dm'(y) = 1$$

for all x.   Thus

$$\int_{M'} \left[\int_M f(x,y) \, dm(x) \right] dm'(y) = 0$$

and

$$\int_M \left[\int_{M'} f(x,y) \, dm'(y) \right] dm(x) = +\infty.$$

Here all that may be deduced from the Fubini theorem is the fact that f is not measurable on $M \otimes M'$.   This of course is obvious since f fails to vanish outside a countable union of basic rectangles.   Note, however, that f is locally measurable on $M \otimes M'$.

SCHOLIUM 3.9   *If g and h are measurable extended-real-valued functions on the basic measure spaces M and M', where $M = (S,\mathfrak{R},m)$ and $M' = (S',\mathfrak{R}',m')$, then the function f defined by*

$$f(x,y) = g(x)h(y), \qquad x \in S, \, y \in S',$$

is measurable on $M \otimes M'$.   The σ-ring of measurable sets on $M \otimes M'$ contains all sets of the form $A \times B$, where A is a measurable subset of M, and B is a measurable subset of M'.

Proof If g and h are any extended-real-valued functions on S and S', respectively, let $g \otimes h$ denote the function on $S \times S'$, defined by

$$(g \otimes h)(x,y) = g(x)h(y), \qquad x \in S, y \in S'.$$

If g is a fixed step function on M, let **C** be the collection of all measurable functions h on M' such that $g \otimes h$ is measurable on $M \otimes M'$. Then **C** contains all step functions on M' and is obviously closed under pointwise convergence of sequences. Thus $g \otimes h$ is measurable on $M \otimes M'$ for all measurable h. Now fix h, and consider the class **D** of all measurable functions g on M such that $g \otimes h$ is measurable on $M \otimes M'$. Then **D** contains all step functions and is also closed under pointwise convergence of sequences, so that $g \otimes h$ is measurable on $M \otimes M'$ whenever g and h are measurable on M and M'. The final statement follows from the relation

$$\varphi_{A \times B} = \varphi_A \otimes \varphi_B.$$

REMARK The equivalence class (as defined in the preceding section) of the direct product of two basic spaces depends only on their equivalence classes. Thus the concept of direct product of measure spaces is a legitimate one, corollary to that of the direct product of basic spaces. To see this it is only necessary to show that the direct product of two basic spaces is equivalent to the direct product of the equivalent spaces in which the respective rings are replaced by the rings of all measurable sets of finite measure. But this follows readily from Scholium 3.9 and the Hahn-Kolmogoroff theorem (Corollary 3.5.1).

EXERCISES

1 Construct an example of a nonnegative essentially measurable function f on the direct product $M \otimes M'$ of the basic spaces M and M' such that $x \to f(x,y)$ is not essentially measurable on M for all y.

2 If f is an essentially measurable real- or vector-valued function on the direct product $M \otimes M'$ of the basic spaces M and M', show that $f(x,y)$ is essentially measurable as a function of y for almost all x.

3 If f is a Baire function on the product of two topological spaces S and S', show that $f(x,y)$ as a function of x is a Baire function on S for each y in S' and that as a function of y it is a Baire function on S' for each x in S.

4 Construct a nonnegative real-valued Baire function f on the unit square such that

$$\iint f(x,y)\, dx\, dy = 0,$$

which has the property that

$$\int_0^1 f(x,y)\, dx = \infty$$

for every rational y in $[0,1]$.

5 Give an example of a Borel-measurable function f in the euclidean plane for which the two iterated integrals

$$\int\left[\int f(x,y)\,dx\right]dy \qquad \text{and} \qquad \int\left[\int f(x,y)\,dy\right]dx$$

exist and are unequal.

6 Give an example of a Borel-measurable function f in the euclidean plane that is not integrable for which the two iterated integrals exist and are both equal to 0.

7 Show that the σ-ring generated by the products of Borel subsets of two euclidean spaces consists of the ring of all Borel subsets of the product of the two spaces.

8 *a.* Let $K(x,y)$ be an integrable function on the product $M \otimes M$ of a basic measure space with itself such that $\int |K(x,y)|\,dx$ is essentially bounded. Show that for every measurable and integrable function f on M, the integral $\int K(x,y)f(y)\,dy$ exists almost everywhere and defines an essentially integrable function $g(y)$. In addition, show that the mapping $K\!:\!f \to g$ is linear and bounded in the sense that $I(|Kf|) \le \text{const } I(|f|)$ for a fixed constant independent of f.

  b. Show that if $L(x,y)$ is another such function, with corresponding map L, then $K + L$ is the map corresponding to $K(x,y) + L(x,y)$, while KL is the map corresponding to $\int K(x,z)L(z,y)\,dz$, which integral exists for almost all (x,y) and defines an essentially integrable function on $M \otimes M$. (Note that this extends the usual rule for matrix multiplication.)

9 Let f be a nonnegative measurable function on the measure space M. Let M' denote the space with set $[0, \infty]$, the Borel subsets as the ring of measurable sets, the measure being Lebesgue measure. Let H_f denote the "ordinate set" of f, defined as the set of all pairs (x,y) in $M \otimes M'$ with $f(x) \ge y$. Show that H_f is a measurable set in the product space of measure equal to $I(f)$. (*Hint:* First establish the conclusion for the case in which f is a step function. Consider, then, the general character of the set of all functions f for which the conclusion is valid.)

10 *a.* Show that the mapping $(x,y) \to (x, y - x)$ of the euclidean plane into itself carries Borel sets into Borel sets. (The same is true of any homeomorphism.)

  b. Show that if E is any Borel subset of the plane, then it has the same measure as its transform E' under the foregoing transformation. [(*Hint:*

$$\text{meas } E' = \iint \varphi_{E'}(x,y)\,dx\,dy = \iint \varphi_E(x, y + x)\,dx\,dy;$$

by the Fubini theorem this is $\int[\int \varphi_E(x, y + x)\,dy]\,dx$; by the invariance of Lebesgue measure under translations, $\int f(y + x)\,dy = \int f(y)\,dy$ for any Lebesgue-integrable function y, so that meas $E' = \int[\int \varphi_E(x,y)\,dy]\,dx = $ meas E).]

   c. Show that (*a*) and (*b*) hold also for Lebesgue-measurable sets. (Recall that these differ from Borel sets only by subsets of Borel subsets of measure zero.)

   d. Conclude that if f and g are Lebesgue-measurable functions on the line, then $f(x - y)g(y)$ is a measurable function on the plane. [$f(x)g(y)$ is measurable on the plane, and is carried by the transformation $x \to x - y$, $y \to y$, into the indicated function.]

   e. Show that if f and g are essentially integrable functions on the line, then $\int f(x - y)g(y)\, dy$ exists for almost all values of x, and defines an essentially integrable function of x. (This function is called the *convolution* of f with g, and commonly denoted $f * g$.)

   f. Show that if f, g, and h are essentially integrable on the line, then $f * g = g * f$ and $f * (g * h) = (f * g) * h$. (For the last result, use the Fubini theorem.)

11 Show that if f is integrable and g is a continuously differentiable function vanishing outside a compact set, then $f * g$ is differentiable with the derivative $f * g'$.

12 Give an example to show that not every Borel subset of the plane of positive measure contains the direct product of two Borel sets of the line of positive measures.

13 *a.* Evaluate the convolution of e^{-ax^2} and e^{-bx^2}, a and b being positive constants, starting from the fact that $\displaystyle\int_{-\infty}^{+\infty} e^{-x^2}\, dx = \sqrt{\pi}$.

   b. Show that if f is any bounded uniformly continuous function on the line, then its convolution f_a with e^{-ax^2} exists and is an infinitely differentiable function.

   c. Find a function of a, $c(a)$ such that $c(a)$ times the convolution in (*b*) converges uniformly to f as $a \to \infty$.

   d. Extend (*b*) and (*c*) to the case when f is only Lebesgue-integrable, replacing uniform convergence by the convergence to zero of the integral of the absolute value of the difference.

14 Show that a countably additive measure on linear Borel sets which is finite on bounded sets and invariant under translations is proportional to Lebesgue measure. [*Hint:* Let m denote Lebesgue measure, and n the measure in question; let B denote a bounded Borel set; and let f be an arbitrary nonnegative Borel-measurable function on the line. By an argument similar to that in Exercise 10, but with the planar measure replaced by $m \otimes n$, show that $n(B) \int f(x)\, dm(x) = m(B) \int f(-x)\, dn(x)$.]

3.7 LARGE SPACES

In most concrete applications of measure theory one deals with finite or σ-finite measure spaces. For such spaces the foregoing material provides an adequate framework for the general theory. On the other hand, certain "large" spaces which are not σ-finite arise quite naturally, for example, in the theory of harmonic analysis on locally compact Abelian groups and in connection with the spectral resolution of Abelian self-adjoint algebras of operators on Hilbert space. For such applications it is necessary to study the class of locally measurable functions in detail.

> DEFINITION If M is a measure space and Y a topological space, a function on M with values in Y is said to be *locally measurable on M* if it is locally measurable with respect to the σ-ring of measurable sets.

If $\mathfrak{R}$ is the ring of all measurable subsets of finite measure on a given measure space $M = (S,\mathcal{S},m)$, every element of $\mathcal{S}$ is a countable union of sets in $\mathfrak{R}$. From this it is readily verified that a function f is locally measurable on M if and only if it is locally measurable with respect to $\mathfrak{R}$. Moreover, if M' is any basic measure space defining M, every function locally measurable on M' is locally measurable on M. We leave the proof of this as an exercise.

The integral on any measure space extends to an integral on the class of nonnegative locally measurable functions in a fashion analogous to the extension of a basic measure given by Scholium 2.4.

> SCHOLIUM 3.10 *For any measure space M let the integral I be defined for arbitrary nonnegative locally measurable functions f by the equation*
>
> $$I(f) = \sup_{E} I(f\varphi_E),$$
>
> *where the supremum is taken over all measurable subsets E. The extended integral is then additive on the given class and sequentially continuous for increasing sequences.*

Proof If f is a nonnegative locally measurable function on M and E is a measurable subset, then $f\varphi_E$ is measurable by Scholium 3.2, and hence $I(f\varphi_E)$ is well defined. Moreover, the identity of notation between the original integral and its extension is justified by their identity on the nonnegative measurable functions. To prove the additivity, note that for any nonnegative locally measurable function f there exists a measurable set E such that $I(f) = I(f\varphi_E)$ and that the same is true for any measurable set F containing E. For if measurable sets E_1, E_2, ... are chosen so that $I(f\varphi_{E_n}) \uparrow I(f)$, then the set $E = \bigcup_n E_n$ has this property. Thus, if f and g are given, there exists a measurable set E such that $I(f + g) = I[(f + g)\varphi_E]$, $I(f) = I(f\varphi_E)$, and $I(g) = I(g\varphi_E)$. Since $(f + g)\varphi_E = f\varphi_E + g\varphi_E$, it follows that $I(f + g) = I(f) + I(g)$.

Next, suppose $f_1, f_2, \ldots$ is an increasing sequence of nonnegative locally measurable functions with limit f. Choose measurable sets E_n such that $I(f_n) = I(f_n\varphi_{E_n})$ and $I(f) = I(f\varphi_{E_n})$. If E is the union of the E_n, it is then immediate that $I(f_n\varphi_E) \uparrow I(f\varphi_E)$.

In connection with this result, the *integral of a nonnegative locally measurable function f over a locally measurable set* A is defined as $I(f\varphi_A)$. However, the *indefinite integral* of such an f will be defined as the set function F on the measurable sets (rather than the locally measurable ones), given by

$$F(E) = I(f\varphi_E).$$

The original measure m is then the indefinite integral of the function identically one. Moreover, any indefinite integral is countably additive by the sequential continuity of the integral for increasing sequences.

If f is a nonnegative locally measurable function for which $I(f)$ is finite, then there exists a measurable set E such that $f\varphi_E$ is integrable and $I(f) = I(f\varphi_E)$. The functions f and $f\varphi_E$ differ only on a set which is locally a null set or a local null set in the following sense.

> DEFINITIONS A subset N of a measure space M is said to be a *local null set* if $N \cap E$ is a null set for every measurable set E. A property is said to hold *locally almost everywhere* (abbreviation l.a.e.) if it is valid with the exception of a local null set. Thus, for functions, $f = g$ l.a.e. means that $f(x) = g(x)$ for all x outside a certain local null set.

If g and h are measurable, or essentially measurable, functions on a measure space such that $g = h$ l.a.e., then $g = h$ a.e. For there exist measurable sets, say, A and B, such that g is supported by A, and h by B. If $g = h$ outside the local null set N, then in fact $g = h$ outside the null set $N \cap (A \cup B)$. This remark justifies the following final and rather trivial extension of the integral.

> DEFINITION A function f on a measure space M is *extendedly integrable* if there is an integrable function g such that $f = g$ l.a.e., and the integral $I(f)$ of f is then defined as $I(g)$.

EXERCISES

1 If $\mathbb{S}$ is the σ-ring generated by a ring $\mathcal{R}$ of subsets of a set S, show that every function on S locally measurable with respect to $\mathcal{R}$ is also locally measurable with respect to $\mathbb{S}$.

2 A measure space $M = (S,\mathcal{R},m)$ may be said to be *essentially a direct sum of finite spaces* if S is the disjoint union of measurable subsets S_i $(i \in I)$ of finite measure such that for any measurable set E, $E \cap S_i$ is nonnull for at most countably many i.

    a. Give an example of a direct sum which is not σ-finite.

    b. Show that a complex-valued function on a direct sum of finite measure spaces is locally measurable if and only if its restriction to each finite subspace is measurable.

3 Suppose M is a direct sum of finite measure spaces. Suppose also that, with every measurable set A of finite measure, there is associated a measurable function f_A which vanishes outside A, and that $f_A = f_B$ a.e. on $A \cap B$ for all measurable B of finite measure. Show that there is a locally measurable function f on M such that $f = f_A$ a.e. on A for every A.

4 Show that if I_1 is any additive functional from the nonnegative locally measurable functions f to $[0, \infty]$ which is continuous with respect to increasing sequences and which extends the integral for measurable functions, then $I(f) \leq I_1(f)$ for all f, where I is the extension given by Scholium 3.10. Give an example to show that I_1 may differ from I.

IV

CONVERGENCE
AND
DIFFERENTIATION

4.1 LINEAR SPACES OF MEASURABLE FUNCTIONS

The reader is probably familiar with the extensive and penetrating theory of finite-dimensional linear spaces and operators thereon. One cannot hope to obtain a similarly penetrating theory when the vector space is infinite-dimensional without additional structure in the space. A first step in civilizing a linear space for the purposes of analysis is the introduction of a convenient topology, i.e. one with respect to which relevant operations are continuous. A natural candidate for such a topology in the linear space $\mathbf{M}$ of all complex-valued locally measurable functions is that of sequential pointwise convergence. However, this topology may fail to have cogent structural properties. One is led rather to deal with a modification $\widetilde{\mathbf{M}}$ of $\mathbf{M}$ which is obtained by identifying functions which agree except on local null sets; and which is given the topology it inherits naturally from that of pointwise sequential convergence in $\mathbf{M}$. It will be seen in fact that $\widetilde{\mathbf{M}}$ is metrizable in this topology, in the key case of a finite measure space; and that it becomes a topological linear space in the following sense.

DEFINITION A *topological linear space* $\mathbf{X}$ over a topological field F is the system composed of a linear space $\mathbf{X}$ over F together with a topological

93

structure in **X**, having the property that the maps $(x,y) \to x - y$ from **X** × **X** into **X**, and $(a,x) \to ax$ from F × **X** into **X**, are continuous.

Here, *topological field* refers to a field F with a given topology such that the operations $(x,y) \to x - y$ on $F \times F$ to F, and $(x,y) \to xy^{-1}$ on $F' \times F'$ to F', where $F' = F - \{0\}$, are continuous. Actually, only the real and complex fields will be involved in this book, although the case of p-adic fields is interesting in connection with number theory, and some aspects of measurability theory extend to such cases. There are no known parallels to many of the central results of integration theory except in the cases of the real and complex fields.

In the theory of integrals *per se*, as distinguished from the theory of measurability, there will be seen to be no material distinction between functions which differ only on a local null set. Their integrals, indefinite integrals, distribution functions, etc., are identical. From a physical or statistical standpoint the distinction seems unobservable, since actual measurements or expectations correspond to integrals rather than point evaluations.

Therefore, for a variety of reasons, it is frequently desirable to replace the algebra of locally measurable functions by the quotient algebra obtained by identifying functions equal l.a.e. In order to describe this situation more precisely, it is helpful to take explicit note of the negligibility of local null sets, with regard to many aspects of integration theory.

DEFINITIONS   Two functions on a measure space M are called *equivalent* if they differ only on a local null set. The equivalence class $\tilde{f}$, determined by an extended-real- or complex-valued locally measurable function f, will be called a *function class*. From the properties of local null sets, it follows that for function classes $\tilde{f}$ and $\tilde{g}$, we may say $\tilde{f} \leq \tilde{g}$ if $f \leq g$ l.a.e. Similarly, $\tilde{f}_n \to \tilde{f}$ means $f_n \to f$ l.a.e. If λ is any Baire function, $\lambda \circ \tilde{f}$, or simply $\lambda(\tilde{f})$, may be defined as $(\lambda \circ f)^{\sim}$. A function class $\tilde{f}$ is (extendedly) *integrable* provided the same is true of f, and $I(\tilde{f})$ is then defined as $I(f)$ [the integral $I(\tilde{f})$ is clearly single-valued; i.e., $I(f)$ depends only on the equivalence class of f]. For function classes $\tilde{f}$ and $\tilde{g}$, in which f and g are extended-real-valued functions, $\max(\tilde{f},\tilde{g})$ and $\min(\tilde{f},\tilde{g})$ are defined as the classes corresponding to $\max(f,g)$ and $\min(f,g)$, respectively. The collection of function classes determined by the extended-real-valued locally measurable functions is evidently a lattice. If $\tilde{f} + \tilde{g}$, $\tilde{f}\tilde{g}$ and $c\tilde{f}$ are defined as $(f + g)^{\sim}$, $(fg)^{\sim}$, and $(cf)^{\sim}$, the set $\tilde{\mathbf{M}}$ of all function classes defined by complex-valued locally measurable functions is a complex linear algebra. Of course, $\tilde{\mathbf{M}}$ may also be described as the quotient of **M** by the ideal of function classes determined by the complex local null functions. We shall say that a subset of $\tilde{\mathbf{M}}$ is *closed* if and only if its complete inverse image under the map $f \to \tilde{f}$ is closed in **M**. It is easy to see that a subset $\tilde{\mathbf{K}}$ of $\tilde{\mathbf{M}}$ is closed if and only if whenever $\tilde{f}_n \in \tilde{\mathbf{K}}$ and $\tilde{f}_n \to \tilde{f}$, then $\tilde{f}$ is also in $\tilde{\mathbf{K}}$. The given class of closed sets defines a topology in $\tilde{\mathbf{M}}$, which we shall call the *topology of sequential convergence*. In this topology, $\tilde{\mathbf{M}}$ is a topological linear algebra.

The topologies in **M** and $\tilde{\mathbf{M}}$ have been defined in terms of sequential convergence; such definitions are generally convenient for analytic purposes, but leave relatively obscure the topological properties of the resulting spaces. As a result of the theory developed in Chap. 3, simple alternative characterizations in terms of neighborhoods which are topologically illuminating may be given. For example, when M is σ-finite, the topology on $\tilde{\mathbf{M}}$ may be defined by a metric, relative to which $\tilde{\mathbf{M}}$ is complete. This result, which is essentially a slight strengthening of a classical one, usually called the "Riesz-Fischer theorem," will emerge from several preliminary results of independent utility.

Since the convergence of a sequence $\tilde{f}_1, \tilde{f}_2, \ldots$ of function classes to a function class $\tilde{f}$ is equivalent to the l.a.e. convergence of the corresponding sequence of functions, it is appropriate in studying the topology of $\tilde{\mathbf{M}}$ to study in particular what can be said about the a.e. convergence of sequences of measurable functions. The following standard result is useful in a number of connections.

THEOREM 4.1   (EGOROFF)   *If f_1, f_2, $\ldots$ is a sequence of complex-valued measurable functions on a finite measure space $(S,\mathfrak{R},m)$ that converges a.e. to such a function f, then for every $\epsilon > 0$ there exists a measurable set A such that $m(A) < \epsilon$ and $f_n \to f$ uniformly on $S - A$.*

Proof   First observe as a matter of general set theory that if f_n and f are any complex functions on S, then the set

$$C = \bigcap_{i \geq 1} \bigcup_{k \geq 1} \bigcap_{j \geq k} [x : |f(x) - f_j(x)| < 1/i]$$

is precisely the subset on which $f_n \to f$ pointwise. Suppose f_n and f are measurable complex functions such that $f_n \to f$ a.e. Then C differs from S by a null set, and if

$$A_{ij} = [x : |f(x) - f_j(x)| < 1/i]$$

and $\epsilon > 0$, it follows that there is an index $k(i)$ such that

$$m\left(S - \bigcap_{j \geq k(i)} A_{ij}\right) < \epsilon/2^i.$$

Then, setting

$$A = \bigcup_{i \geq 1}\left(S - \bigcap_{j \geq k(i)} A_{ij}\right)$$

we see that $m(A) < \epsilon$.   The uniform convergence of $f_1, f_2, \ldots$ to f on $S - A$ follows from the relation

$$S - A = \bigcap_{i \geq 1} \left(\bigcap_{j \geq k(i)} A_{ij} \right).$$

Egoroff's theorem shows that a sequence of measurable functions on a finite measure space which converges a.e. converges almost uniformly in the following sense.

DEFINITION   A sequence $f_1, f_2, \ldots$ of complex-valued functions on a measure space *converges almost uniformly* if for every $\epsilon > 0$ there exists a locally measurable set A of extended measure less than ϵ with the property that the sequence converges uniformly on the complement of A.

It is easy to prove that a sequence which converges almost uniformly also converges l.a.e.

SCHOLIUM 4.1   *If $f_1, f_2, \ldots$ is a sequence of complex-valued locally measurable functions on a measure space which converges almost uniformly to a complex-valued locally measurable function f, then the extended measure of $[x : |f(x) - f_n(x)| \geq \epsilon]$ tends to 0 as n tends to ∞.*

Proof   Denote the measure space by $(S, \mathfrak{R}, m)$, and suppose ϵ and δ are positive.   Then there is a locally measurable set A and an integer j such that $m(A) < \delta$ and $|f(x) - f_i(x)| < \epsilon$ for all x in $S - A$ and all $i \geq j$.   Then, for $i \geq j$

$$[x : |f(x) - f_i(x)| \geq \epsilon] \subset A$$

so that $m[x : |f(x) - f_i(x)| \geq \epsilon] < \delta$.

Thus the sequence $f_1, f_2, \ldots$ in the statement above converges in measure to f in accordance with the following.

DEFINITION   A sequence $f_1, f_2, \ldots$ of functions on a measure space *converges in measure to a function f* in the case where, for every $\epsilon > 0$, $[x : |f(x) - f_n(x)| \geq \epsilon]$ is a locally measurable set whose extended measure tends to 0 as n tends to ∞.   The sequence *converges locally in measure* to f if, for every $\epsilon > 0$ and every measurable set E of finite measure, $[x \in E : |f(x) - f_n(x)| \geq \epsilon]$ is a measurable set whose measure tends to 0 as n tends to ∞.

The notion of convergence just defined is sometimes called *asymptotic convergence*, and is one of the oldest types of convergence in mathematics, having been involved implicitly in early work in probability.   It is usually called *convergence in probability*, or *stochastic convergence*, in the literature on probability and statistics.

The extent to which convergence in measure implies pointwise convergence is given in the following important result.

THEOREM 4.2 (RIESZ-FISCHER) *A sequence f_1, f_2, ... of complex-valued measurable functions on a measure space converges in measure to some complex-valued measurable function f if and only if the sequence is Cauchy in measure. In this event there is a subsequence f_{n_1}, f_{n_2}, ... which converges almost uniformly to f.*

Proof Suppose $f_1, f_2, \ldots$ converges in measure to f on the measure space $(S, \mathcal{R}, m)$. Then it is elementary to verify that for $\epsilon > 0$ and $\delta > 0$ there exists an integer $k(\epsilon, \delta)$ such that

$$m[x : |f_i(x) - f_j(x)| \geq \epsilon] < \delta$$

for all $i, j \geq k(\epsilon, \delta)$; and this is what is meant by the statement that the sequence is Cauchy in measure.

Conversely, suppose this condition is satisfied, and let $n_1 = k(2^{-1}, 2^{-1})$. Define n_i inductively by the equation

$$n_{i+1} = \max\, [n_i + 1,\ k(2^{-(i+1)},\ 2^{-(i+1)})].$$

Then $n_1 < n_2 < \cdots$, and

$$m[x : |f_{n_{i+1}}(x) - f_{n_i}(x)| \geq 2^{-i}] < 2^{-i}.$$

Let $S_i = [x : |f_{n_{i+1}}(x) - f_{n_i}(x)| \geq 2^{-i}]$. Then $m(S_i) < 2^{-i}$, and

$$m\left(\bigcup_{i \geq j} S_i \right) \leq 2^{-j+1}.$$

Setting $E = \bigcap_{j \geq 1} \bigcup_{i \geq j} S_i$, we find that

$$m(E) \leq 2^{-j+1}$$

for $j = 1, 2, \ldots$; hence $m(E) = 0$. For x outside E, there is an integer $j(x) > 0$ such that

$$x \in S - \bigcup_{i \geq j} S_i$$

and hence such that

$$|f_{n_{i+1}}(x) - f_{n_i}(x)| \leq 2^{-i}$$

for all $i \geq j(x)$. Thus, for any integer $k \geq j(x)$.

$$\sum_{i \geq k} |f_{n_{i+1}}(x) - f_{n_i}(x)| \leq 2^{-k+1}.$$

Hence the series

$$f_{n_1}(x) + \sum_{i \geq 1} [f_{n_{i+1}}(x) - f_{n_i}(x)]$$

is absolutely convergent for x outside E and uniformly convergent on $S - \bigcup_{i \geq j} S_i$. Thus the sequence $f_{n_1}, f_{n_2}, \ldots$ converges a.e. to a finite-valued measurable function f, and the convergence is uniform on the complement of a set of arbitrarily small measure. Now, by Scholium 4.1 this implies $f_{n_i} \to f$ in measure. Moreover, since the original sequence $f_1, f_2, \ldots$ is Cauchy in measure, it follows easily that $f_1, f_2, \ldots$ also converges to f in measure, and this completes the proof.

Next observe that for any measure space M, the notion of convergence in measure or of local convergence in measure makes sense in $\widetilde{M}$ when defined in the obvious fashion. Having made this remark, we shall now introduce a new topology on $\widetilde{M}$ in which convergence of a sequence is the same as the local convergence in measure of the sequence.

SCHOLIUM 4.2 *For any measure space $M = (S,\mathfrak{R},m)$ there is a neighborhood basis for a topology in $\widetilde{M}$ in which a basic neighborhood of an element $\tilde{f}$ depends on two positive numbers ϵ, δ and a measurable set E of finite measure and consists of all $\tilde{g}$ such that*

$$m[x \in E : |f(x) - g(x)| \geq \epsilon] < \delta.$$

The same topology is obtained if m is replaced by any nonnegative measure which is σ-finite on $\mathfrak{R}$ and has the same null sets as m.

Proof First we must show that the indicated system forms a neighborhood basis for a topology in $\widetilde{M}$. For this denote the basic neighborhood of $\tilde{f}$ defined above by $N(\tilde{f},E,\epsilon,\delta)$. Then $N(\tilde{f},E,\epsilon,\delta)$ contains $\tilde{f}$, and if $N(\tilde{f},E',\epsilon',\delta')$ is another such neighborhood, it is straightforward to check that their intersection contains the neighborhood $N(\tilde{f},E'',\epsilon'',\delta'')$, in which $E'' = E \cup E'$, $\epsilon'' = \min(\epsilon,\epsilon')$, and $\delta'' = \min(\delta,\delta')$. Next it must be shown that $N(\tilde{f},E,\epsilon,\delta)$ contains a neighborhood of each of its elements. Suppose, then, that $g \in N(\tilde{f},E,\epsilon,\delta)$. Let $A = [x \in E : |g(x) - f(x)| \geq \epsilon]$ and $A_i = [x \in E : |g(x) - f(x)|, \geq (1 - 1/i)\epsilon]$. Then $A_i \supset A_{i+1}$ and $A = \bigcap_i A_i$. Hence there exists an i such that $m(A_i) < \delta$. Let $\epsilon' = \epsilon/i$ and $\delta' = \delta - m(A_i)$. We shall show that $N(\tilde{f},E,\epsilon,\delta)$ contains $N(\tilde{g},E,\epsilon',\delta')$. For this purpose suppose $\tilde{h} \in N(\tilde{g},E,\epsilon',\delta')$. Then

$$m[x \in E : |h(x) - f(x)| \geq \epsilon] \leq m(A_i) + m[x \in E - A_i : |h(x) - f(x)| \geq \epsilon].$$

Now for x in $E - A_i$

$$|h(x) - f(x)| \leq |h(x) - g(x)| + |g(x) - f(x)|$$
$$< |h(x) - g(x)| + \epsilon - \epsilon/i.$$

Thus $[x \in E - A_i : |h(x) - f(x)| \geq \epsilon]$ is contained in $[x \in E : |h(x) - g(x)| \geq \epsilon/i]$, and since this set has measure less than $\delta - m(A_i)$, it follows that $h \in N(\tilde{f},E,\epsilon,\delta)$. Hence the indicated system is indeed a neighborhood basis for a topology in $\widetilde{M}$.

Next suppose m' is a nonnegative measure on $\mathfrak{R}$ such that $(S,\mathfrak{R},m')$ is a measure space, and suppose m and m' have the same null sets. To show that the neighborhood system defined by m' is equivalent to the given one defined by m, it is sufficient by symmetry to show that each neighborhood $N(\tilde{f},E,\epsilon,\delta)$ contains a corresponding neighborhood of $\tilde{f}$ defined by m'. For this, write

E as a countable union of sets E_1, E_2, ... in $\mathfrak{R}$, for which $m'(E_i)$ is finite, and set

$$F_i = \bigcup_{j \leq i} E_j, \qquad i = 1, 2, \ldots .$$

Then choose an i such that $m(E - F_i) < \delta/2$, and set $F = F_i$. Then $m'(F)$ is finite, and there exists a constant $\delta' > 0$ such that $m(A) < \delta/2$ whenever A is a measurable subset of F for which $m'(A) < \delta'$. For if not, there exists a sequence A_1, A_2, ... of measurable sets contained in F such that $m'(A_i) < 2^{-i}$, while $m(A_i) \geq \delta/2$. Setting $A = \limsup_i A_i$, we then see that

$$m'(A) \leq \sum_{i \geq j} m'(A_i) \leq 2^{-j+1}.$$

Thus $m'(A) = 0$, but since

$$m(A) = \lim_j m\left(\bigcup_{i \geq j} A_i \right) \geq \liminf_j m(A_j) \geq \delta/2$$

this contradicts the assumption that m and m' have the same null sets. If $N'(\tilde{f},F,\epsilon,\delta')$ is the set of elements $\tilde{g}$ for which

$$m'[x \in F : |f(x) - g(x)| \geq \epsilon] < \delta',$$

it follows that $N'(\tilde{f},F,\epsilon,\delta') \subset N(\tilde{f},E,\epsilon,\delta)$. Since $m'(F)$ is finite, this completes the proof.

DEFINITION The topology in $\widetilde{\mathbf{M}}$ which is characterized by the preceding result will be called the *topology of local convergence in measure*.

THEOREM 4.3 *For a σ-finite measure space M, the topology of sequential convergence in $\widetilde{\mathbf{M}}$ is the same as the topology of local convergence in measure; moreover, the topology may be defined by means of a metric relative to which $\widetilde{\mathbf{M}}$ is complete.*

Proof Let $M = (S,\mathfrak{R},m)$ and $\mathbf{M}$ the class of complex-valued measurable functions on M (since M is σ-finite there is no distinction between measurability and local measurability). To construct a metric on $\widetilde{\mathbf{M}}$, we shall first define a pseudometric on $\mathbf{M}$. For this purpose, let S_1, S_2, ... be any sequence of measurable sets such that $0 < m(S_i) < +\infty$ and $S = \bigcup_i S_i$. Then the equation

$$m'(A) = \sum_i \frac{1}{2^i} \frac{m(A \cap S_i)}{m(S_i)}$$

defines a finite countably additive measure on $\mathfrak{R}$ with the same null sets as m. Let I' denote the integral on the measure space $(S,\mathfrak{R},m')$, and for any f and g in $\mathbf{M}$, set

$$d(f,g) = I'\left(\frac{|f - g|}{1 + |f - g|} \right).$$

Since the integrand is a bounded measurable function on the finite measure space $(S,\mathcal{R},m')$, $d(f,g)$ is well defined and finite. From the elementary inequality

$$\frac{|a - c|}{1 + |a - c|} \leq \frac{|a - b|}{1 + |a - b|} + \frac{|b - c|}{1 + |b - c|}$$

it is readily verified that d is a pseudometric on $\mathbf{M}$; that is, d satisfies the usual axioms for a metric except that $d(f,g) = 0$ does not necessarily imply $f = g$. In fact, $d(f,g) = 0$ if and only if $f = g$ a.e.   From this it is easy to check that the definition $d(\tilde{f},\tilde{g}) = d(f,g)$ for function classes $\tilde{f}$ and $\tilde{g}$ is unambiguous and provides a metric on $\widetilde{\mathbf{M}}$.

Next, we shall show that the metric topology in $\widetilde{\mathbf{M}}$ defined by d is the same as the topology of local convergence in measure.   For this, let $\tilde{f}$ be a given function class, and $N_a(\tilde{f})$ the open sphere of radius a about $\tilde{f}$, that is, the set of all elements g such that $d(\tilde{f},\tilde{g}) < a$.   For any E in $\mathcal{R}$ and positive ϵ and δ, let $N(\tilde{f},E,\epsilon,\delta)$ denote the basic neighborhood for the topology of local convergence in measure consisting of all $\tilde{g}$ such that

$$m'[x \in E : |f(x) - g(x)| \geq \epsilon] < \delta.$$

If ϵ and δ are small enough, then $N_a(\tilde{f})$ contains $N(\tilde{f},S,\epsilon,\delta)$.   For if $\tilde{g} \in N(\tilde{f},S,\epsilon,\delta)$ and $A = [x : |f(x) - g(x)| \geq \epsilon]$, then $d(\tilde{f},\tilde{g}) \leq m'(A) + \epsilon m'(S - A) < \delta + \epsilon m'(S)$.   Conversely, a given neighborhood of the form $N(\tilde{f},E,\epsilon,\delta)$ contains a neighborhood of the form $N_a(\tilde{f})$ for sufficiently small a.   For suppose $d(\tilde{f},\tilde{g}) < a$ and $b > 0$, and let

$$B = \left[x : \frac{|f(x) - g(x)|}{1 + |f(x) - g(x)|} \geq b \right].$$

Then $m'(B) < a/b$, and if $0 < b < 1$,

$$B = \left[x : |f(x) - g(x)| \geq \frac{b}{1 - b} \right].$$

Now b can be chosen so that $0 < b < 1$ and $b/(1 - b) < \epsilon$.   Then a positive a can be chosen so that $a/b < \delta$, and for such an a, $N_a(\tilde{f})$ will be contained in $N(\tilde{f},E,\epsilon,\delta)$.   Thus the metric topology on $\widetilde{\mathbf{M}}$ coincides with the topology of local convergence in measure.

To show that $\widetilde{\mathbf{M}}$ is complete as a metric space, suppose that $\tilde{f}_1, \tilde{f}_2, \ldots$ is a Cauchy sequence, i.e., that $d(\tilde{f}_i,\tilde{f}_j) \to 0$ as $i, j \to +\infty$.   Since

$$m'\left[x : |f_i(x) - f_j(x)| \geq \frac{b}{1 - b} \right] < \frac{a}{b}$$

if $d(\tilde{f}_i,\tilde{f}_j) < a$ and $0 < b < 1$, the sequence $f_1, f_2, \ldots$ is Cauchy in measure on the space $(S,\mathcal{R},m')$.   By Theorem 4.2 there exists a complex-valued measurable function f on $(S,\Re,m')$, which is then also an element of $\mathbf{M}$, and a subsequence $f_{n_1}, f_{n_2}, \ldots$ which converges almost uniformly to f. By the dominated convergence theorem, $d(\tilde{f}_{n_i}, \tilde{f}) \to 0$ as well. It now follows, by a general argument valid in any metric space, that the entire sequence $\tilde{f}_1, \tilde{f}_2, \ldots$ converges to $\tilde{f}$ in the metric.

To prove the identity of the metric topology with the topology of sequential convergence, let $\tilde{\mathbf{K}}$ be a subset of $\widetilde{\mathbf{M}}$ which is closed under sequential convergence. Then $\tilde{\mathbf{K}}$ is also closed relative to the metric. For if $\tilde{f}_n \in \tilde{\mathbf{K}}$ and $d(\tilde{f}_n, \tilde{f}) \to 0$, then as just seen, there exists a subsequence n_i of the integers such that $f_{n_i} \to f$ a.e., and this implies that $\tilde{f} \in \tilde{\mathbf{K}}$. On the other hand, the dominated convergence theorem implies, as above, that a set closed under convergence in the metric is also closed under sequential convergence. This completes the proof.

Although the space $\mathbf{M}$ of complex-valued locally measurable functions on M is quite useful, it is nevertheless inappropriate for many applications, being in a sense too large and insufficiently structured; and certain submanifolds endowed with richer structures play more important roles.

The most important of these is the subspace $L_1(M)$, consisting of all integrable functions in $\mathbf{M}$, with the pseudometric $d(f,g) = I(|f - g|)$. In the topology defined by d, $L_1(M)$ is a linear topological space on which the integral is a continuous linear functional, by virtue of the inequality

$$|I(f) - I(g)| \le d(f,g).$$

Moreover, L_1 is essentially a complete metric space. More precisely, as we shall soon see, the space $\tilde{L}_1(M)$ consisting of function classes defined by elements of $L_1(M)$ is actually a complete metric space with the metric $d(\tilde{f}, \tilde{g}) = I(|f - g|)$. It will also be seen that the identity map of $\tilde{L}_1(M)$ into $\widetilde{\mathbf{M}}$ is continuous, but that in general its inverse is not. In other words, the topology on $\tilde{L}_1(M)$ derived from the metric above is generally stronger than that induced by the topology of $\widetilde{\mathbf{M}}$. In addition, $\tilde{L}_1$ provides a typical example of a Banach space.

DEFINITIONS   A *seminormed linear space* is a system consisting of a linear space $\mathbf{L}$ over the real or complex field, together with a nonnegative real functional (the seminorm) $x \to \|x\|$ on $\mathbf{L}$ such that

(*1*) $\|x + y\| \le \|x\| + \|y\|$
(*2*) $\|cx\| = |c|\, \|x\|$

where x and y are arbitrary in $\mathbf{L}$, and c is any scalar. The seminorm is called a *norm*, and $\mathbf{L}$ a *normed linear space* if $\|x\| > 0$ when $x \ne 0$. A normed linear space is a *Banach space* in case it is complete as a metric space with regard to the metric $d(x,y) = \|x - y\|$.

For a σ-finite measure space M, the metric in $\widetilde{M}$ constructed in the proof of Theorem 4.3 is not given by a norm; hence $\widetilde{M}$ is not a Banach space (nor can it be given the structure of a Banach space by an equivalent metric).

In proving that L_1 is essentially a Banach space, we shall show that the same is true for all of a useful similar class of spaces, called the L_p-spaces.

DEFINITIONS The L_p-*norm* of a locally measurable function f on a measure space M is defined as

$$[I(|f|^p)]^{1/p}, \qquad 1 \leq p < \infty,$$

and for $p = \infty$ it is defined as the greatest lower bound of all extended-real numbers a such that $|f(x)| \leq a$ locally almost everywhere. The L_p-norm is denoted by $\|f\|_p$, and the subscript p is occasionally omitted when it is implied by the context. If $1 \leq p < \infty$, $L_p(M)$ is defined as the set of all numerical locally measurable functions on M whose L_p-norm is finite. The space $L_\infty(M)$ is defined as the collection of all locally measurable f for which $\|f\|_\infty$ is finite. The L_p distance between two complex-valued locally measurable functions f and g is defined as $\|f - g\|_p$. For function classes the concepts of L_p-norm, the space $\tilde{L}_p(M)$, and L_p distance $\|\tilde{f} - \tilde{g}\|_p$ are defined similarly.

It is not a priori clear that the L_p-spaces are linear spaces or that the L_p-norm has any of the basic properties of a norm. The proof of this involves the following fundamental result, which is basic in a number of connections.

SCHOLIUM 4.3 (HÖLDER'S INEQUALITY) *If f and g are locally measurable functions on a measure space and $1/p + 1/q = 1$ where $1 < p < \infty$, then*

$$I(|fg|) \leq \|f\|_p \|g\|_q$$

and the inequality is also valid in the limiting case in which $p = 1$ and $q = \infty$.

Proof The case $p = 1$, $q = \infty$, follows from the observation that fg is locally measurable and $|fg| \leq \|g\|_\infty \|f\|$ l.a.e. Next suppose that $1 < p < \infty$. Then the inequality is trivial unless both $\|f\|_p$ and $\|g\|_q$ are finite and nonzero. In fact, as is easily seen by homogeneity, there is no loss of generality if it is assumed that $\|f\|_p = \|g\|_q = 1$. The proof then follows immediately by integration from the inequality

$$|f(x)g(x)| \leq \frac{|f(x)|^p}{p} + \frac{|g(x)|^q}{q}.$$

This in turn follows when $f(x)g(x) \neq 0$ upon appropriate substitution from the inequality

$$a^{1/p}b^{1/q} \leq \frac{a}{p} + \frac{b}{q}$$

for positive real numbers a and b.   To verify this it is enough to consider the case in which $a \geq b$.   By means of the substitutions $x = a/b$ and $t = 1/p$, the problem is then reduced to showing that

$$x^t - 1 \leq t(x - 1)$$

for $x \geq 1$, and this is elementary.

THEOREM 4.4   *If M is a measure space and $1 \leq p \leq \infty$, then $\tilde{L}_p(M)$ is a Banach space in the L_p-norm.*

Proof   We shall show first that $L_p(M)$ is a seminormed linear space over the complex field.   For this purpose suppose that f and g are elements of $L_p(M)$.   If $p = 1$ or $p = \infty$, the inequality $|f + g| \leq |f| + |g|$ shows that $f + g \in L_p(M)$, and it also follows without difficulty that $\|f + g\|_p \leq \|f\|_p + \|g\|_p$.   To show that the same is true when $1 < p < \infty$, first observe that $|f + g| \leq 2 \max (|f|,|g|)$.   This implies

$$|f + g|^p \leq 2^p \max (|f|^p,|g|^p) \leq 2^p(|f|^p + |g|^p).$$

Hence $f + g \in L_p(M)$.   Furthermore,

$$|f + g|^p = |f + g| \, |f + g|^{p-1} \leq |f| \, |f + g|^{p-1} + |g| \, |f + g|^{p-1}.$$

Hence $I(|f + g|^p) \leq I(|f| \, |f + g|^{p-1}) + I(|g| \, |f + g|^{p-1})$.   On   the   other hand, by Hölder's inequality,

$$I(|f| \, |f + g|^{p-1}) \leq \|f\|_p \, I(|f + g|^{q(p-1)})^{1/q},$$

and since $q(p - 1) = p$,

$$I(|f| \, |f + g|^{p-1}) \leq \|f\|_p \, \|f + g\|_p^{p/q} < +\infty.$$

Now this inequality is also valid with f and g interchanged, and since $p/q = p - 1$, it follows that

$$I(|f + g|^p) \leq \|f\|_p \, \|f + g\|_p^{p-1} + \|g\|_p \, \|f + g\|_p^{p-1}.$$

The triangle inequality

$$\|f + g\|_p \leq \|f\|_p + \|g\|_p$$

is trivial when $\|f + g\|_p = 0$, and otherwise follows upon division by $\|f + g\|_p^{p-1}$.   If c is a complex number, it is easy to verify that $\|cf\|_p = |c| \, \|f\|_p$.   Hence $L_p(M)$ is indeed a normed linear space, except for the fact that $\|f\|_p = 0$ only implies $f = 0$ a.e. or $f = 0$ l.a.e., in the case $p = \infty$. Thus $\tilde{L}_p(M)$ is a normed linear space.

The completeness of $\tilde{L}_p(M)$, which is the least elementary aspect of the theorem, follows for $1 \leq p < \infty$ from the Riesz-Fischer theorem.   For suppose $1 \leq p < \infty$ and $f_1, f_2, \ldots$ is a sequence of functions in $L_p(M)$ such that

$$I(|f_i - f_j|^p) \to 0$$

as $i, j \to \infty$.   Then the set on which $|f_i(x) - f_j(x)| \geq \epsilon$ is easily seen to have measure less than

$$\epsilon^{-p} I(|f_i - f_j|^p),$$

from which it follows that the sequence is Cauchy relative to convergence in measure.   Thus there is a complex-valued measurable f and subsequence f_{n_i} such that $f_{n_i} \to f$ almost uniformly.   Now $f_j - f = \lim_i (f_j - f_{n_i})$ a.e.; so by Fatou's lemma,

$$I(|f_j - f|^p) \leq \lim_i \inf I(|f_j - f_{n_i}|^p).$$

Hence $\|f_j - f\|_p \to 0$.

If $f_1, f_2, \ldots$ is a Cauchy sequence in $L_\infty(M)$, then there is a local null set N_{ij} such that $\|f_i - f_j\|_\infty = \sup |f_i(x) - f_j(x)|$ over x in the complement of N_{ij}.   The union N of the N_{ij} is a local null set, and the sequence converges uniformly on the complement of N to a bounded locally measurable function f, and this implies $\|f_i - f\|_\infty \to 0$.

There are two related results of a useful but more technical nature whose proofs are similar to that of the theorem.

Lemma 4.4.1   *If $f \in L_p(M)$, $1 \leq p < \infty$, and $f_1, f_2, \ldots$ is a sequence of measurable functions such that $\|f - f_n\|_p < 4^{-n/p}$, then $f_n \in L_p(M)$ and $f_n \to f$ almost uniformly.*

Proof   Let $S_i = [x: |f(x) - f_i(x)|^p > 2^{-i}]$.   Then $2^{-i} m(S_i) \leq \|f - f_i\|_p^p < 4^{-i}$, so that $m(S_i) \leq 2^{-i}$.   Then $f_n \to f$ uniformly on the complement of $\bigcup_{i > k} S_i = B_k$, say, and $m(B_k) \leq \sum_{i > k} m(S_i) \leq \sum_{i > k} 2^{-i} = 2^{-k} \to 0$.

COROLLARY 4.4.1   *If $\|f - f_n\|_p \to 0$, where f and the f_n are in $L_p(M)$, $1 \leq p < \infty$, and $f_n \to g$ a.e., then $g \in L_p(M)$ and $f = g$ a.e.*

Proof   By the preceding lemma, there is a subsequence f_{n_i} such that $f_{n_i} \to f$ a.u.   Since $f_{n_i} \to g$ a.e., it results that $f = g$ a.e.

COROLLARY 4.4.2   *If f is a real-valued function in $L_p(M)$, where $1 \leq p < \infty$, and M is the measure space determined by an integration lattice (S, L, J), then there exists a sequence $g_1, g_2, \ldots$ of bounded functions in $\mathbf{L}$ such that $\|f - g_n\|_p \to 0$.*

Proof   Let $A_n = [x: 1/n < |f(x)| \leq n]$ and $f_n = f\varphi_{A_n}$.   If m denotes the measure on M, then

$$m(A_n) < n^p I(|f|^p) < \infty.$$

Since f_n is bounded by n and supported by A_n, it follows that $f_n \in L_q(M)$ for $1 \leq q \leq \infty$.   At the same time

$$|f - f_n|^p \leq 2^p |f|^p$$

and $I(|f - f_n|^p) \to 0$ by dominated convergence. If g is any measurable function, then

$$
\begin{aligned}
|f_n - g|^p &= |f_n - g|^{p-1} \, |f_n - g| \\
&\le (|f_n| + |g|)^{p-1} \, |f_n - g| \\
&\le (n + \|g\|_\infty)^{p-1} \, |f_n - g|.
\end{aligned}
$$

Thus, to finish the proof, it will suffice to construct a sequence $g_1, g_2, \ldots$ of functions in $\mathbf{L}$ such that g_n is bounded by n and

$$
\|f_n - g_n\|_1 < (2n)^{-p}.
$$

For this choose h_n' in $\mathbf{L}'$ so that $f_n \le h_n'$ and

$$
I(f_n) \le I(h_n') < I(f_n) + (2n)^{-p-1}.
$$

If $h_n = \min(h_n', n)$, it follows that $h_n \in \mathbf{L}'$ and that

$$
\|f_n - h_n\|_1 = I(h_n) - I(f_n) < (2n)^{-p-1}.
$$

Now by the definition of $\mathbf{L}'$ there exists a function g_n' in $\mathbf{L}$ such that $g_n' \le h_n$ and

$$
I(h_n) - (2n)^{-p-1} < I(g_n') \le I(h_n).
$$

If $g_n = \max(g_n', -n)$, then $g_n \in \mathbf{L}$, $|g_n(x)| \le n$ for all x, and

$$
\|h_n - g_n\|_1 = I(h_n) - I(g_n) < (2n)^{-p-1}.
$$

Thus $\|f_n - g_n\|_1 < (2n)^{-p}$.

EXERCISES

1 Give examples of sequences of integrable functions on $(0,1)$ relative to Lebesgue measure which

    a. Converge in L_1 but converge pointwise nowhere.

    b. Converge pointwise everywhere to an integrable limit but do not converge in L_1.

2 For which values of a and p is x^{-a} in $L_p(0,1)$ or $L_p(1, \infty)$ relative to Lebesgue measure?

3 If M is a finite measure space, show that a function in $L_p(M)$, $1 \le p \le \infty$, is integrable. More generally, show that $L_{p_1} \subset L_{p_2}$ if $p_1 > p_2$.

4 If S is the unit disk in euclidean 2-space E_2, for which a and p is $|x|^{-a}$ in $L_p(S)$ or $L_p(E_2 - S)$? Generalize to euclidean n-space.

5 Show that, in general, on infinite measure spaces no L_p-space contains any $L_{p'}$-space for any distinct values of p and p'

6 If M is a measure space and $f \in L_p(M)$ for some $p \geq 1$, show that $\|f\|_\infty = \lim\limits_{p \to \infty} \|f\|_p$.

7 Construct an example in which $\|f\|_p$ is finite for $1 \leq p < \infty$ but $\|f\|_\infty = \infty$.

8 A *Banach algebra* is a linear topological algebra with the additional structure of a Banach space (as the source of its linear topological structure). Show that $\tilde{L}_\infty(M)$ is a Banach algebra for any measure space M and that it has the property $\|\tilde{f}^2\|_\infty = \|\tilde{f}\|_\infty^2$ for all f in $L_\infty(M)$.

9 An essential value of a locally measurable function f on a measure space may be defined as a value c such that for every neighborhood N of $c, f^{-1}(N)$ has positive (extended) measure; the set of all essential values is sometimes called the "essential range" of f.

   a. Show that the essential range of f is a closed set which depends only on $\tilde{f}$ (it is sometimes called the "spectrum" of $\tilde{f}$, or of f).
   b. Show that if f is a continuous function on $[0,1]$, then its essential range relative to $[0,1]$ as a measure space under Lebesgue measure is identical with its usual range.
   c. Show that, in general, the essential range is contained in the closure of the ordinary range.
   d. Show that for a locally measurable function f with $\|f\|_\infty > 0$ there exists a point in the essential range of absolute value $\|f\|_\infty$.
   e. Show that a complex scalar c is not in the essential range of f if and only if $(f - c)^{-1}$ exists as a bounded function l.a.e.

10 Show that if f and g are in L_p and L_q, respectively, over euclidean space relative to Lebesgue measure, and if $p^{-1} + q^{-1} = r^{-1} + 1$ with $r \geq 1$, then the convolution

$$(f * g)(x) = \int f(x - y)g(y)\, dy$$

exists for almost all x and is essentially in L_r (i.e., differs only on a null set from an element of L_r).   (*Hint:* Use Hölder's inequality.)

11 Let $\{f_n\}$ be a sequence of complex-valued measurable functions on a measure space M which converges in measure to the measurable function f.   Then

   a. Show that if ψ is any continuous function of a complex variable, then $\psi \circ f_n \to \psi \circ f$ in measure.
   b. Show that if the f_n are uniformly dominated by an integrable function (i.e., there exists an integrable extended-real function g such that $|f(x)| \leq g(x)$ for all x), then f is integrable and $I(f_n) \to I(f)$.
   c.* Show that if M is finite, then the distribution function for f_n converges pointwise to that for f as $n \to \infty$, at the latter's points of continuity.

12 Show that the system of all extended-real locally measurable function classes on a finite measure space is complete as a partially ordered set.   (*Hint:* Consider first the case of a uniformly bounded collection of function classes; then reduce to this case by an appropriate transformation.)

13 A subset of a partially ordered set with the property that any two elements are comparable is called a chain.   Show that if $\tilde{\mathbf{N}}$ is a chain of nonnegative extended-real locally measurable function classes on a finite measure space, with supremum $\tilde{g}$, then $I(\tilde{g})$ is the supremum of the $I(\tilde{f})$ for $\tilde{f} \in \tilde{\mathbf{N}}$.

14 Extend Exercises 12 and 13 to the case of a σ-finite measure space.   Show that without the σ-finiteness hypothesis, the conclusion of Exercise 12 is in general not valid.

15 The *abstract measure ring* of a measure space may be defined as the Boolean ring of all locally measurable sets modulo the ideal of all such null sets.   Show that the abstract measure ring is complete if and only if the lattice of all extended-real locally measurable function classes is complete.

16 A space with the property of Exercise 15 is sometimes called *localizable*. Show that a space is localizable if and only if there exists a collection E_λ of locally measurable sets of finite measure such that $E_\lambda \cap E_{\lambda'}$ has measure zero for $\lambda \neq \lambda'$ and such that for any locally measurable set A, $m(A) = \sum_\lambda m(A \cap E_\lambda)$, where m denotes the measure.

4.2 SET FUNCTIONS

The indefinite integral in general integration theory cannot be represented by a point function as in calculus, nor can set functions in general be represented by such.   The basic reason for this is the absence, in a general abstract space, of an analog to the intervals on the reals.   Nevertheless, set functions occur in a natural way in many applications; for example, mass, charge probability, and volume are all basically functions of sets.

In this section, we shall treat additive set functions on rings.   The main result to be proved is that the real-valued countably additive set functions on σ-rings are simply the differences of finite measures (essentially the "Jordan" decomposition), which, moreover, may be chosen to have disjoint supporting sets (a "Hahn" decomposition).   Such a decomposition is quite analogous to the representation of a real-valued point function as a difference of non-negative functions having disjoint supports.   Such real-valued countably additive set functions are frequently called *signed measures*.   More generally, one may consider complex-valued countably additive set functions, or functions whose values lie in a finite-dimensional euclidean space.   With any such measure m there is associated a finite nonnegative measure $|m|$, called the *total variation* of m.   The Radon-Nikodym theorem, to be proved in the next section, will show that m is simply the indefinite integral of an integrable function relative to $|m|$.   The results of this section all follow immediately from this, but they are needed for the proofs of the next section, and form a natural independent unit.

The following concept is convenient at this point; in a specialized form it is important in classical analysis.

DEFINITIONS   A real-valued function n on a ring $\Re$ of sets is said to be of *bounded variation* on a set E in $\Re$ in case the quantities

$$n^+(E) = \sup_{F \subset E, F \in \Re} n(F), \quad n^-(E) = \sup_{F \subset E, F \in \Re} [-n(F)]$$

are both finite.   If n is of bounded variation on all sets in $\Re$, it is said simply to be of bounded variation.   In any case, n^+ and n^- are called the *upper* and *lower variation functions* of n, and their sum $n^+ + n^-$ is called the *(total) variation function*, and designated $|n|$.

Example 4.2.1   Let M be a measure space, and s the indefinite integral of a real-valued integrable function f.   Let G denote the set on which f is positive.   For any measurable set E, $s^+(E) = \sup_{F \subset E} s(F)\,(F$ measurable); now $s(F) = s(F \cap G) + s(F - G)$; inasmuch as $s(F - G)$ is the integral of f over $F - G$, it is nonpositive, so that $s(F) \leq s(F \cap G) \leq s(E \cap G)$, showing that $s^+(E) \leq s(E \cap G)$.   On the other hand, choosing F as $E \cap G$ in the foregoing shows that $s^+(E) \geq s(E \cap G)$.   Thus $s^+(E) = s(E \cap G)$.   Similarly, or replacing f by $-f$, it follows that $s^-(E) = -s(E - G)$.   On setting $f^+ = \max\,(f,0)$ and $f^- = \max\,(-f,0)$, $s^\pm$ may be described as the indefinite integral of $f^\pm$, and the total variation function $|s|$ is the indefinite integral of $|f|$.

The importance of functions of bounded variation is evident from

SCHOLIUM 4.4   *Any countably additive real-valued function on a σ-ring of sets is of bounded variation.*

Proof   Let s be real and countably additive on the σ-ring $\Re$ of subsets of the set S.   As the basis of an indirect argument, suppose that s is not of bounded variation, and let A be a set in $\Re$ on which it is not of bounded variation.   A sequence $A_1, A_2, \ldots$ of elements of $\Re$ may now be constructed by induction in the following fashion.   Set $A_1 = A$, and assume that the A_i have been defined for $i \leq k$ in such a fashion that $|s|\,(A_i) = \infty$, $|s(A_i)| \geq i - 1$, and for $i < k$, $A_{i+1} \subset A_i$.   Since $|s|\,(A_k) = \infty$, either $s^+(A_k) = \infty$ or $s^-(A_k) = \infty$; in either case there exists a set B contained in A_k such that $|s(B)|$ exceeds any preassigned number (on which B naturally depends).   In particular, there exists a set B in $\Re$, $B \subset A_k$, such that $|s(B)| \geq |s(A_k)| + k$.   In case $|s|\,(B)$ should be infinite, let A_{k+1} be defined as B; in case $|s|\,(B)$ is finite, let A_{k+1} be defined as $A_k - B$.   In either case, the induction hypothesis is satisfied at the $(k + 1)$st stage; if $|s|\,(B) = \infty$, it is only necessary to verify that $|s(B)| \geq k$, as is clear; if $|s|\,(B) < \infty$, then $|s|\,(A_k - B) = \infty$, since otherwise $|s|\,(A_k)$ could readily be shown to be finite, by virtue of the "subadditivity" of $s^\pm$: $n^\pm(E \cup F) \leq n^\pm(E) + n^\pm(F)$, for any additive function n on a ring $\mathfrak{R}$ containing the sets E and F; in addition, $|s(A_{k+1})| = |s(A_k - B)| = |s(A_k) - s(B)| \geq |s(B)| - |s(A_k)| \geq k$.

At this point we make use of the assumed countable additivity of s, which implies that $s(A_k) \to s\left(\bigcap_k A_k\right)$; this implies in particular that $\{s(A_k)\}$ is bounded, which is incompatible with the inequalities: $|s(A_k)| \geq k - 1$.

Functions of bounded variation may be characterized in terms of measures much as in the example above. The main point is that they are precisely the differences of finite-valued measures; more exactly,

THEOREM 4.5   *If s is a bounded finitely (respectively, countably) additive function on a ring (respectively, σ-ring), then so also are its upper and lower variation functions $s^{\pm}$, and $s = s^+ - s^-$.*

Proof If E and F are elements of $\mathfrak{R}$ such that $F \subset E$, then the finite additivity of s implies that

$$s(F) = s(E) - s(E - F);$$

now taking the supremum of both sides of this equality, with respect to F, it results that

$$s^+(E) = \sup_{F \subset E, F \in \mathfrak{R}} [s(E) - s(E - F)] = s(E) + \sup_{F \subset E, F \in \mathfrak{R}} [-s(E - F)]$$

the last equality being valid because of the constancy of $s(E)$ as a function of F. Now as F ranges over the collection of all subsets of E which are in $\mathfrak{R}$, $E - F$ varies over the same collection, so that

$$\sup_{F \subset E, F \in \mathfrak{R}} [-s(E - F)] = \sup_{F \subset E, F \in \mathfrak{R}} [-s(F)] = s^-(E).$$

Thus $s^+(E) = s(E) + s^-(E)$ for all $E \in \mathfrak{R}$, which means that $s = s^+ - s^-$.

Now let E_1 and E_2 be arbitrary disjoint elements of $\mathfrak{R}$. As F varies over the elements of $\mathfrak{R}$ which are contained in $E_1 \cup E_2$, the sets $F_i = E_i \cap F$ ($i = 1, 2$) vary independently over the elements of $\mathfrak{R}$ which are contained in E_i; for it is evident that the F_i are such sets, and conversely, if F_i is a subset of E_i which is an element of $\mathfrak{R}$, they have the foregoing form, with $F = F_1 \cup F_2$. It follows that

$$s^+(E_1 \cup E_2) = \sup_{F \subset E_1 \cup E_2, F \in \mathfrak{R}} s(F) = \sup_{F_i \subset E_i, F_i \in \mathfrak{R}} [s(F_1) + s(F_2)]$$

$$= s^+(E_1) + s^+(E_2)$$

the last equality resulting from the independent variation of F_1 and F_2 within the indicated collections. Thus s^+ is finitely additive, and since $s^- = (-s)^+$, the same is true of s^-.

Now suppose that s is countably additive; to show that $s^{\pm}$ are likewise countably additive it suffices, for the reason just indicated, to treat the case of s^{+}.   Let A_1, A_2, ... be a sequence of mutually disjoint sets in $\Re$, and let B be an element of $\Re$ which is contained in their union E.   Then

$$s(B) = s\left[\bigcup_i (B \cap A_i)\right] = \sum_i s(B \cap A_i) \le \sum_i s^{+}(A_i),$$

so that $s^{-}(E) \le \sum_i s^{+}(A_i)$.   On the other hand, $E \supset \bigcup_{i=1}^{n} A_i$ and $s^{+}(E) \ge s^{+}\left(\bigcup_{i=1}^{n} A_i\right)$.   Since s^{+} is finitely additive, it follows that .

$$s^{+}(E) \ge \sum_{i=1}^{n} s^{+}(A_i).$$

Because this is true for every n, $s^{+}(E) = \sum_i s^{+}(A_i)$.

The decomposition $s = s^{+} - s^{-}$ is known as that of Jordan.   In the light of Example 4.2.1, it is natural to anticipate that s^{+} and s^{-} may be supported by disjoint sets, under suitable conditions, where a set function F is said to be supported by a set E in case it vanishes on all sets in its domain which are disjoint from E.   This is indeed the case, according to the following result, sometimes referred to as the "decomposition of Hahn."

COROLLARY 4.2.1   *Let s be a countably additive real-valued function on a σ-ring of sets.   Then the representation of s as the difference of its upper and lower variations is the unique decomposition of s as the difference of two measures having disjoint supports.*

Proof   First note that for any finite-valued countably additive measure m on a σ-ring $\Re$, there exists a set $A \in \Re$ which supports m.   For if $\lambda = \sup_{E \in \Re} m(E)$, then λ is finite because of the countable additivity of m, and if E_n is a sequence of sets in $\Re$ such that $m(E_n) \to \lambda$, it is not difficult to verify that the set $A = \bigcup_i E_i$ has the property in question.   In particular, each of s^{+} and s^{-} is supported by a set in $\Re$; let S denote the union of a set supporting s^{+} with a set supporting s^{-}.

To show that s^{+} and s^{-} have disjoint supports, it is no essential loss of generality to assume that $\Re$ consists of subsets of S, since they both vanish outside S.   Now let $\{E_i\}$ be a sequence of sets in $\Re$ such that $s(E_i) > s^{+}(S) - 2^{-i}$.   This means that $s^{+}(E_i) - s^{-}(E_i) > s^{+}(S) - 2^{-i}$; in view of the monotone (-increasing) character of s^{+}, which implies that $s^{+}(E_n) \le s^{+}(S)$, it follows that $s^{+}(S) - s^{-}(E_i) > s^{+}(S) - 2^{-i}$, which implies that $s^{-}(E_i) < 2^{-i}$.   On the other hand,

$$s^{+}(S - E_i) = s^{+}(S) - s^{+}(E_i) = s^{+}(S) - [s(E_i) + s^{-}(E_i)]$$
$$\le s^{+}(S) - s(E_i) < 2^{-i}.$$

Now setting $A = \lim\sup_i E_i$, then $S - A = \lim\inf_i (S - E_i)$, which set is contained in $\bigcup_{i>j} (S - E_i)$ for every fixed index j. From the estimate just made, it follows that $s^+(S - A) \leq \sum_{i>j} 2^{-i} = 2^{-j}$, which implies in turn that $s^+(S - A) = 0$. On the other hand, $\lim\sup_i E_i$ is contained in $\bigcup_{i>j} E_i$ for every fixed index j, which implies that $s^-(A) \leq \sum_{i>j} s^-(E_i) < \sum_{i>j} 2^{-i} = 2^{-j}$. From this it follows in turn that $s^-(A) = 0$, so that s^+ is supported by A while s^- is supported by $S - A$.

To complete the proof, it remains only to show that if $s = s_1 - s_2$ is any composition of s as a difference of finite measures having disjoint supports, then $s_1 = s^+$ and $s_2 = s^-$. Let G_1 and G_2 denote disjoint supports for s^+ and s^-, and let the F_i $(i = 1, 2)$ denote disjoint supports for the s_i. Then s^- vanishes on any subset E of $F_1 \cap G_1$ which is in $\Re$; thus $s(E) \geq 0$. Similarly, $s_1(E) = 0$, so that $s(E) \leq 0$. It follows that all the set functions presently in question are supported by $(F_1 \cap G_1) \cup (F_2 \cap G_2)$. It is easily seen that s^+ and s_1 agree on subsets of $F_1 \cap G_1$ which are in $\Re$, and vanish on $F_2 \cap G_2$; from this it follows that they agree on all sets in $\Re$, and consequently that $s^- = s_2$ also.

The extension of much of the foregoing argument to the case of additive set functions whose values are complex numbers, or more generally, lie in a finite-dimensional unitary space, is largely a straightforward application of the theory of such spaces, especially from the vantage point of the theory of differentiation of set functions treated in the next section. We therefore defer this extension, which is significant for applications, to the end of that section. For brevity, an additive set function whose values are, respectively, real-, complex-, or vector-valued is called, respectively, a *signed, complex, or vector-valued measure.*

The following notation is convenient and often employed. If s is a signed countably additive measure on a σ-ring $\Re$, of subsets of the set S, and if f is integrable on the measure space $(S, \Re, |s|)$, the quantity

$$\int f(x)\, ds^+(x) - \int f(x)\, ds^-(x)$$

is denoted

$$\int f(x)\, ds(x).$$

This evidently conforms with earlier notation when it is applicable, as well as with notation in the calculus. The quantity in question is called the (Stieltjes) *integral of f with respect to s.*

EXERCISES

1 Suppose n is the indefinite integral of a real function g in $L_1(M)$, for a measure space $M = (S,\mathcal{R},m)$.

a. Show that for any function f measurable with respect to $\mathcal{R}$

$$\int |f(x)|\, d|n|\,(x) = \int |f(x)|\,|g(x)|\, dm(x).$$

b. If f is measurable and

$$\int |f(x)|\, d|n|\,(x) < \infty,$$

show that

$$\int f(x)\, dn(x) = \int f(x)g(x)\, dm(x).$$

2 Let $\mathcal{R}$ be a σ-ring of subsets of a set S.

a. Show that all signed measures on $\mathcal{R}$ form a Banach space relative to the obvious linear structure and putative norm

$$\|n\| = \sup_{E \in \mathcal{R}} |n|\,(E).$$

b. Show that each set function in this space is supported by a set in $\mathcal{R}$.

c. If m is any positive measure on $\mathcal{R}$ such that $(S,\mathcal{R},m)$ is a measure space M, show that the indefinite integrals of all functions g in $L_1(M)$ form a closed subspace of the above Banach space.

3 *a.* For a given real-valued function μ on the real interval S, let m be the finitely additive function on the ring $\mathcal{R}$ defined in Lemma 2.2.2. If E is an interval in $\mathcal{R}$ with endpoints a and b, show that

$$|m|\,(E) = \sup \sum_{i \leq n} |\mu(t_{i+1}) - \mu(t_i)|$$

over all partitions $a = t_1 < t_2 < \cdots < t_{n+1} = b$ of E. The point function μ is said to be of *bounded variation on S* if

$$\sup |m|\,(E) < +\infty,$$

where the supremum is taken over all E as above, and the indicated supremum is called the *total variation of μ over S*.

b. If μ is real-valued, show that μ is of bounded variation on S if and only if it is the difference of two bounded monotone functions. (*Hint:* Use Theorem 4.5.)

c. Prove a result along the lines of (*b*) for the case in which μ is complex-valued and of bounded variation on S.

d. Show that a function of bounded variation on S has left- and right-hand limits at interior points of S and finite right- and left-hand limits, respectively, at the left and right endpoints of S.

e. Show that a function of bounded variation on S has at most countably many discontinuities.

f. Show that for any function of bounded variation, there is a unique (left-) right-continuous function of bounded variation to which it is equal, except possibly at its points of discontinuity. A function of bounded variation is said to be *normalized* if it is left- or right-continuous.

4 *a.* If μ is a real-valued function of bounded variation on an open interval S, show that there exists a unique signed measure m on the Borel subsets of S such that

$$m[(a,b)] = \mu(b) - \mu(a)$$

for all points a and b in S with $a < b$ at which μ is continuous.

(*Hint:* Use Exercise 3*d* to define m at single-point subsets of S.)

b. If the function μ in (*a*) is right-continuous, show that m is characterized by countable additivity and the condition

$$m[(a,b)] = \mu(b) - \mu(a)$$

for all points a and b in S with $a < b$.

c. If the function μ in (*a*) is left-continuous, find another characterization of m similar to that given in (*b*).

d. If μ_1 and μ_2 are real-valued right-continuous functions of bounded variation on S, and m_1 and m_2 are the corresponding set functions as in (*a*), show that $m_1 = m_2$ if and only if μ_1 and μ_2 have the same points of discontinuity and differ by a fixed constant at the points of continuity.

5 The *Fourier-Stieltjes transform of* a signed measure n on the reals—by which is meant a signed measure on the Borel subsets of the reals—is defined as the function N given by the equation

$$N(t) = \int_{-\infty}^{\infty} e^{itx}\, dn(x).$$

Show that $N(t)$ actually exists and is a continuous function of the real variable t. If n is supported by a bounded set, show that N can be extended to an entire function in the complex plane.

6 If μ is a real-valued function of bounded variation on a real interval S, and m the corresponding signed measure on the Borel subsets of S, the integral

$$\int_S f(t)\, dm(t)$$

of an integrable function f is frequently written

$$\int f(t)\, d\mu(t)$$

and called the Lebesgue-Stieltjes integral of f relative to μ.   Suppose f is a non-negative measurable function on a measure space and that

$$\mu(t) = \text{meas}\,[x : f(x) > t]$$

for t in the open interval $S = (0, + \infty)$.

    a. Show that μ is monotone-decreasing and right-continuous on S.

    b. If the measure space is finite, show that μ is bounded and that

$$I(f) = - \int_S t\,d\mu(t)$$

whether or not f is integrable.

    c. Show that $I(f) = - \lim_{\epsilon \to 0^+} \int_\epsilon^\infty t\,d\mu(t)$ for any measure space.

7 *Integration by parts.*   Another normalization for functions of bounded variation is the symmetrical one in which $f(x) = \frac{1}{2}[f(x + 0) + f(x - 0)]$.   Show that if f and g are of bounded variation and either one is symmetrically normalized or the other is continuous, at every point of the interval $[a,b]$, then

$$\int_a^b f\,dg + \int_a^b g\,df = f(b + 0)g(b + 0) - f(a - 0)g(a - 0).$$

[*Hint:* Apply Fubini's theorem to the characteristic function of the triangle $[(x,y) : a \le x \le b, y \le x]$ relative to the direct product of the signed measures determined by f and g, to obtain the result that

$$\int_a^b (f(b + 0) - f(x - 0))\,dg(x) = \int_a^b (g(x + 0) - g(a - 0))\,df(x).$$

Combine this with the result of interchanging f and g.]

8   Let m be a countably additive function on a σ-ring of sets to the extended reals exclusive of $-\infty$.   If the entire basic set S is a countable union of sets on which m is finite, show that m has a Hahn-Jordan decomposition.   If it is only assumed that each set of infinite measure contains a subset of finite positive measure, is this result true?

9   Let M denote the measure space indicated in the paragraph preceding the exercises, and let $L(f) = \int f(x)\,ds(x)$, for arbitrary $f \in L_1(M)$.   Show that L is a linear functional on L_1, that $|L(f)| \le \|f\|_1$, and that L is the unique linear functional on L_1 having this last property and such that $L(\varphi_E) = s(E)$ for any set $E \in \mathcal{R}$.

4.3   DIFFERENTIATION OF SET FUNCTIONS

If f is integrable on a measure space $(S,\mathcal{R},m)$, its indefinite integral F is a signed measure on the ring of all measurable sets.   From an intuitive viewpoint, it is reasonable to describe f as a kind of derivative of F with respect to the given measure m.   It is naturally possible to define f as this derivative, but this leaves open the question of when a given signed measure will have a derivative relative to a given measure m.   This section is devoted to a study of the question.

THEOREM 4.6   (LEBESGUE DECOMPOSITION)   *If n is a signed measure on the measurable subsets of a measure space M, then there exists a function f in $L_1(M)$, unique a.e., such that n is the sum of the indefinite integral of f and a signed measure supported by a null set.*

Proof   To see that such an f is unique a.e., suppose f and g both have the indicated property.   Then the indefinite integral of $f - g$ is supported by a null set, and hence identically 0.   Thus $f - g = 0$ a.e.

By Theorem 4.5, the problem of establishing the existence of f reduces readily to the special case in which n has only finite nonnegative values.

Suppose, then, that n is nonnegative and finite-valued on the σ-ring $\mathcal{R}$ of measurable sets.   Then, by the proof of Theorem 4.5, there exists a set A in $\mathcal{R}$ which supports n.   Let $\mathbf{C}$ be the collection of all nonnegative f in $L_1(M)$ vanishing outside A such that $n - F \geq 0$, where F is the indefinite integral of f.   Then $\mathbf{C}$ contains max (f,g) whenever it contains both f and g.   For if $E \in \mathcal{R}$ and

$$B = [x \in E : f(x) \geq g(x)],$$

then B is a measurable set, and

$$\int_E \max{(f,g)} = \int_B f + \int_{E-B} g \leq n(B) + n(E - B) = n(E).$$

Thus there exists a monotone-increasing sequence $f_1, f_2, \ldots$ of functions in $\mathbf{C}$ such that

$$\lim_k I(f_k) = \sup_{g \in \mathbf{C}} I(g).$$

If $f' = \lim_k f_k$, it follows that

$$I(f') = \sup_{g \in \mathbf{C}} I(g) \leq n(A).$$

Let $f = f'$ on the set, where f' is finite, and set $f = 0$ on the complementary null set.   Then $f \in \mathbf{C}$, and the set function $s = n - F$ is a finite nonnegative measure on $\mathcal{R}$ supported by A.   Moreover, s has the property that if g is any nonnegative function in $L_1(M)$ such that $s(E) \geq \int_E g$ for all measurable sets E, then $g = 0$ a.e.

To complete the proof, it suffices to show that any such s is supported by a null set in $\mathcal{R}$.   For this, observe that as A is a countable union of sets of finite measure, there exists a function g in $L_1(M)$ which is strictly positive on A and vanishes outside A.   Let s_i be the countably additive function on $\mathfrak{R}$ defined by

$$s_i(E) = s(E) - \int_E i^{-1}g, \qquad i = 1, 2, 3, \ldots.$$

Then it follows readily from Corollary 4.2.1 that there exist measurable sets B_i such that $B_i \subset A$, s_i^+ is supported on B_i, and s_i^- is supported on $A - B_i$. If E is any measurable set, it follows that

$$0 \leq s_i(E \cap B_i) = s(E \cap B_i) - \int_{E \cap B_i} i^{-1}g \leq s(E) - \int_E i^{-1}g\varphi_{B_i}.$$

Hence B_i is a null set, by our assumption about s, and the union B of the B_i is also a null set.   Moreover, $A - B \subset A - B_i$, so that

$$s_i(A - B) = -s_i^-(A - B) \leq 0.$$

This implies

$$s(A - B) \leq i^{-1}\int_{A-B} g, \qquad i = 1, 2, \ldots,$$

and hence $s(A - B) = 0$, i.e., s is supported by B.

DEFINITIONS   An extended-real function n on a σ-ring $\mathfrak{R}$ of sets is said to be *absolutely continuous with respect to a signed measure m* on $\mathfrak{R}$ if

(1) $n(E) = 0$ whenever E is a set in $\mathfrak{R}$ for which $|m|(E) = 0$.

(2) Whenever E is a set in $\mathfrak{R}$ which is a countable union of subsets in $\mathfrak{R}$ on which m is finite, it is likewise a countable union of subsets in $\mathfrak{R}$ on which n is finite.

On the other hand, n is said to be *singular* with respect to m if n is supported by a set N such that $|m|(E \cap N) = 0$ for all E in $\mathfrak{R}$.   Two signed measures m and n are called *equivalent* if each is absolutely continuous with respect to the other.

COROLLARY 4.3.1   (RADON-NIKODYM)   *If n is a signed measure on the measurable subsets of a measure space $M = (S, \mathfrak{R}, m)$, then n is the indefinite integral of a function f in $L_1(M)$ if and only if it is absolutely continuous with respect to m.*

Proof   If $f \in L_1(M)$ and $n(E) = \int_E f(x)\, dm(x)$ for all E in $\mathfrak{R}$, it is evident that $n(E) = 0$ whenever $E \in \mathfrak{R}$ and $m(E) = 0$.   Since n is finite-valued, it follows that n is absolutely continuous with respect to m.

Conversely, suppose n is absolutely continuous with respect to m.   Then, by the theorem, $n = F + s$, where F is the indefinite integral of a function f in $L_1(M)$, and s is supported by a null set in $\mathfrak{R}$.   Since n and F are both absolutely continuous with respect to m, their difference s also has this property.   But any such function supported by a null set is obviously 0, and therefore $n = F$.

There is a variant of the corollary which is applicable to general nonnegative measures.

DEFINITION   A nonnegative measure n on the measurable subsets of a measure space $M = (S,\mathcal{R},m)$ is *locally an indefinite integral* if for each set A in $\mathcal{R}$ there exists a nonnegative finite-valued measurable function $f = f_A$ such that

$$n(E \cap A) = \int_E f(x)\, dm(x)$$

for all E in $\mathcal{R}$.

COROLLARY 4.3.2   (RADON-NIKODYM)   *Suppose n is a nonnegative measure on the measurable subsets of a measure space $M = (S,\mathcal{R},m)$. Then n is locally an indefinite integral if and only if n is absolutely continuous with respect to m.*

Proof   Suppose n is locally an indefinite integral, and let A be a set in $\mathcal{R}$. Then $n(A) = \int_A f(x)\, dm(x)$, where f is a nonnegative finite-valued measurable function on M.

To prove that A is a countable union of sets in $\mathcal{R}$ on which n has finite values, observe first that A is a countable union of sets A_i in $\mathcal{R}$ on which m assumes finite values.   For each set A_i, let

$$A_{ij} = [x \in A_i : f(x) \leq j], \qquad j = 1, 2, \ldots .$$

Then $A_{ij} \in \mathcal{R}$, A_i is the union of the A_{ij}, and $n(A_{ij})$ is evidently finite.   Thus A is a countable union of sets in $\mathcal{R}$ on which n is finite.   Moreover, in the event that $m(A) = 0$, it is obvious that $n(A) = 0$ as well, so that n is absolutely continuous with respect to m.

Conversely, suppose n is absolutely continuous with respect to m, and let A be a set in $\mathcal{R}$.   Then A is a countable disjoint union of sets A_i in $\mathcal{R}$ on which n assumes finite values.   Let n_i be the function on $\mathcal{R}$ defined by

$$n_i(E) = n(E \cap A_i).$$

Then n_i is a finite positive measure on $\mathcal{R}$ absolutely continuous with respect to m.   By Corollary 4.3.1, n_i is the indefinite integral of a nonnegative function f_i in $L_1(M)$.   Since n_i is supported by A_i, we may assume that f_i vanishes outside A_i.   It then follows that $f = \sum_i f_i$ is a nonnegative finite-valued measurable function on M.   Moreover, if $E \in \mathcal{R}$, then

$$\int_E f(x)\, dm(x) = \sum_i \int_E f_i(x)\, dm(x) = \sum_i n(E \cap A_i) = n(E \cap A).$$

Thus n is locally an indefinite integral.

For σ-finite measure spaces $M = (S,\mathcal{R},m)$, that is, ones for which $S \in \mathcal{R}$, the preceding result reduces to the following.

COROLLARY 4.3.3 (RADON-NIKODYM)   *A nonnegative measure n on the measurable subsets of a σ-finite measure space $M = (S,\mathcal{R},m)$ is the indefinite integral of a nonnegative finite-valued measurable function f if and only if it is absolutely continuous with respect to m.*

DEFINITIONS   A function n on a σ-ring $\mathcal{R}$ of subsets of a set S is said to be *differentiable* with respect to a measure m on $\mathcal{R}$ in case $(S,\mathcal{R},m)$ is a measure space and n is the indefinite integral of a finite-valued, integrable, or nonnegative locally measurable function f on $(S,\mathcal{R},m)$. When n is differentiable with respect to m, the function f, which is unique within equivalence, is called the (Radon-Nikodym) *derivative* of n with respect to m, and f is denoted by dn/dm.

A nonnegative measure n on the measurable subsets of a measure space $M = (S,\mathcal{R},m)$ which is absolutely continuous with respect to m need not be differentiable with respect to m. All that can be said, in general, is that n has *local derivatives* in the obvious sense, specified by Corollary 4.3.2. For some measure spaces, called *localizable* (see the exercises), the family of local derivatives may always be pieced together to give a locally measurable function extending each of the local derivatives. For such a space, the result of Corollary 4.3.2 has an obvious reformulation. On the other hand, not all measure spaces are localizable; however, one which is not has never arisen in analytical practice, although an example may be given to show that such exist.

SCHOLIUM 4.5   *If a function p on a σ-ring $\mathcal{R}$ is differentiable with respect to a measure n, and n with respect to a measure m, then p is differentiable with respect to m and*

$$\frac{dp}{dm} = \frac{dp}{dn}\frac{dn}{dm}.$$

Proof The hypothesis is equivalent to the assumptions that there exist finite-valued locally measurable functions f and g with $g \geq 0$ such that

$$p(E) = \int_E f(x)\,dn(x)$$

and
$$n(E) = \int_E g(x)\,dm(x)$$

for all sets E in $\mathcal{R}$.   The conclusion is equivalent to the assertion that

$$p(E) = \int_E f(x)g(x)\,dm(x)$$

for all E in $\Re$.   For the proof of this it is evidently enough to show that

$$\int_E f(x)\, dn(x) = \int_E f(x)g(x)\, dm(x)$$

for any nonnegative locally measurable f.   Now this is clear when f is the characteristic function of a locally measurable set.   By linearity it is also valid for arbitrary step functions based on the ring $\Re^+$ of locally measurable sets.   Since any nonnegative locally measurable function is the limit of an increasing sequence of nonnegative such step functions, the result follows, in general, from the continuity properties of the integral.

The following criterion is sometimes useful.

SCHOLIUM 4.6   *A signed measure n on a σ-ring $\Re$ is absolutely continuous with respect to the nonnegative measure m on $\Re$ if and only if for every $\epsilon > 0$ there is a $\delta > 0$ such that $|n|\,(E) < \epsilon$ whenever $m(E) < \delta$.*

Proof   As the basis of a proof by contradiction of the "only if" part of the scholium, suppose that for some $\epsilon > 0$ there is no δ with the property indicated above.   Then there exists a sequence $E_1, E_2, \ldots$ of sets in $\Re$ such that $m(E_i) < 2^{-i}$, while $|n|\,(E_i) \geq \epsilon$.   Setting $E = \lim\sup_i E_i$, we see that

$$m(E) \leq \sum_{i \geq j} m(E_i) \leq 2^{-j+1}.$$

Thus $m(E) = 0$, but since $|n|$ is finite and countably additive,

$$|n|\,(E) = \lim_j |n| \left(\bigcup_{i \geq j} E_i \right) \geq \lim_j \inf |n|\,(E_j) \geq \epsilon,$$

which contradicts the absolute continuity of $|n|$ with respect to m.

The proof of the "if" part of the scholium is virtually immediate.

We turn now to the question deferred at the end of the preceding section, namely, the consideration of vector-valued measures.   For simplicity, only the case of a countably additive measure on a σ-ring is treated.

DEFINITIONS   A measure s on a σ-ring $\Re$ having values in a finite-dimensional linear euclidean space **H** is, respectively, of *bounded variation/ absolutely continuous* with respect to a nonnegative measure m on $\Re$ in case each component of s is of bounded variation/absolutely continuous, in the sense previously defined.   A component of s is here a signed measure of the form $(\gamma \circ s)(E)$, where γ is an arbitrary linear functional on **H**.

If f is a measurable function having values in **H**, the integral $I = \int f(x)\, ds(x)$ with respect to a numerically valued measure s is defined provided $\gamma \circ f$ is integrable for every linear functional γ on **H**, by the condition that

$$\gamma \left[\int f(x)\, ds(x) \right] = \int (\gamma \circ f)(x)\, ds(x),$$

where the integral on the right is defined as earlier; since it exists and is a linear functional of γ, it defines an element of the dual of the dual of $\mathbf{H}$, which means it has the form $\gamma(I)$ for some vector I in $\mathbf{H}$ by the canonical isomorphism of the second dual with the original space, in the case of a finite-dimensional space.   Thus the indicated integral exists and is unique under the indicated assumptions.

The concept of total variation may be extended in various ways, of which one of the simpler is as follows: Relative to a choice of basis in $\mathbf{H}$, say, e_1, e_2, $\ldots$, e_n, s has the form $s(E) = s_1(E)e_1 + s_2(E)e_2 + \cdots + s_n(E)e_n$, where each $s_i(E)$ is a signed measure.   These measures are countably additive, assuming that s is such, in particular of bounded variation, showing that the countable additivity of s implies that it is of bounded variation [note that $(\gamma \circ s)(E)$ is a finite linear combination of the $s_i(E)$, for any linear functional γ]. It follows also that there exists a nonnegative countably additive measure m on $\mathcal{R}$ relative to which s is absolutely continuous and which is absolutely continuous relative to $|s_1| + |s_2| + \cdots + |s_n|$, for example, this sum itself. It follows by consideration of components in the fashion just indicated that s is the indefinite integral of a measurable vector-valued function f in the sense that for any linear functional γ,

$$(\gamma \circ s)(E) = \int_E \gamma \circ f(x)\, dm(x);$$

this is equivalent to the relations $f(x) = \sum_i f_i(x)e_i$, where $f_i = ds_i/dm$.   The total variation function $|s|$ for s is now definable by the equation

$$|s|\,(E) = \int_E |f(x)|\, dm(x),$$

where $|\cdot|$ indicates lengths in H; if a different measure m' on $\mathcal{R}$ is employed, so that $s(E) = \int_E f'(x)\, dm'(x)$, then the chain rule (Scholium 4.5) shows that $f'(x) = f(x)(dm/dm')$, from which it follows that

$$\int_E |f(x)|\, dm(x) = \int_E |f'(x)|\, dm'(x),$$

showing that $|s|$ is uniquely defined.   It is easily inferred from the definition that

$$\left| \int_E g(x)\, ds(x) \right| \leq \sup_x |g(x)|\, |s|\,(E)$$

for any real-valued integrable function g.

A particularly important case is that of a complex measure, which by the isomorphism of the complex numbers with their usual norm to a two-dimensional linear euclidean space, falls under the foregoing treatment.

To some extent the developments just presented can be extended to the case of measurable functions with values in Banach spaces, but some essential complications intervene; aspects of this further development will be treated later, as they become relevant.

EXERCISES

1   If m and n are countably additive measures on a σ-ring and n is singular with respect to m, show that m is singular with respect to n.

2   Construct a monotone-increasing function on the reals such that the corresponding measure is not absolutely continuous with respect to Lebesgue measure.

3   Construct a complex measure n on the reals which is absolutely continuous with respect to Lebesgue measure m and has the property that dn/dm is not in L_∞.

4   Give an example of a finite measure space $(S,\mathcal{R},m)$ and a nonnegative measure n on $\mathcal{R}$ such that $n(E) = 0$ whenever $E \in \mathcal{R}$ and $m(E) = 0$ but which is not differentiable with respect to m.

5   *a.* Let $(S,\mathcal{R},m)$ be a measure space, $\mathcal{Q}$ a sub-σ-ring of $\mathcal{R}$, and n the restriction of m to $\mathcal{Q}$. Show that, for any f in $L_1((S,\mathcal{R},m))$, there exists an f' in $L_1((S,\mathcal{Q},n))$ such that

$$\int_E f'(x)\, dn(x) = \int_E f(x)\, dm(x)$$

for all E in $\mathcal{Q}$, and that f' is unique a.e.

b. Let T denote the map $f \to f'$. Show that (i) T is linear, (ii) $T^2 = T$, and (iii) $\|Tf\|_1 \leq \|f\|_1$, $T(fg) = (Tf)g$, when g is bounded and measurable with respect to $\mathcal{Q}$ (f' is called the *conditional expectation* of f with respect to $\mathcal{Q}$).

6   Let $\mathcal{C}$ denote the collection of all finite subrings of the σ-ring of sets $\mathcal{R}$, ordered by inclusion. Let m and n be finite countably additive measures on $\mathcal{R}$, and assume that n is absolutely continuous with respect to m. Show that dn/dm is the limit in $L_1(\mathcal{R},m)$ of the $d(n \mid \mathcal{F})/d(m \mid \mathcal{F})$, with respect to $\mathcal{F}$ as a general element of the directed system $\mathcal{C}$. (*Hint:* Let g be integrable with respect to m and $g_{\mathcal{F}}$ its conditional expectation with respect to $\mathcal{F}$, and show that $\lim_{\mathcal{F}} g_{\mathcal{F}} = g$ in the L_1-topology.)

7   Show that the total variation over a measurable set E of a vector measure s may also be defined as the supremum of $|s(E_1)| + |s(E_2)| + \cdots + |S(E_n)|$ over all finite collections $E_1, E_2, \ldots, E_n$ of mutually disjoint measurable subsets of E.

8   Suppose M is a measure space $(S,\mathcal{R},m)$ on which the lattice of extended-real function classes is complete as a partially ordered set. If n is a nonnegative measure on $\mathcal{R}$ which is absolutely continuous with respect to m, prove that n is differentiable with respect to m.

9   Construct an example of a measure space in which the lattice of extended-real function class is not complete as a partially ordered set. (A space for which this lattice is complete, as in Ex. 10, is called *localizable*.)

10   The quotient ring of the ring of all locally measurable sets in a measure space modulo the ideal of all local null sets is sometimes called the "measure ring" of the space.   Show that the measure ring of a direct sum of finite spaces is complete as a partially ordered set and that the induced measure function $\tilde{m}$ is completely additive; i.e.,

$$\tilde{m}\left(\bigcup_i \tilde{E}_i\right) = \sum_i m(E_i)$$

for any mutually disjoint collection $\{\tilde{E}_i\}$ in the measure ring, where i ranges over a not necessarily countable index set.

11   Two measure spaces may be called *weakly equivalent* in case their measure rings are algebraically isomorphic.   Show that $[0,1]$ and $[-\infty,\infty]$, both under Lebesgue measure, are weakly equivalent.

12   If $(S,\mathcal{R},m)$ is a σ-finite measure space, show that there is a finite nonnegative measure n on $\mathcal{R}$ which is equivalent to m.

13   Let f be a nonnegative finite locally measurable function on a measure space $(S,\mathcal{R},m)$ and n its indefinite integral.

    a. Show that n is supported by the set A, where $f(x) > 0$.

    b. Show that $(S,\mathcal{R},n)$ is a measure space (which is not necessarily a direct sum of finite spaces) and that on this space $S - A$ is a local null set.

    c. Show that m has the Lebesgue decomposition $m = m_1 + m_2$ on $(S,\mathcal{R},n)$, where for E in $\mathcal{R}$,

$$m_1(E) = \int_{E \cap A} \frac{dn(x)}{f(x)}$$

    and m_2 is supported by $S - A$.

    d. Conclude that m and n are equivalent measures on $\mathcal{R}$ if and only if the set where $f(x) = 0$ is a local null set on $(S,\mathcal{R},m)$, and that in this event $dm/dn = 1/f$.

V

LOCALLY COMPACT
AND EUCLIDEAN
SPACES

5.1 FUNCTIONS ON LOCALLY COMPACT SPACES

Except in the examples, the set S on which the measures have been defined (more precisely, on certain subsets of which it has been defined) has been an abstract set, devoid of any special structure. In the particular case in which S has additionally the structure of a topological space, e.g., when S is euclidean space, it is natural to consider the relations between the topological and measure-theoretic features of S. For example, elementary analysis suggests that continuous functions should be measurable; this depends, however, on the existence of a suitable relationship between the measure and the topology.

Although results connecting integration theory and topology exist for a variety of types of topological spaces—metric, completely regular, etc.—the case which above all others occurs frequently in practice and for which there is a simple and effective theory is that of a *locally compact space* (understood as always to be Hausdorff). This section is devoted to two results in general topology which relate to functions on locally compact spaces. Although these results are rather commonly useful, because they involve analytical as well as topological aspects they are not always included in the type of introductory course in topology which this book presupposes.

123

First we show the existence of what is known as a "local partition of unity"; this makes possible the localization of some integration-theoretic and other questions on locally compact spaces. There are many refinements and variants, providing differentiable functions in the case of a manifold, but the following result is adequate for our purposes and is quite representative of the general idea of these other results.

It will be convenient to employ the notation $C_0(S)$ for the collection of all continuous complex-valued functions on a given topological space S, each of which vanishes outside of some compact set .

SCHOLIUM 5.1   *Let K be a nonempty compact subset of a locally compact space S, and $\{V_i\}$ a finite covering of K by open subsets of S. Then there exist corresponding functions f_i in $C_0(S)$ such that*

(1) $0 \leq f_i \leq 1$.
(2) V_i *contains the support of* f_i.
(3) $0 \leq \sum_i f_i \leq 1$.
(4) $\sum_i f_i(x) = 1$ *for all x in K.*

Proof   Each point in K has an open neighborhood whose closure is compact and contained in some V_i. A finite number of such neighborhoods cover K. Thus there exist compact sets K_i such that $K_i \subset V_i$ and

$$K \subset \left(\bigcup_i K_i \right)^0$$

where for any set X, X^0 denotes the interior of X. If K_i is nonempty, let g_i be a nonnegative function in $C_0(S)$, which is strictly positive on K_i and whose support is contained in V_i. If K_i is empty, let $g_i = 0$. Next let h be a continuous function with compact support C such that $0 \leq h \leq 1$, $h(x) = 1$ for x in K, and

$$C \subset \left(\bigcup_i K_i \right)^0.$$

For x in C, let

$$f_i(x) = \frac{g_i(x)h(x)}{\sum_j g_j(x)}$$

and set $f_i(x) = 0$ for x outside C. Then, as is easily verified, the functions f_i satisfy all the conditions stated above.

The next result deals with the approximation of functions on compact spaces. It is applicable to functions on locally compact spaces as well, provided they can be continuously extended to the one-point compactification. When the compact space is a bounded interval and the algebra A in question is that of all polynomials on the interval, the theorem reduces to the classical

Weierstrass approximation theorem. The proof can be shortened by the use of this classical result; the proof given below is, however, self-contained.

A collection $\mathbf{C}$ of functions f on a set S is said to be *separating*, or to *separate the points of* S, if for any two points x and y in S there exists a function f in $\mathbf{C}$ such that $f(x) \neq f(y)$. A collection $\mathbf{C}$ of complex-valued functions is called *self-conjugate* if for every function $f \in \mathbf{C}$, the function $\bar{f}$, where $\bar{f}(x) = \overline{f(x)}$, is again in $\mathbf{C}$.

The theorem deals with uniform approximation throughout the space in question. Throughout its treatment, the uniform topology in function space is employed, i.e., the metric topology in which $d(f,g) = \sup_{x \in S} |f(x) - g(x)|$.

THEOREM 5.1 (STONE-WEIERSTRASS) *Let $\mathbf{A}$ be a self-conjugate complex algebra of continuous functions on a compact space S which separates the points of S, and $\mathbf{B}$ its uniform closure. Then $\mathbf{B}$ is either the algebra of all continuous functions on S or consists of all continuous functions vanishing at some fixed point of S.*

Lemma 5.1.1 *Let f be a continuous complex-valued function on a compact space S. Then there exists a sequence of real polynomials p_n with 0 constant term such that $p_n \circ |f|^2$ converges uniformly to $|f|$.*

Proof Since f is continuous and S compact, there exists a positive constant c such that $|f(x)| \leq c$ for all x in S. Let

$$g(x) = (1/c)f(x), \qquad h_0(x) = 0,$$

and

$$h_{n+1}(x) = h_n(x) + \frac{|g(x)|^2 - h_n{}^2(x)}{2}$$

for $n \geq 0$. Then, since $|g(x)| \leq 1$,

$$0 \leq h_1 \leq |g|^2 \leq |g|.$$

Suppose we have shown that $h_k - h_{k-1} \geq 0$ and $h_k \leq |g|$ for $1 \leq k \leq n$. Then

$$h_{n+1} - h_n \geq \tfrac{1}{2}(|g|^2 - h_n{}^2) \geq 0,$$

and

$$h_{n+1} = h_n + \tfrac{1}{2}(|g| + h_n)(|g| - h_n)$$

$$\leq h_n + |g| - h_n = |g|.$$

Hence $0 \leq h_1 \leq h_2 \leq \ldots$, and $h = \lim h_n$ exists and is finite-valued. It follows readily from the defining equation that each h_n is polynomial in $|f|^2$ with real coefficients and 0 constant term, and that

$$h = h + \tfrac{1}{2}(|g|^2 - h^2).$$

Thus $|g| = h$ and $ch_n \to |f|$; moreover, the convergence is uniform by Dini's theorem.

Lemma 5.1.2   *Let f be a continuous real-valued function on a compact space S, and L a lattice of real-valued continuous functions on S such that for any pair of points (x,y) in S there exists a function f_{xy} in L which coincides with f on (x,y). Then, for any $\epsilon > 0$, there exists a function g in L such that $\|f - g\| < \epsilon$.*

Proof   Fix an $\epsilon > 0$, and for any pair of points (x,y) in S, let

$$U_{xy} = [z : f_{xy}(z) < f(z) + \epsilon].$$

Then U_{xy} is an open set containing x and y, and for fixed y there exist a finite number of points $x_1, x_2, \ldots, x_n$ such that $U_{x_1 y}, U_{x_2 y}, \ldots, U_{x_n y}$ cover S. Let

$$f_y = \min (f_{x_1 y}, f_{x_2 y}, \ldots, f_{x_n y}).$$

Then $f_y \in L$ and $f_y(z) < f(z) + \epsilon$ for all z in S.   Next let

$$V_{iy} = [z : f(z) - \epsilon < f_{x_i y}(z)].$$

Then V_{iy} is an open set containing y, and the same is true of the set

$$V_y = \bigcap_i V_{iy}.$$

Moreover, for z in V_y,

$$f(z) - \epsilon < f_y(z).$$

A finite number of such sets $V_{y_1}, V_{y_2}, \ldots$ cover S, and setting

$$g = \max (f_{y_1}, f_{y_2}, \ldots),$$

we obtain an element of L such that

$$f(z) - \epsilon < g(z) < f(z) + \epsilon$$

for all z in S.

Proof of Theorem 5.1   It follows by continuity and uniform convergence that $\mathbf{B}$ is a complex algebra of continuous functions on $\mathbf{B}$ which is closed under complex conjugation.   Thus each element f in $\mathbf{B}$ has a unique decomposition, $f = g + ih$, where g and h are real-valued functions in $\mathbf{B}$, and the set $\mathbf{B}_r$ of all real-valued functions in $\mathbf{B}$ is a uniformly closed real algebra of real-valued continuous functions on S.   If x and y are distinct points of S, there exists a function f in $\mathbf{A}$ such that $f(x) \neq f(y)$.   Thus there also exists a real-valued function g in $\mathbf{A}$, and hence in $\mathbf{B}_r$, such that $g(x) \neq g(y)$. Moreover, since $\mathbf{B}_r$ is uniformly closed, it follows from Lemma 5.1.1 that $\mathbf{B}_r$ is also a lattice.

We now consider the case where there exists for each x in S an element g in $\mathbf{B}_r$ such that $g(x) \neq 0$.   Then, whenever x and y are distinct points of S, the linear map $g \to (g(x), g(y))$ of $\mathbf{B}_r$ into R^2 is easily seen to be of rank 2. Thus, as $\mathbf{B}_r$ is uniformly closed, Lemma 5.1.2 shows that $\mathbf{B}_r$ is the algebra of all continuous real-valued functions on S, and hence $\mathbf{B} = \mathbf{C}(S)$.

We next consider the case where all the functions in $\mathbf{B}_r$ vanish at some given point x_0 in S. Then the set $\mathbf{B}'_r$ of all functions of the form $f(x) = c + g(x)$, where c is a real constant and $g \in \mathbf{B}_r$ is a real algebra of continuous functions separating the points of S and having the property that for each x in S there exists a function f in $\mathbf{B}_r$ such that $f(x) \neq 0$. In addition, $\mathbf{B}'_r$ is uniformly closed. For suppose that $f_n(x) = c_n + g_n(x)$, as above, and that f_n converges uniformly to h. Then $f_n(x_0) = c_n$ and $c_n \rightarrow h(x_0)$, so that $g_n(x) \rightarrow h(x) - h(x_0)$ uniformly. Since the function $g(x) = h(x) - h(x_0)$ is an element of $\mathbf{B}_r$ it follows that $\mathbf{B}'_r$ is uniformly closed. Thus, by the preceding paragraph, $\mathbf{B}'_r$ is the algebra of all continuous real-valued functions on S. This implies that $\mathbf{B}_r$ is the algebra of all such functions vanishing at x_0; hence $\mathbf{B}$ is the algebra of all continuous complex-valued functions vanishing at x_0.

Example 5.1.1 Let S be the unit circle in the complex plane, and $\mathbf{A}$ the set of all functions on S of the form

$$f(z) = \sum_n c_n z^n,$$

where c_n is complex and the sum is extended over a finite set of integers, positive, negative, or 0. Then $\mathbf{A}$ is a complex algebra of continuous functions on S which separates the points of S and is closed under complex conjugation. Since $\mathbf{A}$ contains the constant functions, it follows from the theorem that $\mathbf{A}$ is uniformly dense in $\mathbf{C}(S)$.

Theorem 5.1 is often applied in a form which deals with functions on locally compact spaces. A continuous function f on such a space X is said to "vanish at infinity" if for every positive number c, the set on which $f(x)$ exceeds c in absolute value has compact closure. It is easily seen that a function vanishes at ∞ if and only if it is a uniform limit of a sequence of continuous functions of compact support. A different way to characterize these functions, from which the extension of Theorem 5.1 to locally compact spaces will be deduced, is as follows.

Let $\tilde{X}$ denote the "one-point compactification" of X, that is, $\tilde{X} = X$ if X is compact, and otherwise $\tilde{X}$ is a compact space containing X and such that $\tilde{X} - X$ consists of a single point, which may be called the "point at infinity" and designated ∞. Such a space always exists and is unique. Indeed, on adjoining a point ∞ to the noncompact locally compact space X, and defining a neighborhood of ∞ to consist of the complement of any compact subset in X, together with the point ∞ itself, the space $X \cup \{\infty\}$ is readily seen to be compact. This shows the existence of the indicated space; its uniqueness follows from the fact that a one-to-one continuous mapping of one compact space onto another must be a homeomorphism. There is now no difficulty in verifying that the functions vanishing at infinity on X are precisely those which may be continuously extended to $\tilde{X}$ in such a fashion as to vanish on $\infty \, (= \tilde{X} - X)$.

We may now state

> COROLLARY 5.1.1   *A self-conjugate algebra* **A** *of continuous functions vanishing at infinity on a locally compact space S is uniformly dense in the algebra of all such functions if and only if for every ordered pair* (x,y) *of points of S, there exists an element* $f \in$ **A** *such that* $f(x) \neq 0$ *and* $f(y) = 0.$

The "only if" part of the corollary offers no difficulty. Consider then the "if" part, under the indicated hypothesis. Let $\bar{A}$ denote the set of all extensions of the elements of **A** to continuous functions on the one-point compactification $\bar{S}$ of S. Then $\bar{A}$ clearly separates the points of $\bar{S}$, and so by Theorem 5.1 is uniformly dense in the algebra of all continuous functions on $\bar{S}$ which vanish at ∞. On restricting to S, the conclusion of the corollary follows.

EXERCISES

1   Let I be a linear functional on the space $C_0(S)$, where S is locally compact. Suppose that every point of S has a neighborhood such that I vanishes on all functions supported by the neighborhood. Show that $I = 0$.

2   Find explicitly a partition of unity for the covering of the interval $[-1,1]$ by the open subsets $(-1,1]$ and $[-1,1)$.

3   Show that every periodic function of period 1 on the real line can be uniformly approximated arbitrarily closely by finite linear combinations of the functions $e^{2\pi n i x}$.

4   Show that a continuous numerical function f on a locally compact non-compact space can be continuously extended to the one-point compactification of the space so as to vanish at infinity if and only if it has the property that for every $\epsilon > 0$, the set $[x : |f(x)| \geq \epsilon]$ is compact.

5   Show that an algebra of real continuous functions on a compact space X which separates points and contains the function 1 is dense in the space $\mathbf{R}(X)$ of all real-valued continuous functions on X.

6*   Show that every closed subalgebra of $\mathbf{R}(X)$, where X is a compact space, consists of all functions f on X of the form $f = g \circ T$, where T is a continuous mapping of X onto a compact space Y, and $g \in \mathbf{R}(Y)$.

5.2   MEASURES IN LOCALLY COMPACT SPACES

The collection of all continuous complex-valued (respectively real-valued) functions with compact supports on the locally compact space S will be denoted as $C_0(S)$ [respectively, $\mathbf{R}_0(S)$]; *the support* of a function on a topological space is defined as the closure of the set where it assumes nonzero values. The minimal σ-ring of subsets of S with respect to which all elements of $\mathbf{C}_0(S)$ are measurable will be denoted as $\mathcal{B}$ [or $\mathcal{B}(S)$]; the elements of $\mathcal{B}$ will be called *Baire sets*.

The term *Baire measure* on the locally compact space S will refer to a countably additive measure on $\mathcal{B}$ which is finite on compact sets in $\mathcal{B}$. A *regular locally compact measure space* $(S,\mathcal{R},m)$ is defined as one for which S is a locally compact space, $\mathcal{R}$ is a σ-ring containing $\mathcal{B}(S)$, and m is a (countably additive) measure, finite on compact sets, such that for any set E in $\mathcal{R}$,

$$m(E) = \sup_{C \subset E, C \in \mathcal{R}} m(C) = \inf_{E \subset U, U \in \mathcal{R}} m(U),$$

where C is restricted to be compact, and U to be open.

It is useful to note that there exists a natural one-to-one correspondence between Baire measures on a locally compact space S and positive linear functionals on $\mathbf{C}_0(S)$, where the latter is defined as a linear functional which is nonnegative on nonnegative functions. Such a functional is known as an *integral on the locally compact space S*. If λ is any given such functional on $\mathbf{C}_0(S)$, then it follows from Example 3.3.2, together with the general theory provided by Theorem 3.3, that there exists a unique measure m on $\mathcal{B}$ such that

$$\lambda(f) = \int f(x)\, dm(x)$$

for all $f \in \mathbf{C}_0(S)$. From the topological normality of a compact Hausdorff space it is clear that the characteristic function of an arbitrary compact subset of S is majorized by some nonnegative element of $\mathbf{C}_0(S)$, showing that m is finite on compact sets. Conversely, if m is a given Baire function, the foregoing equation defines a positive linear functional on $\mathbf{C}_0(S)$. The measure m and the functional λ will be said to "correspond," or "be associated," with each other, and the justification for the term "integral" for λ is apparent.

SCHOLIUM 5.2 *Any Baire measure is regular.*

In view of the association of the measure space with an integration lattice just indicated, Corollary 3.5.4 may be applied, and shows that for each set E in $\mathcal{B}$ and any positive number ϵ, there exists an open set U in $\mathcal{B}$ such that $E \subset U$ and

$$m(E) \leq m(U) \leq m(E) + \epsilon.$$

Now since $\mathcal{B}$ is generated by sets of the form $[x: f(x) \geq 1]$, $f \in \mathbf{R}_0(S)$, any set E in $\mathcal{B}$ is contained in a countable union of compact sets in $\mathcal{B}$ of this form. It therefore suffices, in order to show the approximability of E from below by compact elements of $\mathcal{B}$, to consider the case in which E is a subset of a compact set K in $\mathcal{B}$. Then

$$m(E) = m(K) - m(K - E).$$

Now if W is an open set in $\mathcal{B}$ which contains $K - E$, then E contains the compact set $K - W$, and

$$m(E) = m(K) - \inf m(W) = \sup\,[m(K) - m(W)] \leq \sup m(K - W)$$

where the sup and inf are taken over the collection of all such W. This means that $m(E) \leq \sup_{C \subset E, C \in \mathcal{B}} m(C)$, and since the inequality in the reverse direction is obvious, the proof is complete.

COROLLARY 5.2.1   *If* $1 \leq p < \infty$, *then* $\mathbf{C}_0(S)$ *is dense in* $L_p(S,\mathcal{B},m)$.

Proof   This follows from Corollary 4.4.2, but a separate proof is instructive. Since the step functions relative to $\mathcal{B}$ are dense in L_p for $p < \infty$, it suffices to show that the characteristic function φ_E of an arbitrary set E of finite measure in $\mathcal{B}$ may be approximated arbitrarily closely by an element of $\mathbf{C}_0(S)$. As seen in the proof of the scholium, E is the union of elements of $\mathcal{B}$ having compact closures; thus φ_E may be approximated in L_p by characteristic functions of sets in $\mathcal{B}$ having compact closures; it is therefore no essential loss of generality to assume that E is itself of compact closure. It follows that there exists an open set V of compact closure such that $E \subset V$. Now let ϵ be an arbitrary positive number, and let C and U be compact and open sets in $\mathcal{B}$ (respectively) such that $U \supset E \supset C$ and

$$m(U) - \epsilon < m(E) < m(C) + \epsilon.$$

From the topological normality of a compact Hausdorff space, it follows that there exists a continuous function f on S having values in the interval $[0,1]$, which has the value 1 on C and 0 outside of $U \cap V$. It follows that $\|\varphi_E - f\|_p \leq m(U \cap V - C)^{1/p} \leq (2\epsilon)^{1/p}$, which can be made arbitrarily small.

COROLLARY 5.2.2   (LUSIN)   *Let* f *be a complex-valued Baire function on a regular measure space, and* E *a Baire set of finite measure. Then for any positive number* ϵ, *there exists a compact Baire subset* $K \subset E$ *such that* (1) $m(E - K) < \epsilon$; *and* (2) *the restriction of* f *to* K *is continuous (in the induced topology on* K).

Proof   Since f is finite-valued, there exists a set A in $\mathcal{B}$ such that $m(A) < \epsilon/3$ and f is integrable on $E - A$. Let $g = f\varphi_{E-A}$; then there exists a sequence of elements of $\mathbf{C}_0(S)$ which converges in L_1 to g, and hence a subsequence, say, $\{g_n\}$, which is convergent to g a.e. By Egoroff's theorem, there exists a Baire set B such that $m(B) < \epsilon/3$ and $g_n \to g$ uniformly on $E - B$. By the scholium, there exists a compact set K in $\mathcal{B}$ such that $K \subset (E - A) - B$, and

$$m[(E - A) - B] < m(K) + \frac{\epsilon}{3}.$$

Then $g_n \to f$ uniformly on K, so that the restriction of f to K is continuous on K. Finally setting $D = [(E - A) - B] - K, m(D) < \epsilon/3$ and

$$E - K \subset A \cup B \cup D,$$

from which it follows that $m(E - K) < \epsilon$.

REMARK   There is no entirely standard definition of the term "regular measure."   The Baire sets are quite sufficient for most concrete analytical purposes, since it may be shown (cf. Exercises 1 and 7) that every set of finite measure in a regular space of any description will differ from a Baire set only by a null set.   However, this null set will not necessarily be contained in a Baire set, and the extensions given in the literature, for example, to the σ-ring generated by the compact sets, or the larger ring generated by all open sets, involve σ-rings which in general are not contained in that of the completion of the Baire sets. The sets in the σ-ring generated by all open sets may be called "Borel" sets (but some authors use this term for the sets in the generally smaller σ-ring generated by the compact sets).   It is a noteworthy fact that a regular measure on the Baire sets can be extended uniquely to a measure on the Borel sets having the same property (sometimes known as "regularity" but automatic with the present definition) as is concluded in Scholium 5.2 (see the exercises below).   On the other hand, when the space S has a countable base for its open sets, the Baire and Borel sets are the same, and the qualitative theory of regular measures on S is not materially more complicated than in the case of Lebesgue-Stieltjes measures (which may now be designated simply as regular measures) in euclidean space.

The association of a regular measure with a positive linear functional on $C(S)$ was first established by F. Riesz in the case when S is a compact interval on the real line.   It was generalized by a number of authors, in particular, by A. Markoff, who obtained a measure on all Borel sets in the case when S is compact, and later by several authors, to a similar measure when S is merely locally compact.

One of the advantages of locally compact regular measure spaces is the facility with which direct products of such spaces, as regular measure spaces, may be formed.   If m and n are regular measures on the locally compact spaces S and T, then $S \times T$ is again locally compact, and the regular measure $m \otimes n$ may be defined as that associated with the linear functional λ on $C_0(S \times T)$, which is uniquely determined by the property that $\lambda(f \otimes g) = \sigma(f)\tau(g)$, if f and g are arbitrary elements of $C_0(S)$ and $C_0(T)$, and σ and τ are the linear functionals associated with m and n, respectively.   More specifically,

SCHOLIUM 5.3   *Let m and n be regular measures on the locally compact spaces S and T, and let σ and τ be the linear functionals on $\mathbf{C}_0(S)$ and $\mathbf{C}_0(T)$ defined by the equations*

$$\sigma(f) = \int_S f(x)\, dm(x), \qquad \tau(g) = \int_T g(x)\, dn(x).$$

There is then a unique positive linear functional λ on $\mathbf{C}_0(S \times T)$ such that $\lambda(f \otimes g) = \sigma(f)\tau(g)$ [where $f \otimes g$ is the function on $S \times T$ whose value at the point (x,y) is $f(x)g(y)$]; and if p denotes the measure associated with λ, the Baire measure space $(S \times T, \mathcal{B}(S \times T), p)$ is the direct product of the spaces $(S, \mathcal{B}(S), m)$ and $(S, \mathcal{B}(T), n)$.

This result is basically a simple consequence of the Stone-Weierstrass approximation theorem, which implies that an algebra $\mathbf{A}$ of real-valued continuous functions on a compact Hausdorff space X is dense in the space $\mathbf{R}(X)$ of all real-valued continuous functions on X (in the topology of uniform convergence) if it has the property that for any two elements x and x' in X there exists an element $f \in \mathbf{A}$ such that $f(x) \neq f(x')$ and if $\mathbf{A}$ contains the constant function.   This will be applied to prove

Lemma 5.2.1   *A subalgebra $\mathbf{A}$ of $\mathbf{R}_0(X)$ is dense in $\mathbf{R}_0(X)$ in the topology of uniform convergence provided that for any two elements x and x' in X there is an element $f \in A$ such that $f(x) \neq f(x')$ and if for every compact subset C of X there exists an element of $\mathbf{A}$ which is 1 on C.*

Proof   Let g be arbitrary in $\mathbf{R}_0(X)$; let C_1 be a compact subset of X containing the support of g; let h be an element of $\mathbf{A}$ which is 1 on C_1, and let C be a compact set which contains the support of h and whose interior contains C_1.   The set of all restrictions of elements of $\mathbf{A}$ to C forms an algebra to which the Stone-Weierstrass theorem is applicable.   It follows that $g \mid C$ is uniformly approximable by elements of $\mathbf{A} \mid C$.   Multiplication by h does not disturb this uniform approximability; but $hg = g$, and the elements of $h\mathbf{A}$ vanish outside of C, so that the uniform approximability becomes valid throughout all of X, and not merely on C.

An immediate consequence is

Lemma 5.2.2   *The set of all finite linear combinations of functions of the form $f \otimes g$, with $f \in \mathbf{C}_0(S)$ and $g \in \mathbf{C}_0(T)$, is dense in $\mathbf{C}_0(S \times T)$, in the uniform topology.*

Proof   This set is an algebra contained in $\mathbf{C}_0(S \times T)$ which is easily seen to "separate points" in the sense of the hypothesis of Lemma 5.2.1.   Moreover, if K is any given compact set in $S \times T$, then K is contained in the direct product $A \times B$ of compact subsets of S and T (e.g., the projections of K on these spaces), and the direct product of functions in $C_0(S)$ and $C_0(T)$ which are 1 on S and T, respectively, is a function in the algebra which is 1 on K.

We now return to the

Proof of Scholium 5.3   First, observe that $\mathcal{B}(S \times T) = \mathcal{B}(S) \otimes \mathcal{B}(T)$. For $\mathcal{B}(S) \otimes \mathcal{B}(T)$ may be characterized a the minimal σ-ring with respect to which all functions of the form $f \otimes g$ $(f \in \mathcal{R}_0(S), g \in \mathcal{R}_0(T))$ are measurable; $\mathcal{B}(S \times T)$ is the minimal σ-ring with respect to which all functions in $\mathcal{R}_0(S \times T)$ are measurable.   But if a set of functions is measurable with respect to a σ-ring, so are all finite linear combinations of them, and all pointwise limits of sequences of such.   Thus every element of $\mathcal{R}_0(S \times T)$ is measurable with respect to $\mathcal{B}(S) \otimes \mathcal{B}(T)$, showing that $\mathcal{B}(S \times T) \subset \mathcal{B}(S) \otimes \mathcal{B}(T)$.   But the opposite inclusion relation is obvious, so the stated equality holds.   The product measure $p = m \otimes n$ is then regular and defines a positive linear functional λ.   It is clear from the Fubini theorem that $\lambda(f \otimes g) = \sigma(f)\tau(g)$.

On the other hand, if λ' is any other positive linear functional on $C_0(S \times T)$ having the latter property, then λ and λ' agree on all finite linear combinations of the $f \otimes g$.   Now note that by dominated convergence λ and λ' agree also on all limits of uniformly convergent sequences of such linear combinations which vanish outside of a fixed compact set, and hence on all functions in $C_0(S \times T)$.

EXERCISES

1   Show that a real-valued measurable function on a regular measure space S differs on a null set from the limit of a monotone-decreasing sequence of functions, each of which is the limit of a monotone-increasing sequence of elements of $C_0(S)$. (*Hint:* See the proof of Theorem 3.3.)

2   Let S be a locally compact space.

   a. Show that a linear functional on $C_0(S)$ is determined by its restriction to $\mathbf{R}_0(S)$.

   b. If J is a linear functional on $C_0(S)$, show that there is a unique pair of real functionals [i.e., functionals that are real on $\mathbf{R}_0(S)$] J_1, J_2 such that $J = J_1 + iJ_2$.

   c. If J is a linear functional on $C_0(S)$ and $J = J_1 + iJ_2$ as in (*b*), prove that

$$\|J_n\|_K \leq \|J\|_K, \qquad n = 1, 2,$$

   for all compact K, where for any linear functional F and subset E of S, $\|F\|_E$ is defined as the supremum of $|F(g)|$ as g varies over the functions, supported by E and bounded by unity in absolute value.

   d. The *total variation* $|J|$ of a linear functional J is initially defined for non-negative f in $C_0(S)$ by the equation

$$|J|\,(f) = \sup_{g \leq f} |J(g)|.$$

J is called a *complex integral* in case $|J|(f)$ is finite-valued.   Show that if J is a complex integral, then $|J|$ extends uniquely to a positive linear functional on $C_0(S)$.   Show also that $\| |J| \|_K = \|J\|_K$ for every compact K.

e. If J and J' are complex integrals, $J \leq J'$ means that $J' - J$ is positive. Suppose J is a complex integral expressed in the form $J_1 + iJ_2$ as in (c). Prove that $|J_1| \leq |J|$, $|J_2| \leq |J|$, and $|J| \leq |J_1| + |J_2|$.

f. If J is a real complex integral on $C_0(S)$, show that there exist unique positive linear functionals J^+ and J^- such that $J = J^+ - J^-$ and $|J| = J^+ + J^-$.

g. Show that every complex integral on $C_0(S)$ is a linear combination of positive linear functionals.

3   An integral J on $C_0(S)$ is said to be *finite*, or *bounded*, in case $\sup_k \|J\|_K < \infty$, where the supremum is taken over all compact sets K.

a. Suppose J is a positive finite integral on $C_0(S)$ and that J' is an integral such that

$$|J'(f)| \leq J(|f|)$$

for all f in $C_0(S)$.   Let $M = (S, \mathcal{B}, m)$ be the regular measure space associated with J in which $\mathcal{B}$ is the ring of Baire sets.   Prove there is a unique complex measure m' on $\mathcal{B}$ such that

$$J'(f) = \int f(x)\, dm'(x)$$

for all f in $C_0(S)$.   Show also that m' is the indefinite integral of a function $g \in L_1(M)$.

b. Let an *unrestricted Baire function* be defined as one in the minimal class containing all continuous functions and closed under pointwise sequential convergence.

c. Show that every unrestricted Baire function is locally measurable with respect to $\mathcal{B}$, and hence that under the conditions of (a), the functional J' may be extended to the class of all bounded unrestricted Baire functions by the definition

$$J'(f) = \int f(x)g(x)\, dm(x).$$

d. If J' is a bounded integral, observe that $J = |J'|$ satisfies the condition in (a).

4   Let J be a complex integral on $C_0(S)$, and V the union of all open sets W with the property that $J(f) = 0$ whenever $f \in C_0(S)$ and W contains the support of F.   If $f \in C_0(S)$ and the support of f is contained in V, show that $J(f) = 0$.   (The complement of V is called the *support* of J.)

5   Show that the set of all bounded integrals J on $C_0(S)$ is a Banach space under the norm

$$\|J\| = \sup_{f \in C_0(S),\, \|f\|=1} |J(f)|.$$

6   Let S be a compact Hausdorff space.   Show that an integral J on $C_0(S)$ is positive if and only if $\|J\| = J(1)$.

7   Let $M = (S,\mathcal{R},m)$ be a regular measure space, and K a compact set in $\mathcal{R}$.   If $\epsilon > 0$, show there exists an open set V in the ring $\mathcal{B}$ of Baire sets such that $K \subset V$ and $m(V) < m(K) + \epsilon$.

8   Let S be a locally compact Hausdorff space, and m a positive measure on the σ-ring $\mathcal{B}$ of Baire sets which assumes finite values on the compact sets in $\mathcal{B}$.

    a. For any compact set K, let $n(K) = \inf m(V)$, where V varies over the collection of open sets in $\mathcal{B}$ which contain K.   Observe that n is a non-negative finite-valued function on the compact sets, and show that it has the following additional properties:

        (i) $n(K) = m(K)$ if $K \in \mathcal{B}$
        (ii) $n(K_1) \leq n(K_2)$ if $K_1 \subset K_2$
        (iii) $n(K_1 \cup K_2) \leq n(K_1) + n(K_2)$

        with equality in case K_1 and K_2 are disjoint.

    b. For an arbitrary subset A of S, let $n(A) = \sup_{K \subset A} n(K)$, where K is compact. Observe that n is a nonnegative monotone set function and prove that

        (i) $n(V) = m(V)$ for any open set V in $\mathcal{B}$.
        (ii) For any sequence $V_1, V_2, \ldots$ of open sets

$$n\left(\bigcup_i V_i\right) \leq \sum_i n(V_i).$$

    c. If K is a compact subset of an open set U, show that

$$n(U) = n(K) + n(U - K).$$

    d. Let $\mathcal{R}_0$ be the collection of all sets A such that for any $\epsilon > 0$ there exists a compact set K and an open set V with $K \subset A \subset V$ and $n(V - K) < \epsilon$. Prove that $\mathcal{R}_0$ is a ring containing the compact sets, the open sets V with $n(V) < \infty$, and all sets in $\mathcal{B}$ of finite measure.   [*Hint*: If $K_i \subset A_i \subset V_i$, $i = 1, 2$, then

$$V_1 \cup V_2 - (K_1 \cap K_2) \subset V_1 - K_1 \cup V_2 - K$$

      and $(V_1 - K_2) - (K_1 - V_2) \subset V_1 - K_1 \cup V_2 - K_2.$]

    e. If $A_1, A_2, \ldots$ is a disjoint sequence of sets in $\mathcal{R}_0$ such that $\sum_i n(A_i) < \infty$, show that $A = \bigcup_i A_i$ is also in $\mathcal{R}_0$ and that $n(A) = \sum_i n(A_i)$.

    f. Let $\mathcal{S}$ be the algebra of all sets locally measurable with respect to $\mathcal{R}_0$. Prove that

        (i) $\mathcal{S}$ is a σ-algebra.
        (ii) $\mathcal{S}$ contains $\mathcal{B}$ and all open sets.
        (iii) n is countably additive on $\mathcal{S}$.
        (iv) n agrees with m on $\mathcal{B}$.

g. If $\mathcal{R}$ is the σ-ring generated by $\mathcal{R}_0$, show that $(S,\mathcal{R},n)$ is a regular measure space and that n is the only measure on $\mathcal{R}$ which extends m and has this property.

h. If $\mathcal{C}$ is the σ-ring generated by the compact sets, show that $\mathcal{C}$ is contained in $\mathcal{R}$ but that in general $\mathcal{C} \neq \mathcal{R}$.

i. Let $\mathcal{S}'$ be the ring of σ-finite elements in $\mathcal{S}$. Show that in general the measure space $(S,\mathcal{S}',n)$ is not regular.

9   Give an example to show that in general a compact subset of a locally compact space will not be measurable with respect to the completion of the Baire sets relative to a regular measure. (*Hint:* Take S as the one-point compactification, by a point ω, of an uncountable discrete set, and the measure as that associated with the linear functional assigning to any function its value at ω.)

5.3   TRANSFORMATION OF LEBESGUE MEASURE

If T is a transformation of a set S into a set S', which is measurable relative to given σ-rings $\mathcal{R}$ and $\mathcal{R}'$ of subsets of S and S', respectively, in the sense that if $E \in \mathcal{R}'$, then $T^{-1}(E) \in \mathcal{R}$, then any measure m on $\mathcal{R}$ is carried into a measure m' on $\mathcal{R}'$ in the following natural fashion:

$$m'(E) = m[T^{-1}(E)].$$

From the fact that T^{-1} is a complete Boolean homeomorphism (i.e., is a Boolean homeomorphism in the elementary algebraic sense, and in addition preserves unions and intersections of arbitrary collections of sets), it follows that m' is countably additive provided m is such. Thus $(S',\mathcal{R}',m')$ is a (possibly improper) measure space. The definition of m' is equivalent to the equation

$$(*) \qquad \int f(Tx)\, dm(x) = \int f(x)\, dm'(x)$$

for the case where f is the characteristic function of a set in $\mathcal{R}'$. By an argument similar to that given in the proof of Scholium 4.5, it follows that equation $(*)$ is valid for all functions f which are integrable with respect to the measure space (determined by) $(S',\mathcal{R}',m')$. In practice, it is often important to compute such measures m' explicitly, when T is given. A particularly important case is that where $S = S'$, $\mathcal{R} = \mathcal{R}'$, and T is one-to-one; but even in this special case relatively little can be said concerning m' in a general way. If, however, m is Lebesgue measure on a euclidean space, m' can be described quite explicitly, provided T is sufficiently smooth; instances of this will be familiar to the reader from the calculus. This section is devoted to the derivation of two basic results in this connection.

First we treat in detail a result already alluded to earlier, namely, the characterization of Lebesgue measure as the unique invariant measure on euclidean space.   In this connection, it is convenient on occasion to work with integrals rather than measures. If T is a homeomorphism of a locally compact space S into another such space S', then for any integral I on S there is an integral I' on S' defined by the equation

$$I'(f) = I(f \circ T).$$

It is clear from equation (*) that the measure associated with I' is the measure indicated above. More generally, if T is any *proper* mapping of S into S'— by which is meant that for every $f \in C_0(S')$, $f \circ T$ is integrable on the measure space determined by the integral I on S—an integral I' is definable on S' by the same equation.

Example 5.3.1   For any real a and b with $a \neq 0$, it follows from equation (*) that

$$\int_{-\infty}^{\infty} f(y)\, dy = |a| \int_{-\infty}^{\infty} f(ax + b)\, dx$$

for all Lebesgue-integrable functions f, as noted above. Thus the Lebesgue integral I on the nonnegative Borel functions transforms under the group of *affine transformations*

$$A_{a,b}\colon \quad x \to ax + b, \qquad a \neq 0,$$

according to the relation

$$A_{a,b}(I) = |a|^{-1}I.$$

Hence it is invariant under the (normal) subgroup of *translations*

$$T_b\colon \quad x \to x + b.$$

In particular, if E is a Borel set, then its *translates*

$$E + b = [x + b : x \in E]$$

are also Borel sets, and each translate has the same measure as E.

Example 5.3.2   Let T be a continuous real-valued function mapping the open interval S on the reals into the interval S', and suppose T has a continuous derivative ∂T on S. Let $M = (S, \mathcal{B}, m)$ be the regular measure space on S in which $\mathcal{B}$ is the ring of Borel sets and m is defined on $\mathcal{B}$ by

$$m(E) = \int_E |\partial T(x)|\, dx.$$

If $[a,b]$ is a closed bounded interval contained in S, it follows by elementary calculus that

$$(1) \qquad \int_{T(a)}^{T(b)} f(y)\, dy = \int_b^a f(Tx)\, \partial T(x)\, dx$$

for any continuous complex-valued function f on S'. In general, however, T will not be a proper mapping of M into S'. This is shown by any case where S has infinite measure and S' is compact, for example, $S = (0, \infty)$, $T(x) = \sin(1/x)$, and $S' = [-1,1]$.

We shall now make the additional assumption that T is a one-to-one mapping of S onto S'. Then T is strictly monotone, S' is open, and T has a continuous inverse. Thus, by equation (1),

$$(2) \qquad \int_{S'} f(y)\,dy = \int_{S} f(Tx)|\partial T(x)|\,dx,$$

for all f in $C_0(S')$. By equation (*), this implies that Lebesgue measure on the Borel subsets of S' is the transform of the measure m under T. Moreover, equation (2) is valid for any nonnegative Borel function f on S'. In particular, if E is a Borel subset of S', then

$$\int_{E} dy = \int_{T^{-1}(E)} |\partial T(x)|\,dx.$$

Thus T maps the closed set

$$C = [x \in S: \partial T(x) = 0]$$

onto a closed set C' of Lebesgue measure 0. For y in the open set $S' - C'$,

$$\partial T^{-1}(y) = \frac{1}{\partial T(T^{-1}y)}$$

and if $\partial T^{-1}(y)$ is defined as some arbitrarily chosen constant on C', it follows from equation (2) that

$$\int_{S'} f(y)\,|\partial T^{-1}(y)|\,dy = \int_{S-C} f(Tx)\,dx$$

for any Borel function $f \geq 0$. Taking $f(y) = g(T^{-1}y)$, where g is a nonnegative Borel function on S, we then see that

$$(3) \qquad \int_{S-C} g(x)\,dx = \int_{S'} g(T^{-1}y)\,|\partial T^{-1}(y)|\,dy.$$

Thus T^{-1} is a proper mapping of S' into S with respect to the regular "measure" $|\partial T^{-1}(y)|\,dy$. It also follows from equation (3) that Lebesgue measure on S is the transform of $|\partial T^{-1}(y)|\,dy$ under T^{-1} provided C is empty or has Lebesgue measure 0. On the other hand, the Lebesgue measure of C may be arbitrarily close to that of S. For let $S = (0,1)$, $0 < \epsilon < 1$, and C a compact subset whose complement is dense and has measure less than ϵ. Then set

$$T(x) = \int_{0}^{x} d(u,C)\,du, \qquad 0 < x < 1,$$

where $d(u,C)$ is the distance between u and C.

The following result is actually a special case of one to be proved later, but the method of proof is quite different, and is useful in certain situations where the method employed is not readily applicable; the proof is arranged so as to be applicable with very slight modifications to the proof of the uniqueness of Haar measure on an arbitrary locally compact group.

SCHOLIUM 5.4 *Lebesgue measure is, within a constant multiple, the only Baire measure on euclidean space which is invariant under all translations on the space.*

Proof Suppose I and J are invariant integrals on euclidean space E; let m and n be the corresponding Baire measures, and let f and g be elements of $C_0(E)$. Then, by the Fubini theorem and the invariance of m and n,

$$\int f(x)\,dm(x)\int g(y)\,dn(y) = \int dn(y)\int f(x+y)g(y)\,dm(x)$$

$$= \int dm(x)\int f(y)g(y-x)\,dn(y) = \int f(y)\left[\int g(y-x)\,dm(x)\right]dn(y).$$

By interchanging x and y in the last equation, it follows that

$$(*)\qquad \int f(x)\,dm(x)\int g(y)\,dn(y) = \int f(x)\left[\int g(x-y)\,dm(y)\right]dn(x).$$

Now assuming that m is Lebesgue measure, and using its invariance under the transformation $y \to -y$, it follows that

$$\int g(x-y)\,dm(y) = \int g(y)\,dm(y).$$

Substituting in equation $(*)$, it results that I and J are proportional, g being chosen so that at least one of $I(g)$ and $J(g)$ is not zero.

This result implies the "ergodicity" of the translation group on euclidean space, a collection G of measurable transformations being called *ergodic* if the only essentially invariant locally measurable sets are either local null sets or the complements of such. Here "essentially invariant" means that for any transformation T in G, the set and its transform under T differ only by a local null set, which may however depend on T.

To derive the cited ergodicity implication, suppose A is a measurable essentially translation-invariant subset of E. Then the set function n defined by the equation: $n(x) = m(X \cap A)$ is invariant under translations in E, and hence by Scholium 5.4, has the form $n(X) = cm(X)$ for some constant c. Substituting $X = A$ shows that either $c = 1$ or $m(A) = 0$. In the former case, $E - A$ is a null set; in the latter A is itself a null set.

A similar corollary applies to the general case of a locally compact topological group, treated in Chapter VII, and will have use later on.

COROLLARY 5.3.1   *If T is a nonsingular linear transformation in euclidean space and m denotes Lebesgue measure, then*

$$(m \circ T)(A) = m(A) \, |\det T|$$

for all Lebesgue-measurable sets A.

Proof   It suffices to treat the case in which A is a Borel set, since the general Lebesgue-measurable set differs from such a set by a set of Lebesgue measure zero, which is contained in Borel sets of arbitrarily small measure.   Now $m \circ T$ is again an invariant measure under vector translations; so

$$(m \circ T)(A) = cm(A),$$

where c is a constant, independent of A.   To evaluate the constant, recall that it is shown in linear algebra that any nonsingular linear transformation on E is the product of a finite number of "elementary" transformations $(x_1, x_2, \ldots, x_n) \to (y_1, y_2, \ldots, y_n)$ of the form $T': y_i = ax_i + bx_j$ $(a \neq 0,$ $i \neq j)$; $y_k = x_k$ if $k \neq i$, which affect only one coordinate at a time.   For any such transformation, it is evident that $\det T' = a$.   Thus, by the chain rule for Radon-Nikodym derivatives, it suffices to show that if I denotes the Lebesgue integral, then $I(f) = |a| \, I(f \circ T')$.   But this follows from Example 5.3.1 and the Fubini theorem.

This result for linear transformations is the basis for the extension to Lebesgue measure of the usual rule developed in the calculus for nonlinear transformations of integrals in one or more variables.   Although there is a direct analog of the result described in Example 5.3.2, the proof is rather intricate, and we shall consider only the case where the differential or derivative of the mapping is nonsingular at every point.

THEOREM 5.2   *Let S be an open set in R^n, and T a continuously differentiable mapping on S with nonvanishing Jacobian ∂T.   Suppose T is a one-to-one mapping of S onto the open set S' in R^n.   Then, for any nonnegative Baire function f on S',*

$$\int_{S'} f(y) \, dy = \int_S f(Tx) \, |\partial T(x)| \, dx.$$

Proof   By Corollary 5.2.1 it suffices to prove this for functions f in $C_0^+(S')$. On the other hand, it is actually enough to show that for f in $C_0^+(S')$,

$$(1) \qquad \int_{S'} f(y) \, dy \leq \int_S f(Tx) \, |\partial T(x)| \, dx.$$

For once this has been shown, the same applies to T^{-1}.   Since

$$\partial T(T^{-1}y) \, \partial T^{-1}(y) = 1,$$

it therefore follows that

$$\int_S f(Tx)\,|\partial T(x)|\,dx \le \int_{S'} f(y)\,dy,$$

and this combined with equation (1) yields the desired equality.   Using local partitions of unity (Scholium 5.1), we see that it suffices for the proof of equation (1) to consider the case where $f \in C_0^+(S')$ and the support of $f \circ T$ is contained in a closed cube C_0 which is contained in S.   Suppose C is a closed subcube of C_0.   Then if m denotes Lebesgue measure,

$$\int_{T(C)} f(y)\,dy \le \left[\sup_{y \in T(C)} f(y)\right] m[T(C)].$$

For simplicity of notation, let the differential of T at an arbitrary x in S be denoted by $A(x)$.   Suppose u is a point in C such that

$$f(Tu) = \sup_{y \in T(C)} f(y).$$

Then $A(u)^{-1}$ exists, and by Scholium 5.4,

$$m[T(C)] = |\det A(u)|\, m(A(u)^{-1}T(C)).$$

This implies the inequality

(2) $$\int_{T(C)} f(y)\,dy \le f(Tu)\,|\partial T(u)|\, m[A(u)^{-1}T(C)].$$

The basic idea of the rest of the proof is that if C is small enough, the measure of $A(u)^{-1}T(C)$ should be approximately that of C.   To show this we use the fact that T is continuously differentiable.   This, together with the convexity of C_0, guarantees that for any pair of points (x,x') in C_0, there exists a linear transformation $A(x,x')$ such that

(3) $$T(x') - T(x) = A(x)(x' - x) + A(x,x')(x' - x),$$

where

(4) $$\|A(x,x')\| \le \sup_{0 \le t \le 1} \|A(x + t(x' - x)) - A(x)\|,$$

the norms being the operator bounds associated with the vector norm

$$\|(x_1, x_2, \ldots, x_n)\| = \max_i |x_i|.$$

Now fix $\epsilon > 0$, and let

$$s = \sup_{x \in C_0} \|A(x)^{-1}\|.$$

Then $0 < s < \infty$, and there exists a number $\delta > 0$ such that

$$\|A(x') - A(x)\| < \frac{\epsilon}{s}$$

whenever x and x' are points in C_0 and $\|x' - x\| < \delta$. Suppose C is a closed cube in C_0 with center $x = (x_1, x_2, \ldots, x_n)$. Then C is the cartesian product of n closed real intervals,

$$[x_i - r, x_i + r], \qquad 1 \le i \le n,$$

and if $r < \delta$, it follows that

$$(5) \qquad \|A(x') - A(x)\| < \frac{\epsilon}{s}$$

for all x' in C. Suppose $r < \delta$ and u is a point in C. Then it follows from equation (3) that

$$A(u)^{-1}T(x') - A(u)^{-1}T(x) = (x' - x) + A(u)^{-1}(A(x) - A(u))(x' - x)$$
$$+ A(u)^{-1}A(x,x')(x' - x).$$

Hence, by equations (4) and (5),

$$\|A(u)^{-1}T(x') - A(u)^{-1}T(x)\| \le (1 + 2\epsilon)\,\|x' - x\|$$

for all x' in C. Thus $A(u)^{-1}T(C)$ is contained in a closed cube C' with center $A(u)^{-1}T(x)$ and the property that the length of each of its edges does not exceed $(1 + 2\epsilon)(2r)$. Hence

$$m(C') \le (1 + 2\epsilon)^n (2r)^n,$$

and setting $M(\epsilon) = (1 + 2\epsilon)^n$, we obtain the inequality

$$m[A(u)^{-1}T(C)] \le M(\epsilon)m(C).$$

If u is now chosen as in equation (2), it follows that

$$(6) \qquad \int_{T(C)} f(y)\, dy \le M(\epsilon)f(Tu)\, |\partial T(u)|\, m(C).$$

For any $\gamma < \delta$, we can divide C_0 into a finite number of closed cubes C_i with disjoint interiors such that

$$m(C_i) \le (2\gamma)^n,$$

and in each C_i there is a point u_i such that equation (6) is valid with $C = C_i$ and $u = u_i$. Since the support of f is contained in $T(C_0)$, it follows that

$$\int_{S'} f(y)\, dy \le \sum_i \int_{T(C_i)} f(y)\, dy \le M(\epsilon) \sum_i f(Tu_i)\, |\partial T(u_i)|\, m(C_i).$$

The sum on the right approximates the integral of $f(Tx)\, |\partial T(x)|$ within any desired degree of accuracy. Thus we have

$$\int_{S'} f(y)\, dy \le M(\epsilon)\int_S f(Tx)\, |\partial T(x)|\, dx,$$

and letting ϵ tend to 0, we obtain the inequality which finishes the proof.

EXERCISES

1 Let A be a Lebesgue-measurable subset of euclidean space E such that for every vector x in E, $x + A$ and A differ only by a null set. Show that A is either a null set or the complement of a null set. (*Hint:* Use the invariance characterization of Lebesgue measure.)

2 Give an example to show that not every homeomorphism of the line leaves some nonzero Baire measure invariant.

3* Given a set $\mathcal{C}$ of homeomorphisms T of a compact space X, an invariant measure is called *ergodic* (relative to $\mathcal{C}$) if the only Baire subsets A such that $T(A)$ and A differ only by a null set, for each $T \in \mathcal{C}$, are either null sets or complements of null sets. Show that the set of all invariant Baire measures m such that $m(X) = 1$ is a convex set, and that a measure is ergodic if and only if it is an extreme point of this set; an *extreme point* of a convex set S is defined as a point z which is not of the form $z = ax + by$ for any elements x and y of S with $0 < a < 1$ and $0 < b < 1$ unless $z = x = y$. (*Hint:* Use the Radon-Nikodym theorem.)

5.4 SET FUNCTIONS AND DIFFERENTIATION IN EUCLIDEAN SPACE

The existence of Radon-Nikodym derivatives has been established under fairly general conditions by means of an argument which provides no explicit construction for the derivative. In many applications none is needed, but it is nevertheless natural to seek conditions under which the derivative may be expressed in relatively direct terms. In the case of set functions in euclidean space, there is indeed an explicit connection between the Radon-Nikodym derivative and the derivative as defined in the calculus.

One might take the position that in Lebesgue integration theory, the appropriate notion of differentiation is not quite the pointwise concept of the calculus, but rather in the sense of the L_1-metric. In other words, a function f_1 on the line should be described as the derivative of a function f in case

$$\int |h^{-1}[f(x + h) - f(x)] - f_1(x)| \, dx \to 0$$

as $h \to 0$. On investigation it is found without much difficulty that the fundamental theorem of the calculus is indeed valid in the more general setting of Lebesgue integration theory with this definition of so-called differentiation "in mean." That is, if f is a Lebesgue-integrable function on the line, and if $F(x)$ is its indefinite integral in the sense of calculus, say,

$$F(x) = \int_{-\infty}^{x} f(u) \, du, \text{ then } h^{-1}[F(x + h) - F(x)] \text{ is convergent in mean to}$$

$f(x)$ as $h \to 0$ on every finite interval.

Such results are quite sufficient for most applications, but it is natural to inquire whether the indicated convergence holds also almost everywhere, or in dominated fashion (i.e., so that the difference quotients are uniformly majorized by a fixed function which is integrable or has other suitably designated finiteness properties).   This is clearly the best that could be hoped for; an arbitrary integrable function f is defined only within its values on a null set, so that convergence of the difference quotient to $f(x)$ is out of the question.   One of the most beautiful chapters in Lebesgue integration theory—if for external applications, one of the relatively expendable—deals with these questions in a quite precise and effective way.

The results in question are due basically to Lebesgue, apart from the results on domination due originally to Hardy and Littlewood.   They have been modified and re-proved in many ways, but all proofs have in common a certain intricacy.   Here the argument will be based on a simplified form of the Vitali covering theorem introduced by N. Wiener in connection with ergodic theory.

DEFINITION   If I is an integral on a measure space $(S,\mathcal{R},m)$ and E an arbitrary subset of S, the *outer measure* of E, denoted $m^*(E)$, is defined by the equation $m^*(E) = I^+(\varphi_E)$.

From the properties of the "upper" integral I^+, it is evident that m^* is nonnegative, and readily verified that

$$m^*\left(\bigcup_i E_i\right) \le \sum_i m^*(E_i)$$

for any sequence $E_1, E_2, \ldots$ of subsets of S.

SCHOLIUM 5.5   (WIENER)   *Suppose A is a set in k-dimensional eucliaean space and $\mathcal{C}$ is a collection of open spheres with the property that each point of A is the center of some sphere in $\mathcal{C}$.   Let m denote k-dimensional Lebesgue measure, and suppose that $m^*(A)$ is finite, or that*

$$\sup_{S \in \mathcal{C}} m(S) < \infty.$$

Then there exists a countable collection of disjoint spheres $S_1, S_2, \ldots$ in $\mathcal{C}$ such that

$$m^*(A) \le 3^k \sum_i m(S_i).$$

Proof   If $m^*(A)$ is finite and

$$\sup_{S \in \mathcal{C}} m(S) = \infty,$$

it is enough to take one sufficiently large sphere in $\mathcal{C}$.   Thus we shall suppose

$$\sup_{S \in \mathcal{C}} m(S) < \infty.$$

Next observe that if $S_1, S_2, \ldots$ is a sequence of disjoint spheres in $\mathcal{C}$ such that

$$\sum_i m(S_i) = \infty,$$

then we evidently have a sequence such that

$$m^*(A) \le 3^k \sum_i m(S_i).$$

Thus we shall also suppose that no such sequence exists.

It will next be shown that there exists a finite or infinite sequence $\{S_i\}$ of disjoint spheres in $\mathcal{C}$ such that

(1) $$m(S) < 2m(S_i)$$

whenever $S \in \mathcal{C}$ and S is disjoint from all S_j with $j < i$; and

(2) No element of $\mathcal{C}$ is disjoint from all S_j.

The sequence will be constructed by induction.   First choose S_1 in $\mathcal{C}$ so that

$$m(S_1) > \tfrac{1}{2} \sup_{S \in \mathcal{C}} m(S).$$

Now suppose $n > 1$ and the S_i are given so that equation (1) is satisfied for $i < n$.   If every sphere in $\mathcal{C}$ intersects some S_i with $i < n$, we are through. If not, let $\mathcal{C}_n$ be the collection of all spheres in $\mathcal{C}$ disjoint from all the S_i with $i < n$.   Choose S_n in $\mathcal{C}_n$ such that

$$m(S_n) > \tfrac{1}{2} \sup_{S \in \mathcal{C}_n} m(S);$$

then equation (1) is satisfied for $i \le n$.

If the construction terminates after a finite number of steps, we have a finite sequence of disjoint spheres satisfying equations (1) and (2).   On the other hand, if the construction yields an infinite sequence satisfying equation (1), then

$$\sum_i m(S_i) < \infty,$$

and it follows that $m(S_i) \to 0$.   In view of equation (1), this implies that no element of $\mathcal{C}$ is disjoint from all S_j.

Fixing now any such sequence S_i, for a given S in $\mathcal{C}$ there exists a least $i > 0$ such that $S \cap S_i \ne \emptyset$.   Let r_i be the radius of S_i, and r the radius of S.   Since the Lebesgue measure of an open sphere in k-dimensional euclidean space is proportional to the kth power of its radius, it follows that $r < 2^{1/k} r_i \le 2r_i$.   Let x denote the center of S, and x_i the center of S_i.   Since S and S_i have a point in common, x lies in the sphere S_i^* of radius $3r_i$ about x_i.   Thus

$$A \subset \bigcup_i S_i^*$$

and $$m^*(A) \le \sum_i m(S_i^*) = 3^k \sum_i m(S_i).$$

With the result just obtained it is easy to prove another estimate which is important in classical real variable theory.

SCHOLIUM 5.6   (HARDY-LITTLEWOOD)   *Suppose f is a Lebesgue-integrable function in k-dimensional euclidean space and that*

$$f^*(x) = \sup_S \frac{1}{m(S)} \left| \int_S f(x+y)\, dy \right|,$$

where the supremum is taken over all open spheres with centers at 0, and m denotes Lebesgue measure. Then, for each $t > 0$,

$$m^*[x : f^*(x) > t] < \frac{3^k}{t}\, \|f\|_1.$$

REMARK   It follows from results proved later that the integral

$$\frac{1}{m(S)} \int_S f(x+y)\, dy$$

is continuous as a function of x; in fact, uniformly so. This implies that f^* is lower-semicontinuous, i.e., that $[x : f^*(x) > t]$ is always an open set. Assuming this, we may write $m[x : f^*(x) > t]$ instead of $m^*[x : f^*(x) > t]$ in the above equation.

Proof of Scholium 5.6   Let $t > 0$ and $A = [x : f^*(x) > t]$. Then for each x in A there exists an open sphere S_x with center 0 such that

$$m(S_x) < \frac{1}{t} \left| \int_{S_x} f(x+y)\, dy \right| \le \frac{1}{t} \int_{S_x} |f(x+y)|\, dy$$

$$= \frac{1}{t} \int_{x+S_x} |f(y)|\, dy \le \frac{\|f\|_1}{t}.$$

The sphere $x + S_x$ has center x, and

$$m(x + S_x) = m(S_x).$$

Thus by Scholium 5.5 taking $\mathcal{C}$ as the set of all $x + S_x$, there exists an at most countable disjoint sequence $S_i = x_i + S_{x_i}$ such that

$$m^*(A) \le 3^k \sum_i m(S_i) < \frac{3^k}{t} \sum_i \int_{S_i} |f(y)|\, dy \le \frac{3^k}{t}\, \|f\|_1.$$

The function f^* is called the Hardy-Littlewood *maximal function*. The particular property of f^* just established will be used on the proof of the theorem of Lebesgue on the differentiation of integrals.

THEOREM 5.3   (LEBESGUE)   *If f is a complex-valued Lebesgue-integrable function in k-dimensional euclidean space, then for almost all x,*

$$\lim_{m(S) \to 0} \frac{1}{m(S)} \int_S |f(x+y) - f(x)|\, dy = 0,$$

where the limit is taken over the collection of all open spheres with centers at 0, and $m(S)$ is the Lebesgue measure of S.

Proof Let f be a given Lebesgue-integrable function in k-dimensional euclidean space. Then for the points x at which f is continuous, there is obviously no difficulty in showing that the indicated limit is 0. (On the other hand, it should be kept in mind that f need not be continuous anywhere.)

Let i and j be fixed positive integers. Then by Corollary 5.2.1, there exists a continuous Lebesgue-integrable function g such that

$$\|f - g\|_1 < \frac{1}{ij3^k}.$$

Let A be the set of all points x such that

$$|f - g|^*(x) > \frac{1}{i}.$$

It follows from Scholium 5.6 that

$$m^*(A) \le \frac{1}{j}.$$

Now let $B = [x : |f(x) - g(x)| \ge 1/i]$. Then

$$\frac{1}{i}\, m(B) \le \int_B |f(x) - g(x)|\, dx < \frac{1}{ij3^k},$$

so that $m(B) \le 1/j$.

If S is any open sphere with center 0, then

$$\int_S |f(x + y) - f(x)|\, dy \le \int_S |f(x + y) - g(x + y)|\, dy$$
$$+ \int_S |g(x + y) - g(x)|\, dy + m(S)\, |g(x) - f(x)|.$$

Suppose x is not in $A \cup B$. Then, from the definitions of A and B, it follows that

$$\int |f(x + y) - f(x)|\, dy \le 2\,\frac{m(S)}{i} + \int_S |g(x + y) - g(x)|\, dy.$$

Since g is continuous, this implies

$$\limsup_{m(S) \to 0} \frac{1}{m(S)} \int_S |f(x + y) - f(x)|\, dy \le \frac{2}{i}$$

for x in the complement of $A \cup B$. We now take into account the dependence of A and B on i and j by setting $E_{ij} = A \cup B$. Then, by the estimates for the outer measures of A and B,

$$m^*(E_{ij}) \le \frac{2}{j}.$$

It follows that the set

$$E = \bigcup_i \left(\bigcap_j E_{ij} \right)$$

is a null set.   If x is not in E, then

$$\limsup_{m(s) \to 0} \frac{1}{m(S)} \int_S |f(x+y) - f(x)| \, dy \leq \frac{2}{i}$$

for every i.   Thus for x in the complement of E, the limit indicated in the statement of the theorem exists and equals 0.

The set of all points x such that

$$\lim_{m(S) \to 0} \frac{1}{m(S)} \int_S |f(x+y) - f(x)| \, dy = 0$$

is called the *Lebesgue set for f*; it includes all points of continuity, and is useful in extending various results holding at such points, especially in classical Fourier analysis.

The following result concerning the differentiation of indefinite integrals is an immediate consequence of the theorem.

COROLLARY 5.4.1   *If F is the indefinite integral of a Lebesgue-integrable function f in k-dimensional euclidean space, then*

$$\lim_{m(S) \to 0} \frac{F(S)}{m(S)} = f(x) \qquad a.e.,$$

where the limit is taken over the collection of open spheres S with centers at x.

The second theorem of Lebesgue on differentiation may be stated as follows:

THEOREM 5.4   (LEBESGUE)   *Let n be a complex measure on the Borel subsets of k-dimensional euclidean space, m Lebesgue measure, and $n = F + s$ the Lebesgue decomposition of n with respect to m.   If F is the indefinite integral of the integrable function f, then*

$$\lim_{m(S) \to 0} \frac{n(S)}{m(S)} = f(x) \qquad a.e.,$$

where the limit is taken over the collection of all open spheres S with centers at x.

Proof   In view of Corollary 5.4.1 it is sufficient to show that for almost all x

$$\lim_{m(S) \to 0} \frac{s(S)}{m(S)} = 0,$$

where the limit is taken over the collection of all open spheres with centers at x.   Since the total variation of s is also singular with respect to Lebesgue measure, it is sufficient to consider the case where s is nonnegative and singular with respect to m.

Let D be the set of all points x at which the indicated limit fails to exist or has a value different from 0.   Then, since s is nonnegative,

$$D = \bigcup_{j \geq 1} A_j,$$

where A_j is the set of all points x for which there exist arbitrarily small open spheres S with centers at x such that

$$m(S) < js(S).$$

We shall show that $m^*(A_j) = 0$.   For this let E be a set of Lebesgue measure 0 which supports s, and ϵ and δ arbitrary positive numbers.   Then, by Scholium 5.2, there exists an open set F containing E such that $m(F) < \epsilon$, and a compact set K contained in F such that $s(F - K) < \delta$.   Let $G = A_j - K$, and for each point x in G, choose a sphere S disjoint from K with center x of radius less than 1, and such that $m(S) < js(S)$.   Let $\mathcal{C}$ be the collection of all such S.   Since $A_j - F$ is contained in $A_j - K$, and since

$$A_j = (A_j \cap F) \cup (A_j - F),$$

it follows that A_j is contained in $F \cup G$.   Now $m^*(F \cup G) \leq \epsilon + m^*(G)$, and by Scholium 5.5, there exists a countable disjoint family $\{S_i\}$ of spheres in $\mathcal{C}$ such that

$$m^*(G) \leq 3^k \sum_i m(S_i) < 3^k j \sum_i s(S_i).$$

Since s is supported by E, and E is contained in F, we also have

$$s(S_i) = s(S_i \cap F) = s[S_i \cap (F - K)].$$

Thus $m^*(G) < 3^k js(F - K) < 3^k j \, \delta$.   From this we see that

$$m^*(A_j) < \epsilon + 3^k j \, \delta,$$

and since ϵ and δ are arbitrary, it results that $m^*(A_j) = 0$.   Hence A_j is a null set, and so the set D is a null set as well.

EXERCISES

1 If f is a complex-valued Lebesgue-integrable function on a real interval, show that

$$\lim_{h \to 0} \frac{1}{h} \int_x^{x+h} |f(y) - f(x)| \, dy = 0$$

for almost all x in the interval.

2   Let f be a complex-valued Lebesgue-integrable function on a real interval S, and a a point in S.   Define F on S by

$$F(x) = \int_a^x f(y)\, dy$$

and show that $F'(x) = f(x)$ a.e. in S, where $F'(x)$ denotes the usual pointwise derivative of F at x.

3   A numerical function F on an interval $[a,b]$ is called *absolutely continuous* if for every $\epsilon > 0$ there exists $\delta > 0$ such that for any finite collection of disjoint subintervals $[a_i,b_i]$

$$\sum_i |F(b_i) - F(a_i)| < \epsilon$$

whenever $\sum_i (b_i - a_i) < \delta$.   Show that this is the case if and only if F is a continuous function of bounded variation on $[a,b]$ for which the corresponding set function on the Borel subsets of $[a,b]$ is absolutely continuous with respect to Lebesgue measure.

4   Show that a complex-valued function of bounded variation on an interval is the sum of an absolutely continuous function, a jump function, and a continuous singular function (i.e., such that the associated complex measure is singular with respect to Lebesgue measure).   Show also that this (Lebesgue) decomposition is unique within additive constants.

5   Show that a complex-valued function of bounded variation on an interval is differentiable a.e. in the interval.   If the function is decomposed as in Exercise 4, show that its jump part and continuous singular part have derivatives equal to 0 a.e.

6   *a.* Show that the Cantor set may be defined as the subset of $[0,1]$ consisting of all points x having a ternary expansion $x = \Sigma a_n 3^{-n}$, with $a_n = 0$ or 2.
   b. For any point x of the Cantor set, define $f(x)$ as $\sum_n (a_{n/2})2^{-n}$, and otherwise define $f(x)$ as the value $f(y)$ at the point y of the Cantor set which is nearest to x among the points such that $y \le x$.   Show that the same value is obtained at the point z in the Cantor set which is nearest among the points for which $x \le z$.
   c. Show that f is an increasing function which is constant on a collection of subintervals of $[0,1]$ of total measure 1.
   d. Show that f is a continuous singular function of bounded variation.†

7   The *density* of a Lebesgue-measurable set E on the real line at the point x is defined as the limit as $h \to 0$ of

$$\frac{1}{2h} m\, [E \cap (x - h, x + h)]$$

† For a more complete account of this example, see E. Hille and J. D. Tamarkin, *Am. Math. Monthly*, vol. 36, pp. 255–264, 1929.

when the limit exists. Show that the density is 1 a.e. on E and 0 a.e. in the complement of E.

8 If $f_1, f_2, \ldots$ is a sequence of monotone-increasing functions such that $\sum_i f_i(x)$ is convergent to a finite-valued function $f(x)$, then $f'(x)$ is almost everywhere

$$\sum_i f_i'(x).$$

9 If f is Lebesgue-measurable and has a finite derivative at every point in $[a,b]$ and if f' is integrable on $[a,b]$, then

$$\int_a^b f'(x)\,dx = f(b) - f(a).$$

10 Show that if f is an integrable function on the line (with respect to Lebesgue measure) of compact support, and if $F(x) = \int_{-\infty}^x f(u)\,du$, then $h^{-1}[F(x+h) - F(x)]$ converges in L_1 to f as $h \to 0$. (*Hint:* First approximate to f by a continuous function of compact support.)

VI

FUNCTION
SPACES

6.1 LINEAR DUALITY

In the case of a finite-dimensional linear space **L**, the *dual*, or *conjugate*, space **L*** consisting of all linear functionals on **L** plays an important role in the general theory.   The primary result in the finite-dimensional case is that the natural injection of **L** into the dual **L**** of **L** is an isomorphism of **L** onto **L****.   This means that if x is an element of **L** and x^{**} denotes the element of **L**** given by the equation $x^{**}(f) = f(x)$ for all f in **L***, then the map $x \to x^{**}$ is one-to-one from **L** onto **L****.

It is impossible to extend this result to infinite-dimensional spaces; the algebraic dual of an infinite-dimensional linear space is simply too large. In any event, such an extension would have very little relevance for analysis. Only the continuous (or other suitably smooth) functionals arise in concrete analytic situations, and in fact the others have been shown to exist only by transfinite induction.   No example of an explicit nature has ever been given of such a discontinuous functional, and quite conceivably this is, for the function spaces of interest, not accidental, but in the nature of things.   When these continuity requirements are taken into account, it is found that the duality theory for infinite-dimensional linear topological spaces is much more complex than that for finite-dimensional spaces.   However, the theory arises in a natural way.   For example, before linear topological spaces were considered as such, the L_p-spaces were closely examined from a standpoint which is now recognizable as implicitly that of linear duality.   It was found that interesting relations hold among L_p-spaces which, in the terms indicated above, amount to the equivalence of the L_p-spaces with their second duals for $1 < p < \infty$.

The theory of duality in linear topological spaces may be looked on as an economical and suggestive way of formulating a variety of concrete analytical results or, on the other hand, from a more algebraic point of view, as a natural extension of the finite-dimensional theory; however, for the purposes of pure algebra, it is the extension to the case of spaces over more general fields than those of the real and complex numbers that is most useful.   But the theory of such algebraic extensions resembles the theory in the real and complex cases in many important respects, and these two central cases are the only ones to be treated here in any detail.

DEFINITION   The *dual* to a topological linear space **L** over a topological field F is the collection **L*** of all continuous linear functionals f on **L**.

In other words, an element f of **L*** is a function on **L** with values in F such that (1) f is continuous, (2) $f(cx) = cf(x)$ for all c in F and all x in **L**, and (3) $f(x + y) = f(x) + f(y)$ for any x and y in **L**.   For any two such elements f and g, $f + g$ is defined in the usual way, $(f + g)(x) = f(x) + g(x)$, as is cf for c in F, namely, $(cf)(x) = cf(x)$.   It is then evident that **L*** is itself a linear space over the field F.

The dual **L*** as just defined is a linear space, but has no topological structure.   Such structures do, however, play an important role, especially where **L** is a Banach space, in which case **L*** is also one in a natural way.   To see this it is first convenient to prove the following general result.

SCHOLIUM 6.1   *A linear transformation T from one normed linear space* **X** *into another* **Y** *is continuous if and only if it is bounded, in the sense that there exists a constant c such that* $\|Tx\| \le c\|x\|$ *for all x in* **X**.

Proof   If such a constant c exists, then $\|Tx - Ty\| \le c\|x - y\|$ for all x and y, so that T is certainly continuous.   Conversely, if T is linear and continuous at 0, then there exists a constant $b > 0$ such that $\|Tx\| \le 1$ when $\|x\| \le b$.   Since $T(cx) = cT(x)$ for all x, this implies the relation $\|Tx\| \le b^{-1}\|x\|$ for all x.

DEFINITION   For a continuous linear transformation T, the greatest lower bound of all numbers c such that $\|Tx\| \le c\|x\|$ for all x is called the *bound* of T and is frequently designated by $\|T\|$.

It is easily seen that

$$\|T\| = \sup_{\|x\|=1} \|Tx\|.$$

The *Banach dual* to a Banach space **B** is its dual **B*** as defined above, with the additional structure of a Banach space derived from the norm

$$\|f\| = \sup_{\|x\|=1} |f(x)|$$

There is no difficulty in showing that **B*** is indeed a Banach space. In fact, the following more general result is true.

SCHOLIUM 6.2   *Let* **X** *be a normed linear space and* **Y** *a Banach space. Then the set* **L** *of all continuous linear transformations* T *of* **X** *into* **Y** *is a Banach space relative to the norm given by the operator bound.*

Proof   It is evident that **L** is a normed linear space, and the only problem is to verify that **L** is complete.   For this suppose that $\{T_n\}$ is a Cauchy sequence in **L**.   Then the sequence $\{T_n x\}$ is Cauchy in Y for any vector x in X.   Since Y is complete, it results that there is a unique element y in Y such that $T_n x \to y$.   The map $T : x \to y$ is single-valued and obviously linear. On the other hand, for any $\epsilon > 0$, there exists a positive N_ϵ such that $\|T_k x - T_n x\| \leq \epsilon \|x\|$ for all $k, n \geq N_\epsilon$ and any x.   Thus $\|Tx - T_n x\| \leq \epsilon \|x\|$ for $n \geq N_\epsilon$ and any x, which shows that $\|T - T_n\| \leq \epsilon$ for $n \geq N_\epsilon$. Since $T = (T - T_n) + T_n$, it follows that $T \in$ **L**, and also that $T_n \to T$.

THEOREM 6.1   (F. RIESZ)   *For any measure space* M, *the Banach dual to* $\tilde{L}_p(M)$ *for* $1 < p < \infty$ *is naturally isomorphic to* $\tilde{L}_q(M)$, *where* $q = p/(p-1)$: *every continuous linear functional* F *on* $\tilde{L}_p(M)$ *has the form*

$$F(\tilde{f}) = I(fk)$$

for a unique element $\tilde{k}$ *in* $\tilde{L}_q(M)$; *and* $\|F\| = \|k\|_q$.

Proof   For k in L_q the functional F which assigns the value $I(fk)$ to the element $\tilde{f}$ in $\tilde{L}_p$ is well defined, and $\|F\| \leq \|k\|_q$ by Hölder's inequality.   To show that $\|F\| = \|k\|_q$, it suffices by homogeneity to consider the case $\|k\|_q = 1$.   Let sgn z be defined for complex z as $z/|z|$ when $z \neq 0$ and as 0 otherwise, and set

$$f = |k|^{q-1} \operatorname{sgn} \bar{k}.$$

Then $|f|^p = |k|^{p(q-1)} = |k|^q$, $\|f\|_p = 1$, and $F(\tilde{f}) = I(|k|^q) = 1$; hence $\|F\| = 1$.   Therefore the mapping $\tilde{k} \to F$ is an isometric isomorphism of $\tilde{L}_q$ into $(\tilde{L}_p)^*$.   All that remains is to show that this map is onto.

For this suppose F is a given continuous linear functional on L_p, and assume also that the measure space is finite.   Then the characteristic function φ_E of every measurable set E is an element of L_p, and the equation $n(E) = F(\tilde{\varphi}_E)$ defines a finite complex-valued function n on the σ-ring of measurable sets.   If $E_1, E_2, \ldots$ is an increasing sequence of such sets with union E, then

$$|\varphi_E - \varphi_{E_n}|^p \leq \varphi_E$$

and $\varphi_E - \varphi_{E_n} \to 0$, so that $\|\varphi_E - \varphi_{E_n}\|_p \to 0$.   Since F is continuous, it follows that $n(E_n) \to n(E)$.   Thus n is countably additive, and since $\tilde{\varphi}_E = 0$ when E has measure 0, n is absolutely continuous with respect to the measure m on M.   Hence there exists an integrable function k on M such that

$$n(E) = \int_E k(x)\, dm(x)$$

for all measurable sets E.   It follows by linearity that $F(\tilde{f}) = I(fk)$ for all simple functions f.   Next we shall show that $k \in L_q$.

For this let c_n denote the characteristic function of $[x : |k(x)| \le n]$, $n = 1$, $2, \ldots$, and set $k_n = kc_n$.   Then $k_n \in L_q$, and the corresponding functional F_n is continuous on $\tilde{L}_p$ with norm $\|k_n\|_q$.   Moreover, $\|k_n\|_q \le \|F\|$.   For if f is any simple function, then $F_n(\tilde{f}) = F(\tilde{f}\tilde{c}_n)$, and consequently

$$|F_n(\tilde{f})| \le \|F\|\, \|fc_n\|_p \le \|F\|\, \|f\|_p.$$

A linear functional or linear transformation which is bounded by a given constant on a (linear) subspace of a normed linear space is also bounded by the same constant on the closure of the subspace.   Since the simple functions form a dense subspace of L_p, it follows that $\|F_n\| \le \|F\|$ for all n.   In addition, $|k_n| \uparrow |k|$ and $I(|k_n|^q) \uparrow I(|k|^q)$.   Therefore k is indeed an element of L_q.

Now F and the continuous functional induced by k agree on the simple functions, and hence they agree, by continuity, on all of $\tilde{L}_p$.   Thus F has the stated form when the measure space is finite.

The proof for the general case is derived by relativization in the following way.   For any measurable set A of finite measure, there exists an element k_A in L_q such that $F(\tilde{f}) = I(fk_A)$ for all functions f in L_p supported by A.   Moreover, $\|k_A\|_q \le \|F\|$.   Let $A_1, A_2, \ldots$ be any sequence of measurable sets of finite measure such that

$$\|k_{A_n}\|_q \uparrow \sup \|k_B\|_q$$

and set $E = \bigcup_n A_n$.   Then $F(\tilde{f}) = 0$ for any f in L_p which vanishes on E. For otherwise there would exist a measurable set C of finite measure disjoint from E such that $\|k_C\|_q > 0$.   But since

$$k_{A_n \cup C} = k_{A_n} + k_C \qquad \text{a.e.,}$$

this would imply the contradiction

$$\lim_n \|k_{A_n \cup C}\|_q > \lim_n \|k_{A_n}\|_q.$$

Write E as a disjoint union of measurable subsets $E_1, E_2, \ldots$ of finite measure, and set $k_n = k_{E_n}$.   Let k be the measurable function which agrees with k_n on E_n and vanishes outside E.   Then, since

$$\sum_{j \le n} \|k_j\|_q^q = \left\| \sum_{j \le n} k_j \right\|_q^q \le \|F\|^q,$$

it follows by dominated convergence that the series $\sum_{n} k_n$ converges to k in L_q. If f is a function in L_p supported by E, let $f_n = f\varphi_{E_n}$. Then the series $\sum_{n} f_n$ converges to f in L_p, and

$$F(\tilde{f}) = \sum_{n} F(\tilde{f}_n) = \sum_{n} I(f_n k_n) = I(fk).$$

This completes the proof, inasmuch as any function in L_p may be written as the sum of two L_p functions, one supported by E and one by the complement of E.

The foregoing argument applies in many respects to the case $p = 1$, $q = \infty$; however, the L_∞-norm is not given by an integral, and it cannot be concluded, as above, that a continuous linear functional F on $\tilde{L}_1$ is supported by a measurable set E. A simple example of this is given by the integral I, which is supported by a measurable set if and only if M is σ-finite.

COROLLARY 6.1.1   (F. RIESZ)   *For any measure space M there is a natural isomorphism of $\tilde{L}_\infty(M)$ into the Banach dual of $\tilde{L}_1(M)$. If M is σ-finite, this isomorphism is onto, and every continuous linear functional F on $\tilde{L}_1(M)$ has the form*

$$F(\tilde{f}) = I(fk)$$

for a unique element $\tilde{k}$ of $\tilde{L}_\infty(M)$, and $\|F\| = \|k\|_\infty$.

Proof If $k \in L_\infty$, it is clear that the indicated functional is well defined and bounded by $\|k\|_\infty$. To show that $\|F\| = \|k\|_\infty$, it is enough by homogeneity to consider the case where $\|k\|_\infty = 1$. Then, for $0 < \epsilon < 1$, there exists a measurable set E of finite positive measure $m(E)$ such that $|k(x)| > 1 - \epsilon$ for x in E. Set $f = c\,\mathrm{sgn}\,(\bar{k})\varphi_E$, where $c = 1/m(E)$. Then $|f| = c\varphi_E$, $I(|f|) = 1$, and

$$fk = c|k|\varphi_E > c(1 - \epsilon)\varphi_E.$$

Hence $I(fk) > 1 - \epsilon$, and $\|F\| > 1 - \epsilon$. Since this is true for all ϵ such that $0 < \epsilon < 1$, it follows that $\|F\| = 1$. Thus $\tilde{k} \to F$ is an isometric isomorphism of $\tilde{L}_\infty$ into $(\tilde{L}_1)^*$.

Now suppose F is a given continuous linear functional on $\tilde{L}_1$. Then, for an arbitrary measurable set A of finite measure, there exists, by the same argument used in the preceding theorem, a function k_A in $L_1(M)$ vanishing outside A such that $F(\tilde{f}) = I(fk_A)$ for all simple functions f supported by A. Defining c_n, k_n, and F_n as before, we see that $\|k_n\|_\infty = \|F_n\| \leq \|F\|$, and that $k_n \uparrow k_A$. Thus $\|k_A\|_\infty \leq \|F\|$, and it follows readily that $F(\tilde{f}) = I(fk_A)$ for all f in L_1 supported by A. In the event that there exists an L_∞ function k which agrees with k_A a.e. on A for each A, we see that $F(\tilde{f}) = I(fk)$ for all f in L_1. On a σ-finite measure space such a function is readily constructed. Simply write the entire space as a countable disjoint union of measurable sets A_i of finite measure and add the corresponding k_{A_i}.

The case $p = \infty$ was omitted for good reason: L_∞^* is materially greater than L_1 even in simple cases such as that of Lebesgue measure on the reals. More precisely, the natural map $f \to f^{**}$ of L_1 into $(L_1^*)^* = L_\infty^*$ is an isomorphism into, but in general not onto, L_∞^*. In essence this is a general fact about Banach spaces. To establish this, as well as to explore the case $p = \infty$ more readily, we shall prove a basic result concerning continuous extensions of linear functionals, which plays the role of a cornerstone in the general theory of duality in linear topological spaces. It should perhaps be noted that it is an immediate deduction from transfinite induction that a linear functional on a subspace of a linear space may be extended to a linear functional on the entire space. The substance of the result to be proved is that if the functional on the subspace is bounded in a certain sense, then there is an extension to the entire space which is correspondingly bounded.

THEOREM 6.2 (HAHN-BANACH) *Let p be a real-valued function on a real linear space* **L** *with the properties*

$$p(x + y) \le p(x) + p(y), \qquad p(ax) = ap(x),$$

for arbitrary x and y in **L** *and real $a > 0$. Let f be a linear functional on a linear subspace* **M** *of* **L** *with the property that*

$$f(x) \le p(x), \qquad x \in \mathbf{M}.$$

Then there exists a linear functional F on all of **L** *which agrees with f on* **M** *and has the property that*

$$F(x) \le p(x), \qquad x \in \mathbf{L}.$$

Proof The theorem is proved by transfinite induction, the key step being the extension of f to a suitable functional on the subspace spanned by **M** and one additional element. If u is any element of **L** which is not in **M**, the subspace **M′** spanned by u and **M** consists of all vectors of the form $x + bu$, with x in **M** and b real. If $f(u)$ is defined as some arbitrary real number, the definition

$$F(x + bu) = f(x) + bf(u)$$

then provides a well-defined functional on **M′** extending f. For this functional to have the required property it is necessary and sufficient that

$$f(x) + bf(u) \le p(x + bu)$$

for all x in **M** and all real b. If $b = 0$, this is true by assumption. For $b > 0$ the inequality is equivalent to the condition

$$f(u) \le b^{-1}[p(x + bu) - f(x)],$$

while for $b < 0$, say, $b = -c$ with $c > 0$, it is equivalent to the inequality

$$c^{-1}[f(x) - p(x - cu)] \le f(u).$$

If $t = \inf\limits_{x,b>0} b^{-1}[p(x+bu) - f(x)]$ and

$$s = \sup\limits_{y,c>0} c^{-1}[f(y) - p(y - cu)],$$

it follows that the functional F has the required property if and only if

$$s \le f(u) \le t.$$

Thus there is an admissible value for $f(u)$ if and only if $s \le t$. To show that this is indeed the case, first observe that

$$t = \inf\limits_{x\in\mathbf{M}} [p(x+u) - f(x)]$$

by the homogeneity of f and that of p relative to positive scalars. For the same reasons

$$s = \sup\limits_{y\in\mathbf{M}} [f(y) - p(y - u)].$$

It is therefore sufficient to show that

$$f(y) - p(y - u) \le p(x+u) - f(x),$$

or equivalently, that

$$f(x) + f(y) \le p(x+u) + p(y-u)$$

for all x and y in $\mathbf{M}$. But this is immediate from the linearity of f, the subadditivity of p, and the assumed relation between f and p on $\mathbf{M}$.

A formal proof may now be given along the standard lines by transfinite induction. Specifically, let $\mathcal{C}$ be the collection of all pairs $(\mathbf{N},g)$, where $\mathbf{N}$ is a linear submanifold of $\mathbf{L}$ containing $\mathbf{M}$, and g is a linear functional on $\mathbf{N}$ agreeing on $\mathbf{M}$ with f and such that $g(x) \le p(x)$ for all x in $\mathbf{N}$. Let $\mathcal{C}$ be partially ordered by inclusion; that is, $(\mathbf{N}',g') \subset (\mathbf{N},g)$ means that $\mathbf{N}' \subset \mathbf{N}$ and that $g'(x) = g(x)$ for x in $\mathbf{N}'$. Then $\mathcal{C}$ has the property that any simply ordered subset has a least upper bound. For suppose $\mathcal{D}$ is such a subset of elements $(\mathbf{N},g)$, and let $\mathbf{Q}$ be the union of the subspaces $\mathbf{N}$. Then $\mathbf{Q}$ is a linear subspace of $\mathbf{L}$, and there is a unique linear functional h on $\mathbf{Q}$ such that h agrees with g on $\mathbf{N}$ for any $(\mathbf{N},g)$ in $\mathcal{D}$. The pair $(\mathbf{Q},h)$ is an element of $\mathcal{C}$ which, by construction, is an upper bound for $\mathcal{D}$, and is easily seen to be minimal among such elements of $\mathcal{C}$. By the Zorn formulation of transfinite induction, it follows that $\mathcal{C}$ contains a maximal element, say, $(\mathbf{R},k)$. This pair satisfies exactly the same conditions as $(\mathbf{M},f)$, which was shown to be appropriately extendable to the subspace spanned by $\mathbf{M}$ and any element of $\mathbf{L}$ outside $\mathbf{M}$. Consequently, $(\mathbf{R},k)$ can be maximal only if $\mathbf{R}$ is the entire space $\mathbf{L}$, and k then the desired extension F of f.

> COROLLARY 6.1.2   *A linear functional on a subspace of a normed linear space which is bounded by a given constant has a continuous linear extension to the whole space which is bounded by the same constant.*

Proof Assume first that **L** is a normed linear space over the reals and that f is a linear functional on a subspace **M** such that $|f(x)| \le a \, \|x\|$ for all x in **M**. Setting $p(x) = a \, \|x\|$, we see immediately from the theorem that f has an extension F such that $|F(x)| \le a \, \|x\|$ for all x in **L**.

Now consider the case where **L** is a normed linear space over the complex field. Then $\operatorname{Im} f(x) = \operatorname{Re} [-if(x)] = -\operatorname{Re} [f(ix)]$ for all x in **M**. Thus f can be expressed in terms of its real part g by the equation

$$f(x) = g(x) - ig(ix).$$

Now g is a real linear functional on **M**, and $|g(x)| \le |f(x)| \le a\|x\|$ for all x in **M**. Hence, by the first paragraph, there is a real linear functional G on **L** which extends g and satisfies the relation $|G(x)| \le a \, \|x\|$ for all x in **L**. Let F be the functional defined on **L** by

$$F(x) = G(x) - iG(ix), \qquad x \in \mathbf{L}.$$

Then F is obviously real linear and extends f. Moreover, $iF(x) = G(ix) + iG(x) = G(ix) - iG(-x) = F(ix)$, so that F is complex linear as well. To prove that $|F(x)| \le a \, \|x\|$ for all x in **L**, observe that this is trivial if $F(x) = 0$. When $F(x) \ne 0$, there is a complex number c such that $|c| = 1$ and $|F(x)| = cF(x)$. Thus $|F(x)| = F(cx) = G(cx) \le a \, \|x\|$.

COROLLARY 6.1.3 *If* **M** *is a closed subspace of a normed linear space* **L** *and* x *is a vector not in* **M**, *there exists a continuous linear functional* f *on* **L** *of norm* 1 *such that* f *annihilates* **M** *and* $f(x) = d$, *where* d *is the distance between* x *and* **M**.

Proof Let **M**′ be the subspace of all vectors of the form $cx + y$, where c is a scalar and y an element of **M**. Since x is not in **M**, there is a well-defined linear functional f on **M**′ such that $f(cx + y) = cd$. For $c \ne 0$

$$\|cx + y\| = |c| \left\| x + \frac{y}{c} \right\| \ge |c| \, d.$$

Hence $|f(cx + y)| \le \|cx + y\|$, and this shows that $\|f\| \le 1$. On the other hand, for each $\epsilon > 0$, there exists y in **M** such that $d \le \|x - y\| < d + \epsilon$. For any such vector y

$$\frac{d}{d + \epsilon} < \frac{f(x - y)}{\|x - y\|} \le 1,$$

which implies $\|f\| > d/d + \epsilon$. Thus $\|f\| = 1$ on **M**′, and the result now follows by Corollary 6.1.2.

COROLLARY 6.1.4 *If* **L** *is a normed linear space, the natural injection* $x \to x^{**}$, *where* $x^{**}(f) = f(x)$ *for* f *in* **L***, *is a norm-preserving isomorphism of* **L** *into its second dual* **L****.

Proof By Corollary 6.1.3 there exists for each nonzero x in **L** a linear functional f in **L*** of unit norm such that $f(x) = \|x\|$.   Hence $\|x\| = \|x^{**}\|$.

There is a suggestive geometrical reformulation of the Hahn-Banach theorem which is based on the connection between convex functionals and open convex sets.

DEFINITIONS   A *convex set* in a real linear space **L** is a set S with the property that if x and y are elements of S, then so also is $ax + by$ in case $0 \leq a, b \leq 1$, and $a + b = 1$.   A *convex functional* on **L** is a real-valued subadditive function p which is homogeneous relative to positive scalars; that is, $p(x + y) \leq p(x) + p(y)$ and $p(ax) = ap(x)$ for all x and y in **L**, and $a > 0$.

If p is a convex functional, it is easily verified that $[x:p(x) < 1]$ is a convex set, which, moreover, has the property of containing some positive multiple of every vector in the space.   This suggests the following converse.

SCHOLIUM 6.3   *If **P** is a convex set in a real linear space **L** which contains some positive multiple of every vector in the space, than the equation*

$$p(x) = \inf\,[a:a > 0 \quad \text{and} \quad a^{-1}x \in \mathbf{P}]$$

*defines a convex functional p on **L**.*

Proof Suppose x is a given vector in **L** and $b > 0$.   Let a be any positive number such that $a^{-1}x \in \mathbf{P}$.   Then $ab > 0$ and $(ab)^{-1}(bx) \in \mathbf{P}$, so that $p(bx) \leq ab$.   Since this is true for all such a, it follows that $p(bx) \leq bp(x)$. For the same reasons, $p(b^{-1}bx) \leq b^{-1}p(bx)$, hence and $p(bx) = bp(x)$.   If x and y are arbitrary vectors and $\epsilon > 0$, there exist positive numbers a and b such that $ax \in \mathbf{P}$, $by \in \mathbf{P}$, $a^{-1} < p(x) + \epsilon$, and $b^{-1} < p(y) + \epsilon$.   By the convexity of **P**,

$$\frac{b}{a + b}\,(ax) + \frac{a}{a + b}\,by$$

is an element of **P**, and since this element equals

$$\frac{ab}{a + b}\,(x + y),$$

it follows that $p(x + y) \leq (a + b)/ab$.   Thus $p(x + y) \leq p(x) + p(y) + 2\epsilon$, and this implies the subadditivity of p and completes the proof.

The functional p in the preceding result is often called the *Minkowski functional* for **P**.   Of course, the condition that **P** contain some positive multiple of every vector in the space is quite restrictive, and this will not be true for a general convex set.   In fact, the hyperplanes form a rather important class of convex sets for which the condition is never satisfied.

DEFINITIONS Any set in a real linear topological space **L** which consists of all points at which some nonzero continuous linear functional attains a given value is called a *hyperplane*. An *open half-space* in **L** is a set of the form $[x: f(x) < c]$, where f is a continuous nonzero linear functional and c is a given real constant, while a *closed half-space* is a set of the form $[x: f(x) \leq c]$. The boundary of such a closed half-space is readily seen to be the hyperplane $[x: f(x) = c]$. A hyperplane is said to *separate* two subsets in case they lie in complementary half-spaces bounded by the hyperplane, and to *strongly separate* them if they lie in complementary open half-spaces.

From a geometrical point of view the Hahn-Banach theorem asserts that an open convex subset of a real linear topological space and a point outside it can be separated by a hyperplane. In fact, we shall prove the following somewhat more precise statement.

COROLLARY 6.1.5 *If* **P** *is an open convex set in a real linear topological space and u is a point outside* **P**, *then there exists an open half-space which contains* **P** *but not u.*

Proof By making a suitable vector translation, which does not affect convexity or separation by a hyperplane, we may assume that 0 is in the given convex set **P**. Then, since **P** is open and $ax \rightarrow 0$ as $a \rightarrow 0$ for any x, it follows that every vector has a positive multiple lying in **P**, so that the Minkowski functional for **P** is well defined as above. If now u is the given point outside **P**, let **M** be the one-dimensional subspace spanned by u, and define f on **M** by the equation $f(au) = ap(u)$. Then f is obviously linear on **M**, and since p is nonnegative, there is no difficulty in verifying that $f(x) \leq p(x)$ for all x in **M**. Thus the Hahn-Banach theorem guarantees the existence of a linear functional F defined on the entire space **L** which extends f and has the property that $F(x) \leq p(x)$ for all x. This inequality implies the continuity of F. For suppose $\epsilon > 0$. Then since the mapping $x \rightarrow \epsilon^{-1}x$ is continuous on **L** and sends 0 into 0, there exists a neighborhood N of 0 such that $\epsilon^{-1}x \in$ **P** for all x in N. Hence $p(x) \leq \epsilon$ for x in N, and for x in the intersection of N and $-N$, $|F(x)| \leq \epsilon$. Thus F is continuous at 0 and by linearity is continuous at all points. Therefore $[x: F(x) < 1]$ is an open half-space, and we shall show next that this half-space contains **P**. For this suppose that $x \in$ **P**. Then since **P** is open, the continuity of ax as a function of a implies the existence of an open interval S of reals which contains 1 such that $ax \in$ **P** for all a in S. This in turn implies that $p(x) < 1$, and hence that $F(x) < 1$. To complete the proof it must be shown that $F(u) \geq 1$, i.e., that $p(u) \geq 1$, and this follows from the fact that **P** is a convex set containing 0 but not u.

The Hahn-Banach theorem is useful not only for general theoretical purposes but also for avoiding unnecessary constructions and details in specific situations. A reasonably typical example of this of particular relevance to integration theory is given in the next result, which shows that the dual of L_∞ may be larger than L_1.

SCHOLIUM 6.4   *Not every continuous linear functional on $\tilde{L}_\infty[0,1]$ with respect to Lebesgue measure has the form $\tilde{g} \to I(fg)$ for some integrable function f.*

Proof From the fact that nonempty open sets have positive Lebesgue measure, it is elementary to verify that the essential bound of a continuous function on [0,1] is identical with its ordinary least upper bound. The Banach space of all continuous functions h on [0,1] relative to the norm

$$\|h\| = \sup_{0 \le x \le 1} |h(x)|$$

is consequently indentifiable with a closed linear subspace of $\tilde{L}_\infty$. Therefore the Hahn-Banach theorem implies that the continuous linear functional $h \to h(0)$ on this space has an extension to a continuous linear functional F on all of $\tilde{L}_\infty$. Now if F had the form given by the equation $F(\tilde{g}) = I(fg)$ for some fixed integrable function f, this would apply in particular to the case of a continuous function h, so that f would have the property

$$h(0) = \int_0^1 f(x)h(x)\, dx$$

for all continuous functions h. But if $\epsilon > 0$, there exists a continuous function h such that $h(0) = 1$ and $|I(fh)| < \epsilon$, for example, $h(x) = (1 - x)^n$ for sufficiently large n.

A Banach space with the property that its natural injection into its second dual is onto is called *reflexive*. Thus, for $1 < p < \infty$, $\tilde{L}_p(M)$ is reflexive for any measure space M, while in general $\tilde{L}_1(M)$ is not reflexive. It can be shown also that $\tilde{L}_\infty(M)$ is not in general reflexive. On the other hand, it can be shown that all linear topological spaces whose duals are sufficiently nontrivial are reflexive in a certain weaker sense. The development of this positive general result requires the introduction of the useful notions of weak topology and that of a locally convex space.

DEFINITIONS   If **L** is any set, and **C** is any collection of real- or complex-valued functions on **L**, the *weak topology* on **L** defined by **C** is defined as the weakest (i.e., one whose collection of open sets is minimal) for which all the functions in question are continuous. It is easily seen that this weak topology does in fact exist, and that a *basic family* of neighborhoods of a general point y is provided by the sets of the form

$$[x: |f_i(x) - f_i(y)| < \epsilon, \qquad i = 1, 2, \ldots, n],$$

where ϵ is an arbitrary positive number, and $f_1, f_2, \ldots, f_n$ is an arbitrary finite sequence in **C**. **C** is said to be a *separating collection* if $f(x) = f(y)$ for all f in **C** implies $x = y$. When **C** is a collection of linear functionals on **L**, it is a straightforward exercise to verify that **L** in the weak topology defined by **C** is a linear topological space.

SCHOLIUM 6.5 *Every continuous linear functional on a real or complex linear space* **L** *relative to the weak topology defined by a collection* **C** *of linear functionals on* **L** *is a finite linear combination of elements of* **C**.

Proof Suppose g is a continuous linear functional on **L** in the weak topology defined by **C**. Let δ be an arbitrary positive number, and let f_1, $f_2, \ldots, f_n$ and ϵ define a basic neighborhood of 0, as in the preceding paragraph, which is mapped by g into the set $|z| < \delta$. Suppose x is an element of **L** such that $f_i(x) = 0$ for $i = 1, 2, \ldots, n$. Then for any scalar a, $f_i(ax) = af_i(x) = 0$ as well, so that ax is in the above neighborhood of 0. Hence $|g(ax)| < \delta$ for all a, which implies $g(x) = 0$.

Using this fact, we shall show that g is a linear combination of $f_1, f_2, \ldots, f_n$. This follows essentially by finite-dimensional linear algebra. For let **V** be the vector space of all scalar n-tuples, and let F be the mapping of **L** into **V**, defined by the equation

$$F(x) = (f_1(x), f_2(x), \ldots, f_n(x)).$$

Then F is linear, and $F(\mathbf{L})$ is a linear subspace of **V**. Now since $g(x) = 0$ whenever $F(x) = 0$, it follows by linearity that the relation $F(x_1) = F(x_2)$ implies $g(x_1) = g(x_2)$. Thus there is a unique well-defined linear functional G on $F(\mathbf{L})$ such that $G[F(x)] = g(x)$ for all x in **L**. Any linear functional on **V** is a linear combination of the standard coordinate functionals, and if G is extended in an arbitrary fashion to a linear functional on **V**, it follows that the same is true of G. Hence there exist scalars $a_1, a_2, \ldots, a_n$ such that for all x in **L**,

$$G[F(x)] = \sum_i a_i f_i(x),$$

which is precisely the statement

$$g = \sum_i a_i f_i.$$

As we shall see shortly, this result is the key to a kind of reflexivity property of linear topological spaces which are *locally convex* in the sense that there is a basis for open sets consisting of convex sets. The dual of such a (real linear) space is automatically nontrivial by Corollary 6.1.5. Therefore the category of locally convex linear topological spaces is a natural one to consider in connection with linear duality, and experience confirms, moreover, that it is a maximal category for which a linear theory with a certain roundedness is possible. Any Banach space or normed linear space is obviously locally convex. Another general class is described in the following example.

Example 6.1.1   Let L be a linear space with the weak topology defined by a collection C of linear functionals.   Suppose y is an element of L and N is the basic neighborhood of y, specified by the positive number ϵ and the linear functionals $f_1, f_2, \ldots, f_n$ of C.   If x and x' are in N and $0 \le a$, $b \le 1$, with $a + b = 1$, then

$$f_i(ax + bx') - f_i(y) = a[f_i(x) - f_i(y)] + b[f_i(x') - f_i(y)].$$

This implies

$$|f_i(ax + bx') - f_i(y)| < \epsilon, \qquad 1 \le i \le n.$$

Thus $ax + bx' \in N$, and L is locally convex.

If L is a linear topological space with dual L^*, there is a natural weak topology in L^*, namely, that defined by the collection of all linear functionals on L^* of the form x^{**} with x in L.   We shall refer to this topology in L^* as the *weak topology defined by* L.   To avoid confusion it is often called the *weak-star* (abbreviated w^*) topology, and we may on occasion use this terminology.

The reflexivity result referred to above may now be stated as follows:

SCHOLIUM 6.6   *Let L be a locally convex linear topological Hausdorff space and L^* its dual in the weak topology defined by L.   Then the natural injection of L into the dual L^{**} of L is a continuous linear isomorphism of L onto L^{**} in the weak topology defined by L^*.*

Proof   The map $x \to x^{**}$ is clearly linear from L into L^{**}.   The fact that the map is onto is an immediate deduction from the preceding scholium. It remains to prove that $x^{**} \ne 0$ if $x \ne 0$.

For this suppose that x is a nonzero element of L.   Then there is an open convex neighborhood N of 0 which does not contain x.   Therefore, by Corollary 6.1.5 there is a continuous real linear functional f on L such that $f(x) \ne 0$.   If L is a real linear space, this shows that $x^{**} \ne 0$.   For the complex case, let $g(y) = f(y) - if(iy)$ for arbitrary y in L.   Then g is a continuous complex linear functional on L for which $x^{**}(g) = g(x) \ne 0$.

It is not difficult to see that in a finite-dimensional real or complex vector space there is a unique Hausdorff topology in which it is a linear topological space.   This is far from the case for infinite-dimensional spaces.   For example, in an infinite-dimensional Banach space it may be shown that the norm topology and the weak topology induced by the dual never agree. Nevertheless, in any locally convex Hausdorff space there is an important class of subsets in which closure in the given topology of the space is the same as closure in the weak topology defined by the dual, namely, the convex ones and in particular the linear subspaces.   This may be seen with the aid of the following useful result.

SCHOLIUM 6.7 *A closed convex set in a locally convex real linear topological space may be strongly separated from a point outside it by a hyperplane.*

Proof By making a suitable vector translation which does not affect convexity or separation by a hyperplane, we may take the point to be 0. There is then a neighborhood N_0 of 0 disjoint from the closed convex set **C**. By local convexity and the continuity of the map $(x,y) \to x - y$, there exist open convex neighborhoods N_1 and N_2 of 0 such that $N_1 - N_2 \subset N_0$. Then $-N_1$ and $-N_2$ are also open convex neighborhoods of 0. Thus the intersection **N** of N_1, N_2, $-N_1$, and $-N_2$ is an open convex neighborhood of 0 such that $N - N \subset N_0$ and $N = -N$. Then $C + N$ is an open set, being a union of translates of **N**, and is readily seen to be convex and disjoint from **N**. By Corollary 6.1.5 there exists a hyperplane separating 0 from $C + N$. This implies the existence of a continuous linear functional f such that $f(x) < 0$ for all x in $C + N$. Thus $f(c) + f(n) < 0$ for all c in **C** and all n in **N**. Since $N = -N$, there is a vector n_0 in **N** such that $f(n_0) > 0$. Hence $f(c) < -f(n_0) < 0$ for all c in **C**. This of course shows that **C** and 0 are strongly separated by the hyperplane consisting of all x such that $f(x) = -f(n_0)$.

COROLLARY 6.1.6 *A closed convex set in a locally convex real linear topological space is also closed in the weak topology defined by its dual.*

Proof Let **E** be a closed convex set in the locally convex real linear topological space **L**, and **F** its closure in the weak topology. Then, obviously, $E \subset F$. Suppose y is in **F** but not in **E**. Then by what was proved above, there is a continuous linear functional f on **L** and a real constant a such that $f(x) < a$ for all x in **E**, while $f(y) > a$. But then $[z : |f(y) - f(z)| < \epsilon]$, where $0 < \epsilon < f(y) - a$, is a weak neighborhood of y disjoint from E, a contradiction.

EXERCISES

1 Show that if the sequence $\{f_n\}$ converges weakly in L_p to f, that is, converges to f in the weak topology defined by L_q, $1 \le p < \infty$, $p^{-1} + q^{-1} = 1$, and if $f_n(x) \to g(x)$ a.e., then $f(x) = g(x)$ a.e.

2 Give an example of the situation in Exercise 1 where the sequence $\{f_n\}$ is not convergent in the L_p-norm.

3 *a.* Let **B** denote the Banach space of all complex measures on the Borel subsets of $[0,1]$ relative to the total variation as norm. Show that if k is any bounded Baire function on $[0,1]$, then the map $m \to \int k(x)\,dm(x)$ is a continuous linear functional F_k on **B**.

   b. Show that if k_n is a uniformly bounded sequence of Baire functions converging pointwise to the function k, then $F_{k_n} \to F_k$ in the weak topology relative to $\mathbf{B}^*$.

4 Show that the space $C[0,1]$ of all continuous functions on $[0,1]$, with the supremum norm, is not reflexive.

5 Let $\mathbf{B}$ be a Banach space and $\mathbf{C}$ a separating collection of continuous linear functionals on $\mathbf{B}$. Show that if a vector x is in the closure of a subset $\mathbf{Y}$ of $\mathbf{B}$ in the weak topology defined by $\mathbf{C}$, then

$$\|x\| \le \sup_{y \in Y} \|y\|.$$

6 Show that if, in the situation of Exercise 5, $\{x_n\}$ converges weakly to x, then there is a sequence $\{x'_n\}$, where each x'_n is a convex linear combination of a finite subset of the x_n which converges in norm to x.

7 Show that in the situation of Exercise 5, a bounded sequence $\{x_n\}$ is weakly convergent provided $\{f(x_n)\}$ is convergent for every f in some dense subset of the Banach dual $\mathbf{B}^*$.

8   *a.* If f is a complex-valued measurable function on a measure space M and $1 \le p \le \infty$, show that $\|f\|_p = \sup |I(fg)|$, where g ranges over the simple functions on M such that fg is integrable and $\|g\|_q = 1$, where $p^{-1} + q^{-1} = 1$.

   b. If $\|f\|_p = \infty$, show there exists a dense set of functions g in L_q such that fg is integrable and $I(fg) = 0$.

9*   Show that if $\{f_n\}$ converges weakly in L_p to f, $1 < p < \infty$, and if $\|f_n\|_p \to \|f\|_p$, then $\{f_n\}$ converges in norm to f.

10   *a.* Show that a sequence $\{f_n\}$ in $L_p(0,1)$, $1 < p \le \infty$ (under Lebesgue measure), is weakly convergent to f in L_p in the topology induced by $L_q, p^{-1} + q^{-1} = 1$, provided (i) $\|f_n\|_p$ is bounded, and (ii) $\lim\limits_n \int_0^x f_n(t)\,dt = \int_0^x f(t)\,dt$ for all x in $(0,1)$. (Apply Exercise 7.)

   b. Show that this conclusion is false for $p = 1$, but that if (ii) is replaced by the condition $\int_E f_n \to \int_E f$ for all measurable subsets E of $(0,1)$, it then is true.

11   *a.* Show that if $\{f_n\}$ is a bounded sequence in $L_p(0,1)$, $1 < p < \infty$, such that $\lim\limits_n \int_0^x f_n(t)\,dt$ exists for all x, $0 < x < 1$, then there exists an element f of L_p to which the sequence is weakly convergent in the weak topology defined by $L_q(p^{-1} + q^{-1} = 1)$.

   b. Extend this result to the case $p = \infty$, but show it is false for $p = 1$.

12   *a.* Show that it suffices in Exercises 10 and 11 to have convergence for rational x, or any countable dense subset.

b. Conclude that the unit sphere $[f : \|f\|_p \leq 1]$ in $L_p(0,1)$, $1 < p \leq \infty$, is sequentially compact in the weak topology induced by L_q. [*Hint:* Observe that $\int_0^r f_n(t)\,dt$ forms a bounded sequence and so has a convergent subsequence for any fixed rational r. Use the Cantor diagonal process to obtain a subsequence which is convergent for all rational r.]

13 Show that the conclusion of Exercise 12*b* is valid for any measure space M such that $\tilde{L}_1(M)$ is separable.

14 *a.* Show that there are no continuous linear functionals $\neq 0$ on the space of all finite measurable functions modulo null functions on $[0,1]$ in the topology of convergence in measure.

  b. Show that there are no nonzero linear functionals F defined for all complex-valued measurable functions on $[0,1]$ such that $F(f_n) \to F(f)$ whenever $f_n \to f$ a.e.

  c. Extend these results to the case of general measure spaces.

15* Show that a Banach space whose norm topology agrees with the weak topology induced by the dual must be finite-dimensional.

16* Show that a Banach space whose weak topology induced by the dual is first-countable (i.e., there exists a sequence of neighborhoods of any given point such that every open set containing the point contains a member of the sequence) is finite-dimensional.

17 Show that in a reflexive Banach space whose dual is separable (i.e., second-countable) the unit sphere is sequentially compact in the topology induced by the. dual. (*Hint:* Abstract the argument indicated for Exercise 12.)

18* *a.* Let T be a set, and let F be a collection of mappings $t \to f(t)$ of T into topological spaces X_f. Show that there is a topology on T relative to which all the mappings are continuous and that there is a weakest such, i.e., one whose open sets are open also in any other such topology.

  b. Give explicitly a basic for the open sets of this weak topology, generalizing that indicated in the text for the case of numerical functions, and show that if the X_f are uniform spaces, then T also becomes a uniform space relative to this basis, provided F is separating.

  c. Show that if the spaces X_f are totally bounded ($=$ relatively compact), then the same is true of T in the weak topology.

  d. Show that if each X_f is complete, and if whenever $\{f(t_u)\}$ is a convergent net for all f there is an element t of T such that $f(t_u) \to f(t)$, then T is complete.

19 Derive the following corollaries to the results of Exercise 18.

  a. An infinite product of compact (Hausdorff) spaces is compact (Hausdorff) (Tychonoff theorem for Hausdorff spaces).

  b. The unit sphere in the Banach dual $\mathbf{B}^*$ to a Banach space $\mathbf{B}$ is compact in the weak topology induced by $\mathbf{B}$ (Alaoglu's theorem).

  c. The unit sphere in a Banach space $\mathbf{B}$ is totally bounded in the weak uniform structure induced by $\mathbf{B}^*$.

20*   Show that a locally compact Banach space is finite-dimensional.

21*   Show that a Banach space is reflexive if and only if its unit sphere is compact in the w^*-topology, defined as that induced by the dual.

22   *a.* Show that the minimal closed and convex set (= "closed convex hull") containing a given set in a linear topological space is the closure of the minimal convex set containing the given set, but not necessarily the minimal convex set (= "convex hull") containing the closure of the set.

     b. Show that the convex hull of a compact set in a Banach space is compact. (*Hint:* Use the compactness of a continuous image of a compact Hausdorff space.)

23   Show that the dual of $\bar{L}_1(M)$, M being a given measure space, is precisely $L_\infty(M)$ if and only if M is localizable in the sense defined in Chaps. 3 and 4.

24   Show that if **M** is a closed subspace of a normed linear space **L**, and x is any vector in **L**, then the set of all vectors of the form $cx + y$, where c is an arbitrary scalar and y an arbitrary element of **M**, is a closed subspace of **L**.   (*Hint:* Use an argument similar to that used to prove a continuous linear operation bounded; alternatively, use the Hahn-Banach theorem.)

25   An extreme point of a convex set is a point which is not an inner point of any segment wholly contained in the set.   Show that in any locally convex real linear topological space (i.e., one admitting a basis consisting of convex sets), a compact convex subset is the convex closure of its set of extreme points. [This is known as the Krein-Milman theorem.   A geometrical proof using the concept of "face" of a convex set, defined as a nonempty convex subset containing the endpoints of any segment of which it contains an interior point, proceeds in outline as follows: The closed faces of the given set S, ordered by reverse inclusion, have the property that any chain has an upper bound, so that by the Zorn form of transfinite induction a maximal element exists.   This is an extreme point, for if this face contains two distinct elements x and y, then taking a continuous linear functional f such that $f(x) \neq f(y)$, the subset of the face on which f attains its maximum would be properly smaller.   If the convex closure K of the set of all extreme points does not contain all of S, take a linear functional f such that $f(x) > \sup_{y \in K} f(y)$ for some $x \in S - K$, and take a minimal face contained in the closed face of S on which f attains its maximum.]   Note: the segment between the points x and y is defined as $[ax + (1 - a)y : 0 \leq a \leq 1]$.

26   Let T_n be a sequence of continuous linear mappings from one Banach space **B** to another such space, and suppose that $\lim_n T_n x$ exists for all $x \in$ **B**.   Show that $\|T_n\|$ is bounded, and that the operator T defined by the equation $Tx = \lim_n T_n x$ is continuous.   (*Hint:* Use the Baire category theorem, q.v., to show that the set $[x \in \mathbf{B} : \|T_n x\| \leq k, \quad n = 1, 2, \ldots]$ is nondense for some k, by virtue of the completeness of **B**.)

6.2   VECTOR-VALUED FUNCTIONS

There is a large variety of special types of integrals for vector-valued functions, but in practice virtually all of them are likewise integrals in the sense of the

DEFINITION   A measurable map f from a measure space M to a real or complex locally convex linear topological space **L** is said to be *integrable* and to have the (weak) *integral v* in **L** if for every continuous real linear functional λ on **L**, $\lambda \circ f$ is integrable and has the integral $\lambda(v)$.

The fact that the integral is single-valued on integrable functions follows from Corollary 6.1.5, and the definition evidently generalizes the one given earlier for functions with values in finite-dimensional euclidean spaces. Moreover, the characteristic property

$$I(\lambda \circ f) = \lambda(I(f))$$

implies a more general fact which, roughly speaking, asserts that integration commutes with continuous linear maps.

For suppose that f is an integrable function on **M** with values in **L** and that T is a continuous linear map from **L** into another locally convex linear topological Hausdorff space **L'**.   Then $T \circ f$ is a measurable function on **M** with values in **L'**.   If λ' is a continuous real linear functional on **L'**, then $\lambda' \circ (T \circ f) = (\lambda' \circ T) \circ f$, and since $\lambda' \circ T$ is a continuous real linear functional on **L**, it follows that

$$I[\lambda' \circ (T \circ f)] = (\lambda' \circ T)(I(f)) = \lambda'(T(I(f))).$$

Then $T \circ f$ is integrable, and

$$I(T \circ f) = T[I(f)].$$

In particular, in the case that **L** is a complex space, the integral commutes with all continuous complex linear functionals.

Conditions for the integrability of vector-valued functions are somewhat more complicated than those for complex-valued functions, and the particular theories for some of the special integrals are helpful in this connection. Nevertheless, as is indicated by the following result concerning integrability, the basic theory more or less parallels the scalar theory and is partly deducible from it.

SCHOLIUM 6.8   *A measurable function f on a measure space M with values in a Banach space* **B** *is integrable provided its absolute value is integrable, and in this event*

$$\|I(f)\| \leq I(|f|).$$

Proof   Assume first that f is a countably valued measurable function whose absolute value $|f|$ is integrable, $(|f|(x) = \|f(x)\|)$.   If the distinct nonzero values $y_1, y_2, \ldots$ of f are assumed on the measurable sets $E_1, E_2, \ldots$, then

$$I(|f|) = \sum_j \|y_j\|\, m(E_j) < \infty,$$

where m denotes the measure on M. If λ is any continuous linear functional on $\mathbf{B}$, then $\lambda \circ f$ assumes the values $\lambda(y_1)$, $\lambda(y_2)$, ... on E_1, E_2, ..., and

$$I(|\lambda \circ f|) = \sum_j |\lambda(y_j)|\, m(E_j) \le \|\lambda\|\, I(|f|).$$

Thus $\lambda \circ f$ is integrable and

$$I(\lambda \circ f) = \sum_j \lambda(y_j)m(E_j).$$

On the other hand, the series $\sum_j m(E_j)y_j$ converges in $\mathbf{B}$ to an element y, and by continuity $\lambda(y) = I(\lambda \circ f)$. Thus $I(f) = y$, and

$$|I(f)| = \left\| \sum_j m(E_j)y_j \right\| \le \sum_j m(E_j)\, \|y_j\| = I(|f|).$$

Now suppose f is an arbitrary measurable function with values in $\mathbf{B}$. Then there is a countable sequence z_1, z_2, ... of elements in $\mathbf{B}$ which is dense in the range of f. For this is true of the simple functions, and also for the pointwise limit of a sequence of functions for which it is true. Define f_n as the function on M which at the point x has the value z_k, where $|f(x) - z_k| < 1/n$, and k is the least positive integer for which this is true. Since f is measurable, for any sphere S in $\mathbf{B}$ excluding 0, $f^{-1}(S)$ is measurable. Let A_{nj} be the inverse image under f of the open sphere about z_j of radius $1/n$. Then f_n assumes the value z_k on

$$A_{nk} - \bigcup_{j<k} A_{nj}.$$

Thus f_n is measurable, and evidently $f_n(x) \to f(x)$ for all x. Since the map $z \to \|z\|$ is continuous, $|f_n|$ and $|f|$ are also measurable. Therefore there exists a sequence g_1, g_2, ... of simple functions such that $g_n(x) \to \|f(x)\|$ and $0 \le g_n(x) \le \|f(x)\|$ for all x. Let

$$h_n(x) = \frac{g_n(x)}{\|f_n(x)\|}\, f_n(x)$$

if $f_n(x) \ne 0$, and otherwise set $h_n(x) = 0$. Then h_n is countably valued and clearly measurable; moreover, $h_n(x) \to f(x)$, and

$$\|h_n(x)\| \le \|f(x)\|$$

for all x. If $|f|$ is integrable, it follows that $|h_n|$ is also integrable. This implies by the first part of the proof that $I(h_n)$ exists. In addition,

$$\|f(x) - h_j(x)\| \le 2\,\|f(x)\|$$

for all x, so that $I(|f - h_j|) \to 0$ by dominated convergence. Thus

$$\|I(h_n) - I(h_j)\| = \|I(h_n - h_j)\|$$
$$\le I(|h_n - h_j|)$$
$$\le I(|h_n - f|) + I(|f - h_j|).$$

Therefore $I(h_1)$, $I(h_2)$, ... is a Cauchy sequence in **B**, and

$$y = \lim_n I(h_n)$$

exists.   If λ is a continuous linear functional on **B**, it follows that

$$\lambda(y) = \lim_n \lambda(I(h_n)) = \lim_n I(\lambda \circ h_n).$$

On the other hand, $\lambda \circ h_n \to \lambda \circ f$, and

$$|\lambda(h_n(x))| \le \|\lambda\|\, \|h_n(x)\| \le \|\lambda\|\, \|f(x)\|$$

for all x.   Thus $\lambda \circ f$ is integrable, and $I(\lambda \circ h_n) \to I(\lambda \circ f)$ by dominated convergence.   Hence $I(f) = y$ and $\|I(f)\| = \lim_n \|I(h_n)\| \le \lim_n I(|h_n|) = I(|f|)$.

A fairly typical application of this result is described in the following example.

Example 6.2.1   Let G be a measurable mapping from a measure space M to a complex Banach space **B** whose absolute value $|G|$ is in $L_p(M)$ for some p with $1 \le p \le \infty$.   From the easily established fact that the product of a measurable vector-valued function with a measurable scalar-valued one is again measurable, it follows that fG is a measurable function on M with values in **B** for each f in $L_q(M)$.   Since $|fG| = |f|\,|G|$, Hölder's inequality implies the integrability of $|fG|$ (here it is assumed q is the index conjugate to p).   Thus the equation

$$T(f) = \int_M f(x)G(x)\, dm(x)$$

defines a linear map $T = T_G$ from $L_q(M)$ into **B**, and

$$\|T(f)\| \le \int \|f(x)G(x)\|\, dm(x)$$

$$= \int |f(x)|\, \|G(x)\|\, dm(x)$$

$$\le \|\,|G|\,\|_p\, \|f\|_q.$$

Hence T is bounded, and $\|T\| \le \|\,|G|\,\|_p$.

On occasion this is a good way to look at integral operators such as

$$f(x) \to \int K(x,y)f(y)\, dy,$$

where K is a given "kernel," i.e., measurable function of two variables x and y.   A simple instance is that of the convolution map

$$f(x) \to \int K(x - y)f(y)\, dy,$$

where K is a given function in L_r, $1 \leq r < \infty$, and f a general function in L_1 on the real line relative to Lebesgue measure. Here L_r (or more precisely $\tilde{L}_r$) is the appropriate Banach space **B**. For G we take the map whose value at y is the element K_y of L_r, defined by

$$K_y(x) = K(x - y).$$

Then $\|G(y)\|_r = \|K_y\|_r = \|K\|_r$. For

$$\int |K(x - y)|^r \, dx = \int |K(x)|^r \, dx.$$

Thus $|G| \in L_\infty$, and $T = T_G$ is a bounded linear map from L_1 into L_r. For the bound of T we have the estimate

$$\|T\| \leq \|K\|_r.$$

The reader may undertake as an exercise the problem of proving that G is not only measurable but actually continuous, and that T is convolution by K, that is, that

$$K * f = \int f(y) K_y \, dy.$$

The next result to be proved is an elementary but useful geometric property of the integral for vector-valued functions.

SCHOLIUM 6.9 *If f is an integrable function with values in a locally convex real linear topological space, the integral $I(f)$ lies in the closed subspace spanned by the range of f, and if in addition the measure space is of total measure 1, then $I(f)$ lies in the convex closure of the range of f.*

Proof Suppose f is integrable and that $I(f)$ is not a member of the closed subspace spanned by the values of f. Then by Scholium 6.7 there is a continuous linear functional λ such that $\lambda \circ f = 0$ and $\lambda(I(f)) \neq 0$. But this is obviously impossible.

Now suppose the measure space has total measure 1 and that $I(f)$ is outside the closed convex set **C** spanned by the values of f. Then $I(f)$ can be strongly separated from **C** by a hyperplane (Scholium 6.7). If λ is a real linear functional defining this hyperplane, then $\lambda(I(f)) = I(\lambda \circ f)$, while, on the other hand, there exists a positive number ϵ such that $\lambda(f(x)) < \lambda(I(f)) - \epsilon$ for all x, contradicting the fact that the value of a numerical integral over a space of total measure 1 is bounded by the bound of the function.

Next we shall prove a somewhat different but a similar conclusion.

SCHOLIUM 6.10 *Let M be a measure space, f a nonnegative function in $L_1(M) \cap L_q(M)$, $1 \leq q \leq \infty$, with $I(f) = 1$, and G a measurable function on M to a Banach space **B** such that $|G| \in L_p(M)$, p being the index* *conjugate to q. Then fG is integrable and I(fG) lies in the closed convex set spanned by the values of G.*

Proof The integrability of *fG* is shown in Example 6.2.1. Suppose $I(fG)$ is outside the closed convex set spanned by the values of *G*. Then there exists a continuous real linear functional λ and an $\epsilon > 0$ such that

$$\lambda(G(x)) + \epsilon < \lambda[I(fG)]$$

for all *x*. After multiplying by $f(x)$ and integrating, we obtain the contradiction

$$\lambda[I(fG)] + \epsilon < \lambda[I(fG)].$$

Example 6.2.2 Suppose $M = (S,\mathcal{R},m)$ is a finite measure space, and let $T = T_G$ be the continuous linear map from $L_q(M)$ into **B**, defined as in Example 6.2.1. Let **C** be the closed convex set spanned by the values of *G*. Observe that if $f \in L_q(M)$, then

$$\|f\|_1 \le [m(S)]^{1/p}\,\|f\|_q,$$

where $[m(S)]^{1/p} = 1$ in case $p = \infty$. Thus, for a nonnegative *f* in $L_q(M)$, Scholium 6.10 implies that

$$T(f) \in I(f) \cdot \mathbf{C}.$$

If *f* is any function in $L_q(M)$, the positive and negative parts of its real and imaginary parts are also in $L_q(M)$. If $\|f\|_q \le 1$ and **V** is the set of all vectors in **B** of the form $z - w$ with *z* and *w* in **C**, it follows without difficulty that $T(f)$ lies in the set

$$[m(S)]^{1/p}(\mathbf{V} + i\mathbf{V}).$$

Now suppose the range of *G* is relatively compact. Since the convex closure of a compact set in a Banach space is also compact, it follows that **C** is compact. Then **V** is compact as well, being the image of $\mathbf{C} \times \mathbf{C}$ under the continuous map $(z,w) \to zw$. For similar reasons $\mathbf{V} + i\mathbf{V}$ is also compact. Thus *T* has the property of carrying the unit sphere in $L_q(M)$ into a compact set in **B**.

Such operators are called *compact*, or *completely continuous*. They have a highly developed spectral theory, and indeed, apart from normal operators in Hilbert space, are the only general class for which such a theory does exist. They occur frequently in connection with compact domains, manifolds, and groups.

EXERCISES

1 Show that a measurable function on a finite measure space with values in a Banach space is a.e. the limit of a sequence of simple functions.

2 Show that if $K \in L_2(M \times M)$, then the map

$$f(x) \to \int K(x,y)f(y)\,dm(y)$$

of $L_2(M)$ into $L_2(M)$ may be defined as

$$f \to \int f(y)G(y)\, dm(y),$$

where $G(y)$ denotes $K(x,y)$ as a function of x for those values of y for which this function is in L_2 and is otherwise 0.

3   Show that a continuous function on $[0,1]$ to a Banach space **B** is integrable relative to Lebesgue measure and that the integral is a continuous function on the space $C(\mathbf{B})$ of all such f in the topology of uniform convergence.

4   For real t, let T_t be the operation $f(x) \to e^{itx}f(x)$ on $L_2(-\infty,\infty)$.   Show that if $h \in L_1(-\infty,\infty)$, then $\int h(t)T_t\, dt$ exists as the integral of a function with values in the continuous linear operators on $L_2(-\infty,\infty)$, in the weak topology defined by the functionals $A \to \int Afg$, where f and g vary over L_2 (the weak-operator topology). Show that the value of the integral is the operator $f(x) \to \hat{h}(x)f(x)$, where $\hat{h}$ denotes the Fourier transform of h.

5   Show that if $K(x,y)$ is a continuous function on the unit square $0 \le x, y \le 1$, then the map $f(x) \to \displaystyle\int_0^1 K(x,y)f(y)\, dy$ is compact from any L_p into every L_r $(1 \le p, r < \infty)$.

VII

INVARIANT
INTEGRALS

7.1 INTRODUCTION

The measure spaces or integrals which actually arise in analysis frequently
do so with additional elements of structure which are quite significant, and
interact with the integration features, so that the additional features cannot
really be considered separately. The simplest and most common such
additional feature is a topology on the set under consideration, on which the
measure in question is defined. Some indications have already been given
earlier of the relations between the continuous and measurable functions
which are then valid and useful. Another such feature is a group of transformations on the basic set S, leaving invariant the measure m. There are
questions of the uniqueness of m, its existence when the transformations are
given, and of the special relations and group action on the function spaces
over S which then become relevant. The developments which emerge are
extensive, important, and in some respects striking. It is to some of these
developments that this chapter is devoted.

The most familiar example of an integral invariant under a group of
transformations is the Lebesgue integral on the line. More generally, as

175

already shown in Chap. 5, the Lebesgue integral in euclidean space is invariant, not only under translations, but rotations as well.   Moreover, the invariance under translations is a characteristic property; any integral in euclidean space which is invariant under translations is a multiple of the Lebesgue integral.

A somewhat different sort of invariant integral, also familiar, is the usual one for functions on the sphere in three-dimensional euclidean space; this is invariant under rotations of space, and is also essentially uniquely characterized by this property.   This integral may be defined in a classical style for continuous functions, and extended to a countably additive regular measure by the theory of Sec. 5.2.   More specifically, a rotation of the 2-sphere S, consisting of all real vectors (x_1, x_2, x_3) such that $\sum_j x_i^2 = 1$, may be described as a map of S onto S of the form $x \rightarrow xa$, where a denotes a real 3×3 orthogonal matrix of determinant 1.   If the points (x_1, x_2, x_3) on S are represented by their spherical coordinates,

$$x_1 = \sin \theta \cos \phi, \qquad x_2 = \sin \theta \sin \phi, \qquad x_3 = \cos \theta,$$

it may be shown that the equation

$$\lambda(f) = \int_0^{2\pi} \int_0^{\pi} f(\theta, \phi) \sin \theta \, d\theta \, d\phi$$

where $0 \leq \theta \leq \pi$, $0 \leq \phi \leq 2\pi$, defines an integral λ on the continuous functions which is invariant under rotations; and as will follow from a later theorem, any other invariant integral is a multiple of λ.

In these examples, as in a variety of other analytic-geometric situations, there is given a topological space S and a set of homeomorphisms of S, and a central role is played by a measure on suitably smooth subsets of S which is invariant under the actions on sets of the given homeomorphisms.   Apart from certain infinite-dimensional cases arising in abstract probability and quantum field theory, the space S is generally locally compact, and the measure in question is regular; there is a great variety of important examples of this nature, and this is the only case to be treated in this chapter.   In the simplest cases, as in that of Lebesgue measure, S is itself a topological group, and the transformations in question are the *left translations* $x \rightarrow ax$, by a fixed (but arbitrary) element a of the group; naturally, one may also consider the *right translations* $x \rightarrow xa$.

Euclidean space is not only a locally compact group relative to vector addition, but a *Lie group*, which may be defined as a group having a neighborhood of the unit that is homeomorphic to an n-dimensional euclidean cube in such a way that group composition corresponds to a binary operation in the cube, which may be expressed in terms of convergent power series in the coordinates of the two vectors in question.   The theory of such groups is beyond the scope of this book, but it should be noted that for any Lie group an explicit local construction for an invariant integral has been well known since the establishment of the local theory of Lie groups before the turn of the century. With the development of the global theory of Lie and other topological groups in the twenties (by Weyl, Schreier, and others), the local aspect of the Lie-group construction was removed, and the global form of the integral was shown by Weyl and others to be a powerful tool in harmonic analysis. However, the classical invariant-measure construction for a Lie group depends essentially on its differentiability properties, which are wholly absent in the case of a more general topological group, even a locally compact one; and even in the case of a Lie group, a number of important analytical aspects, such as the possibility of characterizing the measure by its invariance, depend on notions quite distinct from those of Lie theory.

The highly original work of Haar, who established the existence of non-trivial regular invariant measures for all separable locally compact groups, has led to the consolidation, clarification, and extension of a variety of analytical theories, of importance in such diverse fields as classical analysis, number theory, and (indirectly) the mathematical analysis of theoretical physics. Haar's work was apparently motivated in part by the hope that it might contribute to the resolution of Hilbert's fifth problem, which in its simplest form was to show that a topological locally euclidean group is a Lie group. He died not long after his work on invariant measures, but von Neumann was quick to apply his result, which permitted the work of Peter and Weyl on compact Lie groups to be directly carried over to compact groups in general, to the establishment of Hilbert's conjecture for the case of a compact group. It was difficult to apply a similar method in the case of noncompact groups, but attempts to pursue the problem by local and topological means did not prove effective, and the full solution of the problem—certainly one of the great triumphs of axiomatic method in this century—was attained only after ways were found to bring the Haar integral to bear on the noncompact case.

In other work not long after Haar showed the existence of his measure, von Neumann showed its uniqueness (within a constant factor) by a method based on the use of the Fubini theorem, indicated earlier. The separability restriction in Haar's and von Neumann's work was eliminated by Weil, who dealt in fact with an invariant integral rather than measure; this served to separate troublesome measurability questions from the problem of the existence and uniqueness of an invariant integral and to isolate them in such general problems, since found to be for the most part academic, of the representation of the integral in terms of measures of various types. Weil made other significant contributions to the subject, but probably his most important work in this connection was his new proof of uniqueness, which paralleled and illuminated the existence proof and, unlike the Fubini-theorem argument, was applicable to transformation groups as well as to groups acting on themselves by left or right translations. The action of the rotation group in three-dimensional space on the 2-sphere provides an example of a transformation group; the existence and uniqueness of an invariant regular measure on the 2-sphere is one of the simple examples of the application of the theory to a transformation group.

7.2   TRANSFORMATION GROUPS

Purely set-theoretically, a *transformation group* is a system (G,S,F), where G is a group, S is a set, and F is a function defined on $G \times S$ and having values in S, satisfying certain conditions; namely, it is required that if T_g denotes the mapping $x \to F(g,x)$ from S into S, then $T_{gg'} = T_g T_{g'}$ for g, $g' \in G$, and $T_e = I$, where e is the group unit and I is the identity map of S into S. When topological structures are added to G and S, it is naturally appropriate to require, further, that operations defining the system be continuous.

Thus it is appropriate to require in particular that G be a *topological group*, which means that the topology on G is such that the mapping $(x,y) \to xy^{-1}$ of $G \otimes G$ into G be continuous. It follows, in particular, that for any element a of G, the mapping

$$L_a : x \to ax,$$

the so-called *left translation by a*, is a homeomorphism of G, and that the same is true of the *right translation by a*, defined as the mapping

$$R_a : x \to xa.$$

It is also natural to require continuity of the group action on the space S. Specifically, a *continuous*, or *topological*, transformation group will be defined as a system (G,S,F), in which G is a topological group and S a topological space, forming a transformation group in the purely algebraic sense (i.e., when the given topologies are removed) and such that F is continuous on $G \otimes S$ to S.

When S has a given measure-theoretic structure, similar definitions are appropriate. Let $\mathfrak{R}$ be a given σ-ring of subsets of S, and n a measure on $\mathfrak{R}$ such that $(S,\mathfrak{R},n)$ is a measure space. We may speak of an *absolutely continuous transformation group* (the nomenclature is not entirely standard) in case the mappings T_g leave $\mathfrak{R}$ invariant (in their induced action on the power set of S) and if they carry null sets into null sets.

Various mixed topologico-measure theoretic types of transformation groups can be defined, and actually arise in concrete situations. Here only the most common type will be treated, that of a *regular topological transformation group*, defined as a system (G,M,F) in which M is a regular locally compact measure space $(S,\mathfrak{R},m)$ and which forms both a topological and a nonsingular measurable transformation group in the obvious fashion. A simple example is the case where $G = S =$ euclidean space, while $F(x,y) = x + y$ and $m =$ Lebesgue measure; in this case, m is invariant under the action of the group; i.e.,

$$m[T_g(E)] = m(E)$$

for all measurable sets E and elements $g \in G$; we speak then of a *group of measure-preserving transformations*. Another example is that in which G is the group of rotations in 3-space, S is the 2-sphere, and m is the measure on S previously indicated.

The interaction between the topological and measure-theoretic features of transformation groups has been the basis for a number of significant mathematical developments. One of the most important is that with which we are presently concerned, the demonstration of the existence of an invariant measure on a locally compact space with respect to a given topological transformation group, under suitable conditions. It is rather clear that some conditions must be imposed; for example, under the group of all affine transformations on the real line, consisting of those of the form

$$x \to ax + b,$$

where a and b are real and $a \neq 0$, no invariant regular measure exists. This is clear from the fact that invariance under the translations characterizes Lebesgue measure, while the similarity transformation $x \to ax$ evidently carries Lebesgue measure m into $|a|^{-1}m$, where the image of m under a nonsingular measurable transformation T is defined as the measure $m' : m'(E) =\cdot m[T^{-1}(E)]$. The same is true of the affine group on any finite-dimensional space, for essentially the same reason.

There are, roughly speaking, two general approaches to the establishment of an invariant measure for a transformation group on a locally compact space. One depends on compactness, either of the space or the group, naturally a strong limitation; in addition, further restrictions are required, and relatively little can be established in a general way about the uniqueness and nontriviality of the resulting invariant measure. The other depends on a certain uniformity in the action of the group on the space; roughly speaking, the result is that topological uniformity implies the existence of a measure-theoretical one, and that topological "transitivity" implies a measure-theoretical type of transitivity (known at one time as "metrical transitivity," but now usually called "ergodicity," and a simple consequence of the uniqueness of the invariant measure).

EXERCISES

1 Let T be a measure-preserving transformation on the measure space $(S,\Re,m)$; let $U(T)$ denote the induced transformation on the function classes, corresponding to the transformation $f(x) \to f(Tx)$ of the measurable functions f. Show that $U(T)$ is continuous in the topology of local convergence in measure and that its restriction to any L_p-space is a nonsingular isometry of the space (i.e., an automorphism of the space as a normed linear space).

2   Let G be a group which acts via the mapping $g \to T_g$ as absolutely continuous transformations on the measure space M.   Show that the mapping $g \to U(T_q^{-1})$ is a representation of G on the space of all measurable function classes, in the usual algebraic sense.

3   Let T be a nonsingular measurable transformation on the measure space $(S,\mathcal{R},m)$, and let $U(T)$ be defined as in Exercise 1.

    a. Show that $U(T)$ is continuous in the topology of local convergence in measure.

    b. Show that for any σ-finite measure space there exists for any p, $1 \leq p \leq \infty$, a nonnegative measurable function k_p such that for $f \in L_p$ the mapping $f \to k_p U(T)f$ is an invertible isometry, say, $V_p(T)$. (*Hint:* Use the Radon-Nikodym theorem.)

    c. In case $p < \infty$, give a formulation of the operator $V_p(T)$ applicable to not necessarily σ-finite spaces.

4   With the notation of Exercise 3, show that if T' is another nonsingular measurable transformation, then so also is $T'T$, and that $V_p(T'T) = V_p(T)V_p(T')$.

5   Suppose that I is an integral on a locally compact group G which is invariant under all left translations: $I(f(ax)) = I(f(x))$ for all $f \in C_0(G)$.   Show that if I' is the transform of I induced by the mapping $x \to x^{-1}$, then I' is an integral which is invariant under all right translations.

    6   *a.* Show that the set of all measure-preserving transformations T on a measure space is a topological group relative to the weakest topology in which $U(T)\tilde{f}$ depends continuously on T for all measurable function classes $\tilde{f}$.

    b. Show that the topology in (*a*) is the same as the weakest in which $V_2(T)\tilde{f}$, as defined in Exercise 3c, depends continuously on T, for all $\tilde{f}$ in L_2, in the L_2-topology.

    *c.** Extend (*a*) and (*b*) to the group of all absolutely continuous transformations, and show that the subset of all measure-preserving transformations forms a closed subgroup.

7   Let G denote the group of all transformations on the real line of the form $x \to ax + b$, where $a > 0$ and b is arbitrary.

    a. Show that G is a locally compact topological group, in the topology obtained by carrying over the euclidean topology on the right half plane and setting this space in one-to-one correspondence with G by associating the indicated transformation with the point (a,b) in the plane.

    b. Show that the measure on G whose element is $a^{-1}\, da\, db$, relative to the foregoing parametrization, is left-invariant but not right-invariant.   [*Hint:* Observe that the group multiplication law is $(a',b')(a,b) = (a'a, a'b + b')$, $= (a'',b'')$, say; compute $(a'')^{-1}\, da''\, db''$ for fixed (a',b').]

    c. Find the form of a right-invariant measure on G.

8 Show that for a regular topological group of measure-preserving transformations on a locally compact space, the induced representation of G on L_2 is continuous in the sense indicated in Exercise 6*b*. [*Hint:* Apply the last part of Exercise 6*b* to the dense subset $C_0(S)$ of $L_2(S)$.]

9* Extend Exercise 8 to the case where the transformations are not necessarily measure-preserving.

10 A transformation T defined on a convex subset of a real linear space may be called linear in case it has the property that $T(\alpha x + \beta y) = \alpha Tx + \beta Ty$ whenever α and β are real nonnegative numbers such that $\alpha + \beta = 1$. It is known that a continuous linear transformation of a compact convex subset K of a locally convex linear topological space leaves fixed some point of K.

   a. Show that for any given set of commuting continuous linear transformations on K, there is some point of K which is fixed under all the transformations. (*Hint:* Show that the set of all fixed points under one such transformation is again compact and convex; apply transfinite induction.)

   *b** Show that any solvable group of continuous linear transformations on K leaves fixed some point of K. (*Hint:* If G is a group with Abelian normal subgroup A such that G/A is Abelian, show that G/A operates naturally on the set of fixed points of A.)

   c. Show that any solvable group of homeomorphisms of a compact Hausdorff space S leaves invariant some nonzero regular measure on the space. [*Hint:* Consider the induced action of the group on the integrals on S which are normalized to have the value 1 on the function identically 1, and topologize the integrals by their correspondence with linear functionals on $C(S)$, taken in the w^*-topology.]

7.3 UNIFORM SPACES

In the case of a metric space, the concept of uniform continuity is readily introduced, but in a general topological space such uniformity notions are not defined. Even if the space is metrizable, uniform continuity will depend on the particular metric employed, and will not in general be topologically invariant.

There are other cases, besides those of metric spaces, where a natural notion of uniformity may be introduced which turns out to be useful. These are notably the cases of topological groups and linear vector spaces. To consider the case of a group, which the case of a vector space closely resembles, a numerical function f on a topological group may be called (*left*) *uniformly continuous* in case for every positive number ϵ there is a neighborhood N of the group unit e such that $|f(x) - f(y)| < \epsilon$ provided $x \in yN$.

It develops that the fundamental theory concerning uniform continuity, completeness, etc., which is familiar in the case of a metric space can be extended equally well to the case of a topological group or analogous structures. In fact, A. Weil's theory of uniform spaces provides a simple general theory which includes both the case of a metric space and of a structure similar to a topological group.   These uniform spaces provide a natural setting for the formulation of a uniformity condition on a transformation group under which the ideas of Haar measure theory are effectively applicable.

For most applications, uniform spaces are conveniently defined in terms of the auxiliary notion of a uniform structure.   To arrive at this notion, let a *covering system* on a set S be defined as a mapping $x \to N(x)$ defined on S which assigns to each point $x \in S$ a set $N(x)$ containing x.   The *product* of two such covering systems N_1 and N_2, denoted $N_1 N_2$, is defined as the covering system which assigns to any point x the union of all $N_1(y)$ as y ranges over $N_2(x)$; it is straightforward to verify that this product is associative.   The *inverse* (terminology which turns out to be suggestive, but is not directly connected with the product just introduced) is, for a covering system N, the system denoted N^{-1}, which assigns to any point x the set of all $y \in S$ such that $x \in N(y)$.   A *neighborhood system* on a topological space is defined as a collection $\mathcal{C}$ of covering systems N, each of which has the property that $N(x)$ is a neighborhood of x, for all x [that is, x is contained in the interior of $N(x)$], and such that for every point x and neighborhood U of x there exists an element $N \in \mathcal{C}$ such that $N(x) \subset U$.

A *uniform structure* is a system $(S,\mathcal{C})$ consisting of a topological Hausdorff space S together with a neighborhood system $\mathcal{C}$ on S satisfying the following two axioms:

(*1*)   If N and N' are any two elements of $\mathcal{C}$, then there exists an element $N'' \in \mathcal{C}$ such that $N''(x) \subset N(x) \cap N'(x)$ for all $x \in S$.

(*2*)   For any element $N \in \mathcal{C}$ there is an element $N' \in \mathcal{C}$ such that $N'N'^{-1}(x) \subset N(x)$ for all $x \in S$.

> *Example 7.3.1*   Let S be a metric space, with the distance from x to y denoted as $d(x,y)$.   Let ϵ denote an arbitrary positive number, and define $N_\epsilon(x)$ as the open sphere with center at x and radius ϵ.   It is easily seen that if $\mathcal{C}$ is the collection of all N_ϵ, then $(S,\mathcal{C})$ is a uniform structure.   Another uniform structure could be obtained by utilizing the closed spheres; still another by defining $\mathcal{C}$ as the set of all N_ϵ for rational values of ϵ.   Note that $N_\epsilon{}^2 \subset N_{2\epsilon}$.

> *Example 7.3.2*   Let G denote a topological group, let V denote an arbitrary neighborhood of the unit e, and define $N_V(x)$ as xV.   It follows directly from the axioms for a topological group that with $\mathcal{C}$ taken as the set of all N_V as V varies, $(G,\mathcal{C})$ forms a uniform structure.   Another uniform structure is obtained by defining $N_V(x)$ as Vx; still another by defining $N_V(x) = Vx \cap xV$.

It is intuitively fairly plausible in connection with Example 7.3.1 that the several uniform structures indicated are more or less equivalent as regards essential features.   The notion of equivalence may be formalized as follows:

Two uniform structures $(S,\mathcal{C})$ and $(S,\mathcal{C}')$ over the same topological space S are *equivalent* in case for every element $N \in \mathcal{C}$ there is an element $N' \in \mathcal{C}'$ such that $N'(x) \subset N(x)$ for all $x \in S$, and vice versa. It is easily seen that this is a bona fide equivalence relation (i.e., symmetric, transitive, and reflexive); the ultimate invariant object, the *uniform space* over the topological space S, is then defined as an equivalence class of uniform structures on S.

The theoretical usefulness of this concept may be indicated by the formulation of the notion of a *uniformly continuous* function from the uniform space defined by the uniform structure $(S,\mathcal{C})$ to that defined by the uniform structure $(T,\mathfrak{D})$; this is a mapping f from S into T, with the property that for any given $P \in \mathfrak{D}$, there exists an $N \in \mathcal{C}$ such that $f(N(x)) \subset P(f(x))$ for all x; it is easily seen that the notion is independent of the choice of uniform structure within the defining equivalence class. A *unimorphism* between two uniform spaces is defined as a mapping from (the set of) one space to (the set of) the other, having the property of being a uniformly continuous homeomorphism, together with its inverse.

The fundamental theory for uniform spaces is closely analogous to that for metric spaces, with similar appropriate generalizations of the concepts in question. Relatively little use will be made in the following discussion of this general theory, and here it will merely be illustrated by the consideration of some representative notions, which will, moreover, be useful later. The notion of *equicontinuity*, for example, is defined for metric spaces as follows: A class $\mathcal{M}$ of continuous mappings f from a metric space S to a metric space S' is said to be *equicontinuous* at a point x in case for every $\epsilon > 0$ there is a $\delta > 0$, in general dependent on x, *but not on f*, such that

$$d(f(y),f(x)) < \epsilon \quad \text{if} \quad d(y,x) < \delta.$$

Such equicontinuity is *uniform* in case δ can be chosen to be independent of x.

These notions can be carried over directly to the case of uniform spaces. To avoid circumlocution, a uniform space will be denoted by any defining uniform structure for it; experience shows that no confusion results from the use of this convention, in the present connection. Now if $(S,\mathcal{C})$ and $(S',\mathcal{C}')$ are uniform spaces, a collection M of mappings from S to S' is *equicontinuous* at a point x in case for every $N' \in \mathcal{C}$ there exists an $N \in \mathcal{C}$ such that for all f in M

$$f(y) \in N'(f(x)) \quad \text{if} \quad y \in N(x).$$

In general, N' is naturally dependent on x as well as N; the equicontinuity is *uniform* in case the indicated relation holds for all values of x, as well as for all $f \in M$, in other terms, if for any given N' there is a fixed N such that $f(N(x)) \subset N'(f(x))$ for all $x \in S$ and $f \in M$.

The idea of uniformity applies also to local properties of a uniform space— local connectedness, local compactness, etc.—in a rather straightforward way.   In general, there is no apparent reason why a topological property P which holds locally in a purely topological sense need hold also uniformly, and generally there is a material difference between the two notions.   For example, a uniform space is *uniformly locally compact* if it has a structure of the form $(S, \mathcal{C})$, where for all $N \in \mathcal{C}$ and $x \in S$, $N(x)$ is compact.   Such a space is of course locally compact topologically, but it does not follow that a uniform space which is topologically locally compact is uniformly so, and in fact this is not necessarily the case.

The central structure theorem concerning uniform spaces, and the only nontrivial result concerning them which we shall use, is to the effect that every uniform space is, in a certain sense, a limit of metric spaces.   To state this result, it is helpful to introduce the notion of a *pseudometric* on a set S; this is defined as a numerical function d on $S \times S$ which satisfies all the axioms for a metric on S, except that the vanishing of $d(x, y)$ does not necessarily imply that $x = y$.   A collection $\mathcal{M}$ of pseudometrics on a set S may be called *separating* in case $d(x, y) = 0$ for all $d \in \mathcal{M}$ only if $x = y$. If now $\mathcal{M}$ is a separating collection of pseudometrics on a set S, then a uniform space $(S, \mathcal{C})$ is obtained by taking $\mathcal{C}$ as the set of all covering systems N of the form $N(x) = [y \in S : d(y, x) < \epsilon, d \in \mathcal{M}_0]$, where ϵ is an arbitrary positive number and $\mathcal{M}_0$ an arbitrary finite subset of $\mathcal{M}$, and topologizing S by taking these sets $N(x)$ as a basis for the open sets.   The verification that the various conditions on a uniform space are satisfied is quite parallel to that for a metric space; we may say that the uniform space in question is defined by the collection $\mathcal{M}$ of pseudometrics.

THEOREM 7.1   (BIRKHOFF-PONTRJAGIN-WEIL) *Any uniform space is unimorphic with one defined by a collection of pseudometrics.*

Proof   Let an admissible covering system for the uniform space $(S, \mathcal{C})$ be defined as a covering system N with the property that there exists an element $N' \in \mathcal{C}$ such that $N'(x) \subset N(x)$ for all $x \in S$.

Lemma 7.3.1   *Let V_1, V_2, . . . be a sequence of admissible covering systems such that $V_k = V_k^{-1}$ and $V_{k+1}^2 \subset V_k$ for all $k = 1, 2, \ldots$.   For any number r of the form $\sum_{j=1}^{p} 2^{-n_j}$, where the n_i are positive integers such that $n_1 < n_2 < \ldots < n_p$, set $U_r = V_{n_p} V_{n_{p-1}} \ldots V_{n_1}$.   Let $F(x, y)$ denote the function on $S \times S$ defined as follows: $F(x, x) = 0$ for all $x \in S$; $F(x, y) = 1$ in case $y \notin U_1 U_1^{-1}(x)$; otherwise $F(x, y) = \sup [r : y \notin U_r U_r^{-1}(x)]$.   Then $F(x, y) = F(y, x)$, and*

(1) $F(x, y)$ is a uniformly continuous function on $S \times S$.
(2) $F(x, y) < 2^{-m-1}$ implies that $y \in V_m(x)$.

Proof   Observe first that $U_r(x)$ is a monotone-increasing function of r; that is, $U_r(x) \subset U_{r'}(x)$ in case $r < r'$.   It suffices to show that if r has the form $k/2^m$ for some integers k and m, and if $r' = (k + 1)/2^m$, then $V_m U_r(x) \subset U_{r'}(x)$. This will be proved by induction on m. It is evident in case $m = 0$. Now assuming that the indicated inclusion is valid for all lower values of m, then if k is even, say, $k = 2k'$, we have $r = k'/2^{m-1}$, $r' = (k'/2^{m-1}) + 2^{-m}$, and $U_{r'}(x) = V_m U_r(x)$ by definition. Assume, therefore, that $k = 2k' + 1$ for a positive integer k'; then $r' = (k' + 1)/2^{m-1}$, and setting $r'' = k'/2^{m-1}$, the induction hypothesis implies that $V_{m-1} U_{r''}(x) \subset U_{r'}(x)$; but $V_m^2(x) \subset V_{m-1}(x)$ for all x, so it follows that $V_m^2 U_{r''}(x) \subset U_{r'}(x)$; observing that $V_m^2 U_{r''} = V_m U_r$ by definition, the last inclusion relation leads to the stated one.

Observe next that $F(x,y)$ is uniformly continuous; indeed,

$$|F(x',y') - F(x,y)| < 2^{-m+1}$$

if $x' \in V_m(x)$ and $y' \in V_m(y)$. To prove this, let us suppose, first, that $F(x,y) < 1$; then $F(x,y) < r = k/2^m < 1$, so that $y \in U_r U_r^{-1}(x)$, which means that there exists a point $z \in S$ such that $y \in U_r(z)$ and $x \in U_r(z)$. It follows that $x' \in V_m U_r(z) = U_{r'}(z)$, where $r' = (k + 1)/2^m$; similarly, $y' \in U_{r'}(z)$; it follows that $y' \in U_{r'} U_{r'}^{-1}(x')$, implying that $F(x',y') < r'$. On the other hand, since $V_m = V_m^{-1}$, the situation is symmetrical between (x,y) and (x',y'), so that the inequality $F(x',y') < r$ implies the inequality $F(x,y) < r'$; now choosing k as minimal, so that $F(x',y') < r$, it results that $F(x,y) - F(x',y') < 2^{-m+1}$. Combining these two one-sided inequalities for $F(x',y') - F(x,y)$ yields the stated inequality on the absolute value. In case now $F(x,y) = 1$, the argument is similar.

As already seen, the relation $y \in U_r U_r^{-1}(x)$ is equivalent to the existence of an element $z \in S$ such that $y \in U_r(z)$ and $x \in U_r(z)$; this is evidently symmetrical between y and x, and it follows that $F(x,y) = F(y,x)$. Finally, the inequality $F(x,y) < 2^{-m-1}$ means that $y \in V_{m+1} V_{m+1}^{-1}(x) \subset V_m(x)$.

Proof of Theorem 7.1   If N is arbitrary in $\mathcal{C}$, there exists an admissible covering system V_1 such that $V_1 = V_1^{-1}$ and $V_1 \subset N$, for example, $V_1(x) = N_1(x) \cap N_1^{-1}(x)$. By induction it follows from the axioms that there exists a sequence $V_1, V_2, \ldots$ of admissible covering systems satisfying the hypothesized conditions above. It follows that there exists a bounded uniformly continuous function $F_N(x,y)$ on $S \otimes S$ such that $F(x,y) = F(y,x)$, such that the inequality $F_N(x,y) < 2^{-m-1}$ implies that $y \in V_m(x)$.

Now let d_N denote the function on $S \otimes S$ defined by the equation

$$d_N(x,y) = \sup_{z \in S} |F_N(x,z) - F_N(y,z)|;$$

it is easily seen that d_N is a pseudometric, and that it is uniformly continuous on $S \otimes S$. Let $\mathcal{M}$ denote the set of all such pseudometrics, obtained by permitting N to vary; then $\mathcal{M}$ is separating, for if $x \neq y$, then $y \notin N(x)$ for some N, so that $F_N(x,y) \geq 2^{-1}$, while $F_N(x,x) = 0$, from which it follows that $d_N(x,y) > 2^{-1}$.

Conversely, if $d_N(y,x) < 2^{-1}$, then $N(x)$ contains the open sphere of radius 2^{-1} centered at x, relative to the pseudometric d_N. To conclude the proof, it is only necessary to observe that by the uniform continuity of F_N, d_N is likewise uniformly continuous; consequently, for any $\epsilon > 0$ there exists an $N' \in \mathcal{C}$ such that if $y \in N'(x)$, then $d_N(y,x) < \epsilon$. Thus N' is contained in the ϵ-sphere system relative to the pseudometric d_N; and the intersection of such spheres over any finite set of N's contains the finite intersection of certain N''s in $\mathcal{C}$, which in turn contains an element of $\mathcal{C}$.

COROLLARY 7.3.1   *For any uniformly equicontinuous group of unimorphisms of a uniform space, there exists a defining set of pseudometrics, each of which is invariant under the group.*

Let S denote the uniform space, and G the group, the unimorphism of S corresponding to the element $g \in G$ being denoted as T_g. Let $\mathcal{M}$ denote a defining set of pseudometrics on S. It is no essential loss of generality to assume that all these metrics are uniformly bounded by 1, since otherwise they may be replaced by equivalent metrics with this property. Let $\mathcal{M}'$ denote the set of all pseudometrics d' of the form

$$d'(x,y) = \sup_{g \in G} d(T_g x, T_g y)$$

for some metric $d \in \mathcal{M}$. To show that $\mathcal{M}'$ defines an equivalent uniform structure, it is only necessary, since obviously $d(x,y) \leq d'(x,y)$, to show that, given any $\epsilon > 0$, there exist $d_1, d_2, \ldots, d_n$ in $\mathcal{M}$ and $\delta > 0$ such that the relations $d_1(y,x) < \delta,\ d_2(y,x) < \delta, \ldots,\ d_n(y,x) < \delta$ imply that $d'(x,y) \leq \epsilon$. Now the uniform equicontinuity of the transformation group means precisely that for any given $\epsilon > 0$, there exist $d_1, d_2, \ldots, d_n$ in $\mathcal{M}$ and $\delta > 0$ such that the inequalities just indicated imply that $d(T_g x, T_g y) < \epsilon$, for all $g \in G$; it follows that these inequalities imply that $d'(x,y) \leq \epsilon$.

Example 7.3.3   Let G be a topological group, with the uniform structure on G in which $N(x) = xV$, for a generic neighborhood V of the unit e. It is evident that $aN(b) = N(ab)$ for arbitrary a and b, which means that the action of G on itself by left translations is uniformly equicontinuous. Consequently, the uniform structure on G may be equally well defined by a set of pseudometrics d, each of which is invariant under left translations by arbitrary elements of G; that is, $d(ax,ay) = d(x,y)$ for all a, x, and y in G.

In case G satisfies the first countability axiom, i.e., there exists a countable set of neighborhoods V_i of e such that any neighborhood contains one of the V_i, then a countable defining set of invariant pseudometrics is obtained, say, $d_1, d_2, \ldots$; these may be taken to be bounded uniformly by 1, and on defining

$$d(x,y) = \sum_n 2^{-n} d_n(x,y),$$

an invariant metric is obtained which defines the same structure. Thus a topological group is metrizable with an invariant metric (relative to left translations) if (and only if, evidently) it satisfies the first countability axiom.

The same argument shows—or it follows through the use of the transformation $a \to a^{-1}$—that "left" may be replaced by "right" in the foregoing with the redefinition $N(x) = Vx$.   In general, however, these two structures on a topological group are not equivalent; it therefore does not follow that an equivalent metric exists which is invariant both under right and left translations, and in fact this is far from being the case, even for simple Lie groups.

EXERCISES

1   Show that, in a uniform space S with admissible covering system N, $\bigcup_k N^k(x)$ is for any $x \in S$ both open and closed ($k = 1, 2, \ldots$).

2   Show that a connected uniformly locally compact space is a countable union of compact sets.   (*Hint:* Use the result of Exercise 1.)

3   *a.* Show that any locally compact topological group G contains a subgroup which is both open and closed.

 b. Show that any such subgroup is invariant ($=$ normal).

4   A quasinorm on a real linear vector space V may be defined as a function $x \to |x|$ from V to the nonnegative reals, which satisfies the relation $|x - y| \leq |x| + |y|$; and if $|a_n x| \to 0$ whenever $a_n \to 0$ (a_n real); a set of quasinorms is complete if $|x| = 0$ for all norms in the set implies that $x = 0$.

 a. Show that if a generic neighborhood $N(y)$ of an element y is defined as the set of all elements x such that $|x - y| < \epsilon$ for a finite set of norms in a given complete collection, then V becomes a linear topological space, and a uniform space.

 b. Show that, conversely, every real linear topological vector space admits a complete set of quasinorms defining its topology.

 c. Show that the topology in a real linear topological space is definable by a single quasinorm if and only if it satisfies the first countability axiom.

5   Show that a compact topological group admits a separating set of pseudometrics each of which is invariant under both right and left translations.   (*Hint:* Show the equivalence of the covering system $x \to xV$ with the system $x \to Vx$.)

6   Show that a continuous function from a compact uniform space to an arbitrary uniform space is necessarily uniformly continuous. (*Hint:* Parallel the usual argument for functions on metric spaces.)

7.4   THE HAAR INTEGRAL

The main content of this section is the proof of

THEOREM 7.2   *Any uniformly equicontinuous group of unimorphisms of a uniformly locally compact space leaves invariant a nonzero invariant integral on the space, which is unique within a multiplicative constant if and only if there is a point in the space whose orbit is dense.*

We recall that the orbit of a point p under the action of a group G is the set of all transforms of p by elements of G. It will simplify notation and cause no nontrivial confusion to denote the action of the group simply by juxtaposition of the group element on the left; thus, instead of $F(g,p)$ for the transform of p by the unimorphism associated with g, we write simply gp. The orbit of the point p is then simply Gp; if for some p, Gp is all of the space, G is transitive, while if Gp is dense, we shall say it is weakly transitive.

The essential idea of the proof of existence is that of Haar, to the effect that the desired invariant measure m may be approximated in the following way: Let E be given set with compact closure, N be a "small" open set, and E_0 be a fixed "standard" set, which is to have unit measure. Then $m(E)$ should be approximately $n(E,N)/n(E_0,N)$, where $n(E,N)$ is the least number of translates of N required to cover E, which number is finite under the indicated assumptions. As N shrinks down to a single point, the approximation should improve, although either a great deal of optimism, or genius, is required to expect that a countably additive measure could really be obtained in this way. Haar supplied the genius, and the remarkable affinity between the theory of groups and integration shown by this result is indeed one of the authentic natural wonders of mathematics.

For technical reasons it is convenient to modify the construction just outlined, as in the work of Weil, to produce a left Haar integral rather than a measure. The measure may then be obtained from the integral in the standard fashion.

Proof of Theorem 7.2   Observe first that it suffices to consider the case where there exists a point whose orbit is dense. For suppose that the existence and uniqueness have been established in this case. Let S denote the given space, G the given group, and p any point of S. The closure P of Gp is then uniformly locally compact in the relative uniform structure, in which the family of neighborhoods $N'(x)$ relative to P is given by the equation $N'(x) = N(x) \cap P$. It is easily seen that P is invariant under G, so that G, when restricted to P, provides a group of unimorphisms of P, which is moreover equicontinuous, as follows directly from the equicontinuity of the original group.

Now an invariant integral I for G on P gives rise to an invariant integral I' for G on S by the definition $I'(f) = I(f \mid P)$; evidently $f \mid P$ will be a continuous function vanishing outside a compact set in P. Thus it suffices as regards the proof of existence to consider the case where a dense orbit exists. This shows, also, that the invariant integral can be essentially unique only if there is a point whose orbit is dense. For otherwise G would act equicontinuously on the uniformly locally compact space $S - P$, and so admit invariant integrals both on P and on $S - P$, arbitrary linear combinations of which would give invariant integrals on S that are not proportional.

For the remainder of the proof it may therefore be assumed that there exists a point p_0 whose orbit Gp_0 is dense.   This implies that the transforms under G of any neighborhood of p_0 will cover S.   It follows that if f and g are any two elements of the collection $\mathbf{C}_0^+$ of all nonnegative continuous functions of compact support on S, and if $g(p_0) \neq 0$, then there exists non-negative numbers $c_1, \ldots, c_n$ and group elements $a_1, \ldots, a_n$ such that

$$f(x) \leq \sum_i c_i g(a_i^{-1}x)$$

for all x in S.   In fact, if C is any compact set supporting f, and if N is any neighborhood of p_0 on which g is bounded away from zero, then C will be covered by a finite number of translates $a_1 N, \ldots, a_n N$ of N, and with c_i chosen as $\sup_{x \in S} f(x)/\inf_{x \in N} g(x)$, the cited inequality is valid.   The greatest lower bound of the sums $\sum_i c_i$ as the c_i and a_i vary over arbitrary finite sets so that the inequality is valid will be denoted $\Lambda(f,g)$, which can vanish only when f does, since $f(x) \leq [\Lambda(f,g) + \epsilon]\,\|g\|$ for all positive ϵ.   It is evident that $\Lambda(f,g)$ is invariant under the induced action of G on f and that $\Lambda(f + f',g) \leq \Lambda(f,g) + \Lambda(f',g)$, while $\Lambda(cf) = c\Lambda(f)$ for $c \geq 0$.   Another key property is that if h is an element of $\mathbf{C}^+$ with $h(p_0) \neq 0$, then

$$\Lambda(f,g) \leq \Lambda(f,h)\Lambda(h,g),$$

for if $f(x) \leq \sum_i c_i h(a_i^{-1}x)$ and $h(x) \leq \sum_j d_j g(b_j^{-1}x)$, then

$$f(x) \leq \sum_{i,j} c_i d_j g[(a_i b_j)^{-1}x],$$

and
$$\sum_{i,j} c_i d_j = (\sum_i c_i)(\sum_j d_j).$$

For any fixed unit function $f_0 \neq 0$, the real functional Λ_g on $\mathbf{C}_0^+$ defined by the equation

$$\Lambda_g(f) = \frac{\Lambda(f,g)}{\Lambda(f_0,g)}$$

has as a result the properties (1) $\Lambda_g(f_a) = \Lambda_g(f)$ for all $a \in G$, where $f_a(x) = f(a^{-1}x)$; (2) $\Lambda_g(cf) = c\Lambda_g(f)$ for any nonnegative real constant c; (3) $\Lambda_g(f + f') \leq \Lambda_g(f) + \Lambda_g(f')$ (for arbitrary f and f' in $\mathbf{C}_0^+$); (4) $\Lambda(f_0,f)^{-1} \leq \Lambda_g(f) \leq \Lambda(f,f_0)$.   It is, moreover, approximately additive, as detailed in

Lemma 7.4.1 *If k and k' are in $\mathbf{C}_0^+$ and $k + k' \leq 1$, then for arbitrary $\epsilon > 0$ there exists a λ such that for all f in $\mathbf{C}_0^+$*

$$\Lambda_g(fk) + \Lambda_g(fk') \leq (1 + \epsilon)\Lambda_g(f)$$

if g vanishes outside $N_\lambda(p_0)$.

By the uniform continuity of any element of C_0, there exists a covering system N_0 such that k and k' both vary by less than $\epsilon/2$ over any $N_0(x)$. If

$$f(x) \le \sum_i c_i g(a_i^{-1}x),$$

then $g(a_i^{-1}x)$ vanishes unless $x \in a_i N_{\gamma}(p_0)$, and by the uniform equicontinuity of G there is an index λ such that $aN_{\gamma}(x) \subset N_0(ax)$ for all a and x. Now, multiplying the inequality by $k(x)$ and observing that $k(x)$ is $\le k(a_i p_0) + \epsilon/2$ on $N_0(a_i p_0)$, it follows that

$$f(x)k(x) \le \sum_i c_i g(a_i^{-1}x)\left[k(a_i p_0) + \frac{\epsilon}{2} \right].$$

It follows that $\Lambda(fk,g) \le \sum_i c_i[k(a_i^{-1}p_0) + \epsilon/2]$.   Replacing k by k', it results that

$$\Lambda(fk,g) + \Lambda(fk',g) \le \sum_i c_i[k(a_i^{-1}p_0) + k'(a_i^{-1}p_0) + \epsilon] \le (1 + \epsilon) \sum_i c_i,$$

and the lemma follows on choosing the c_i so that $\sum_j c_i$ is arbitrarily close to $\Lambda(f,g)$, followed by division by $\Lambda(f_0,g)$.

It could now be proved relatively directly that $\Lambda_g(f)$ has a limit for any fixed f with respect to $N(= N(g))$, and this approach gives the uniqueness and the existence simultaneously (see the exercises). However, it is essentially equivalent and methodologically less special to establish the existence first, using a compactness argument, and then to derive the uniqueness.

To this end, let $\mathbf{A}$ be the collection of all functions ϕ from the elements of C_0^+ to the nonnegative reals such that $\phi(f) \in I_f$, where I_f denotes the interval $[\Lambda(f_0,f)^{-1}, \Lambda(f,f_0)]$; by property (4) above, $\Lambda_g \in \mathbf{A}$. Let $\mathbf{A}$ be topologized by the weakest topology in which $\phi(f)$ is for every fixed f a continuous function on $\mathbf{A}$; there is then an obvious homeomorphism of $\mathbf{A}$ with the product of the spaces I_f as f varies, so that by Tychonoff's theorem, $\mathbf{A}$ is compact. For any covering system N, let Γ_N denote the closure in $\mathbf{A}$ of the set of all Λ_g for those g vanishing outside $N(p_0)$. Evidently, $\Gamma_N \subset \Gamma_{N'}$ in case $N \le N'$, from which it follows that the Γ_N have the finite intersection property, and so contain a common element, say, Λ. This is then an invariant subadditive functional on C_0^+ to the nonnegative reals, not identically zero, and to complete the existence part of the proof it is only necessary to establish that Λ is additive.

This follows from Lemma 7.4.1 in the following way: Let f and f' be arbitrary in C_0^+, let m be any element of C_0^+ bounded away from zero on a set outside of which both f and f' vanish, let $h = f + f' + m$, and set $k = f/h$ (where this is defined and otherwise 0) and $k' = f'/h$ (similarly extended if necessary). Then, if $\epsilon > 0$, for all sufficiently small N,

$$\Lambda_g(kh) + \Lambda_g(k'h) \le (1 + \epsilon)\Lambda_g(h),$$

provided g is supported by $N(p_0)$, which implies that

$$\Lambda_g(f) + \Lambda_g(f') \le (1 + \epsilon)[\Lambda_g(f + f') + \Lambda_g(m)].$$

From the definition of Λ, $\Lambda(F_i)$ is for any finite set of F_i's arbitrarily closely approximable by the $\Lambda_g(F_i)$, and it results that

$$\Lambda(f) + \Lambda(f') \le (1 + \epsilon)\,[\Lambda(f + f') + \Lambda(m)].$$

Since this holds for every x and m of the indicated type—and any positive multiple of m again has the same property as m—it results in turn that

$$\Lambda(f) + \Lambda(f') \le \Lambda(f + f'),$$

completing the proof of the existence part.

To prove the uniqueness, it suffices to show that if Λ' is any integral which is not identically zero, then for any $\epsilon > 0$ there is an N such that, denoting by m any element of $\mathbf{C_0}^+$ which is 1 on the support of the general element f of $\mathbf{C_0}^+$, and m_0 a corresponding function for f_0,

$$(*) \qquad \frac{\Lambda(f,g) - \epsilon\Lambda(m,g)}{\Lambda(f_0,g)} \le \frac{\Lambda(f)}{\Lambda(f_0)} \le \frac{\Lambda(f,g)}{\Lambda(f_0,g) - \epsilon\Lambda(m_0,g)}.$$

For by the inequality (4) above, $\Lambda(m,g)/\Lambda(f_0,g) \le \Lambda(m_0,f_0)$, and so is bounded, while $\Lambda(m,g)/\Lambda(f_0,g) \le \Lambda(m,f_0)$ and is also bounded, so that $\Lambda(f)/\Lambda(f_0)$ may be made to differ by arbitrarily little from the purely geometrically defined quantity $\Lambda(f,g)/\Lambda(f_0,g)$ by choosing N sufficiently small.

This will be shown, not for all admissible g, but for those with the property that $g(a^{-1}p_0) = g(ap)$ for all a in G. To show that there exist such g for arbitrarily large λ, as well as to eliminate extraneous complications in the proof, recall that, by Example 7.3.3, the present uniform structure may be taken to be defined by a family of invariant pseudometrics dN.

Now setting $g(p) = \psi[d_N(p,p_0)]$, where $\psi(x)$ for real x is defined as $1 - x$ for $0 \le x \le 1$ and 0 otherwise, g is evidently continuous, vanishes outside of the $N(p_0)$ defined by d_N but not at p_0, and in addition

$$g(ap_0) = \psi[d_N(ap,p_0)] = \psi[d_N(p,a^{-1}p_0)] = g(a^{-1}p_0).$$

To prove the inequality $(*)$, let P be a covering system such that $|f(x) - f(x')| < \epsilon$ when $x \in P(x')$; let g be an element of $\mathbf{C_0}^+$ which is supported by $P(p_0)$ and such that $g(p_0) \ne 0$. Setting $h(a) = \Lambda'(fg_a)$, where $g_a(p) = g(a^{-1}p)$, then since $g_a(p)$ vanishes unless $p \in N(ap_0)$, in which case $f(p) > f(ap_0) - \epsilon$, it results that

$$h(a) \ge \Lambda'\{[f(ap_0) - \epsilon]g_a\} = [f(ap_0) - \epsilon]\Lambda'(g).$$

Now choosing an arbitrary positive number δ, set N be a covering system such that $|g(p) - g(q)| < \delta$ for $p \in N(q)$. Let C be a compact set outside of which f vanishes; then f is covered by a finite number of G-translates of

$N(p_0)$, say, the $a_i N(p_0)$.   Let the k_i be the corresponding elements of $\mathbf{C}_0{}^+$ such that k_i vanishes outside of $a_i N(p_0)$ and such that for $p \in C$, $\sum_i k_i(p) = 1$; here Scholium 5.1 is used.   Then $h(a)$ may be written

$$h(a) = \Lambda'[f(\sum_i k_i)g_a] = \sum_i \Lambda'(fk_i g_a),$$

and since k_i vanishes outside of $a_i N(p_0)$, on which g, or by the G-invariance of the uniform-neighborhood system any G-translate of g, such as g_a, varies by less than δ, it results that

$$h(a) \le \sum_i \Lambda'\{fk_i[g_a(a_i p_0) + \delta]\}.$$

At this point the property that $g(ap_0) = g(a^{-1}p_0)$ is required and is assumed. Noting that $\Lambda'(g) \ne 0$, inasmuch as the translates of any neighborhood of p_0 cover M, it results from the combination of the two inequalities attained that

$$f(ap_0) - \epsilon \le \sum_i c_i[g(a_i^{-1}ap_0) + \delta],$$

where $c_i = \Lambda'(fk_i)/\Lambda'(g)$.   Since f and g are continuous functions and the orbit of p_0 is dense, it follows that for all $p \in M$,

$$f(p) - \epsilon \le \sum_i c_i[g(a_i^{-1}p) + \delta].$$

Recalling the definition of m, it follows that

$$f(p) \le \sum_i c_i g(a_i^{-1}p) + \left(\epsilon + \delta \sum_i c_i\right)m(p),$$

implying that

$$\Lambda(f,g) \le \sum_i c_i = \left(\epsilon + \delta \sum_i c_i\right)\Lambda(m,g),$$

which by the arbitrary character of δ leads to the inequality

$$\Lambda(f,g) \le \frac{\Lambda'(f)}{\Lambda'(g)} + \epsilon\,\Lambda(m,g).$$

On the other hand, if the e_i are any nonnegative numbers and the b_j any group elements such that

$$f(p) \le \sum_j e_j g(b_j^{-1}p)$$

for all p, then by the linearity and positivity of Λ',

$$\Lambda'(f) \le \left(\sum_j e_j\right)\Lambda'(g),$$

so that

$$\frac{\Lambda'(f)}{\Lambda'(g)} \le \Lambda(f,g).$$

Multiplication by $\Lambda'(g)$ in the two inequalities attained for $\Lambda(f,g)$ gives the result

$$\Lambda'(g)\Lambda(f,g) - \epsilon\Lambda'(g)\Lambda(m,g) \le \Lambda'(f) \le \Lambda'(g)\Lambda(f,g).$$

Replacing f by f_0 and dividing, the required inequality (*) is obtained.

7.5 DEVELOPMENTS FROM UNIQUENESS

A representative application of the uniqueness part of Theorem 7.2 is the following result, of which, for example, the invariance of Lebesgue measure in euclidean space under rotations as well as translations is a quite particular case.

> COROLLARY 7.5.1 *A uniformly locally compact metric space with a weakly transitive group of isometries admits a unique invariant integral (within a constant factor).*

In the statement, the uniformity means, of course, that for some $\epsilon > 0$, all spheres of radius less than ϵ have compact closures. The corollary is an immediate deduction from Theorem 7.2. With slight effort a more general result may be obtained, of which, for example, the transformation properties of Lebesgue measure under linear transformations are a special case.

> COROLLARY 7.5.2 *If in Theorem 7.2 G acts weakly transitively on S, and G' denotes the group of all homeomorphisms u of S such that $uGu^{-1} \subset G$, then there exists a representation r of G' into the positive reals (as a multiplicative group) such that*

$$I(f_u) = r(u)I(f), \qquad f \in C_0(S)[f_u(x) = f(u^{-1}x)].$$

Note first that if $I'(f)$ is defined as $I(f_u)$, then I' is also an invariant integral, for (denoting by x a bound variable)

$$I'(f_a) = I((f_a)_u) = I(f_{ua}) = I(f_{uau^{-1}\cdot u}) = I[(f_u)_{uau^{-1}}] = I(f_u) = I'(f).$$

By the uniqueness, $I(f_u) = r(u)I(f)$ for some positive constant $r(u)$. Now it is easily seen that $(f_u)_{u'} = f_{u'u}$ for any elements u and u' in G' and $f \in C_0(G)$, and applying I to both sides of this equation and using the definition of r, it follows that $r(uu') = r(u)r(u')$.

A particularly important case is that of a group acting on itself by left or right translations. Let G be a topological group acting on itself, and G' the group of all homeomorphisms u of G such that $uGu^{-1} \subset G$. An *automorphism* of G is a homeomorphism A of G with itself such that $A(ab) = A(a)A(b)$ for all a and b in G. The set of all automorphisms of G is easily seen to form a subgroup, *Aut G*, of G'.

COROLLARY 7.5.3   *For any locally compact group G, there exists a representation δ of Aut G into the multiplicative group of positive reals such that for any Haar measure m on G*

$$m[A(E)] = \delta(A)m(E)$$

for all automorphisms A and all Baire sets E.

Proof   It is an immediate deduction from Corollary 7.5.2, applied to the case of the left action of G on itself, and the essential uniqueness of Haar measures that there is a representation δ having the cited property for all left Haar measures.   If m is any left Haar measure, the equation $n(E) = m(E^{-1})$ defines a right Haar measure, and evidently

$$n[A(E)] = m[A(E^{-1})] = \delta(A)n(E).$$

When this corollary is applied to the case of the inner automorphism $\phi_x : a \to x^{-1}ax$, it follows that $m(x^{-1}Ex) = \delta(\phi_x)m(E)$.   Then noting that $\phi_{xy} = \phi_y\phi_x$ and setting $\Delta(x) = \delta(\phi_x)$, one has the algebraic part of the following result.

COROLLARY 7.5.4   *On any locally compact group G, there exists a continuous representation Δ of G into the multiplicative group of positive reals such that*

$$m(Ex) = \Delta(x)m(E)$$

for any left Haar measure m, and

$$n(x^{-1}E) = \Delta(x)n(E)$$

for any right Haar measure n, E being an arbitrary Baire set and x an arbitrary element of G.

Proof   That Δ is continuous is most easily seen in terms of integrals.   Let I be a left Haar integral on G, and f an element of $\mathbf{C}_0(G)$ such that $I(f) \neq 0$. Then, from the right uniform continuity of f, it is easy to see that if $(R_x f)(y) = f(yx)$, then $I(R_x f)$ is continuous as a function of x.   Since the first equation above implies that $I(f) = \Delta(x)I(R_x f)$, it follows that Δ is continuous.

On the other hand, this relation also shows that $f \to I(f\Delta^{-1})$ is a right Haar integral on $\mathbf{C}_0(G)$.   From this one may derive another useful result.

COROLLARY 7.5.5   *If m is a left Haar measure on the locally compact group G, and n is the right Haar measure defined on the Baire sets by $n(E) = m(E^{-1})$, then m is absolutely continuous with respect to n and $dm/dn = \Delta$.*

Proof   Again this is most easily seen in terms of integrals.   Let I and J be the integrals determined by m and n, respectively.   Then there is a constant

$c > 0$ such that $J(f) = cI(f\Delta^{-1})$ for all f in $\mathbf{C}_0(G)$.   Moreover, $J(f) = I(\check{f})$, where $\check{f}(x) = f(x^{-1})$.   Thus $I(f) = cI(\check{f}\Delta^{-1})$; when this relation is applied to $f\Delta^{-1}$, it results that $I(f) = c^2 I(f)$, and hence that $c = 1$.   Therefore $I(f) = J(f\Delta)$ for all f in $\mathbf{C}_0(G)$.

We take explicit note of the following result, obtained in the course of the proof.

COROLLARY 7.5.6   *If m is a left Haar measure on the locally compact G, then*

$$\int_G f(x)\, dm(x) = \int_G f(x^{-1})\, \Delta(x^{-1})\, dm(x)$$

for any nonnegative Baire function f.

It is natural to explore the question of when the Haar measure on a locally compact group G is invariant under translations on both sides; such a group is called *unimodular*.   On any locally compact group G the representation Δ is generally called the *modular function*.   It is evident that G is unimodular if and only if $\Delta(x) = 1$ for all x in G.   This is obviously the case for Abelian groups and for compact groups as well.   For if G is compact, $m(Gx) = \Delta(x)m(G) < \infty$ and $Gx = G$.   Another class of unimodular groups is given by the connected semisimple Lie groups, which may be shown to have no representations into the positive reals other than the identity.

Observe that Corollary 7.5.6 implies

COROLLARY 7.5.7   *On any unimodular group, Haar measure is invariant under the transformation $x \to x^{-1}$.*

Example 7.5.1   The complex $n \times n$ matrices form a real vector space V of dimension $2n^2$.   Let G be the set consisting of all invertible matrices in V.   Then G is an open set in V, and in the topology induced by that in V, G is a locally compact (non-Abelian) group with respect to matrix multiplication.   The group G is usually designated by $GL(n,C)$, and is called the *complex general linear group of degree n*.   Each element a in V defines a linear transformation

$$L_a:\quad x \to ax, \qquad x \in V,$$

which maps V into V, ax being the usual matrix product.   For any a and b, $L_{ab} = L_a L_b$ and L_a is invertible or nonsingular if and only if $a \in G$.   If m is a Haar measure on V (as a group under addition) and a an element of G, it follows from Corollary 5.3.1 that

$$\int_V f(x)\, dm(x) = |\det L_a| \int_V f(ax)\, dm(x)$$

for any f in $\mathbf{C}_0(V)$.   The elements of V also form a complex vector space W of dimension n^2, and the mapping $T_a: x \to ax$ is a linear transformation on W with the property that $\det L_a = |\det T_a|^2$.   Now W is the direct sum of n

subspaces invariant under T_a, in each of which T_a is equivalent to the standard action of a on column vectors.   Thus $\det T_a = (\det a)^n$, so that

$$\det L_a = |\det a|^{2n}.$$

We therefore have the more explicit transformation formula

$$\int_V f(x)\, dm(x) = |\det a|^{2n} \int_V f(ax)\, dm(x).$$

If $f \in C_0(G)$, we can define f as 0 outside G, and then $f \in C_0(V)$; moreover, since $\det x$ is continuous and nonvanishing on G,

$$\int_V f(x)\,|\det x|^{-2n}\, dm(x) = |\det a|^{2n} \int_V f(ax)\,|\det (ax)|^{-2n}\, dm(x)$$

$$= \int_V f(ax)\,|\det x|^{-2n}\, dm(x).$$

Thus the equation

$$I(f) = \int_V f(x)\,|\det x|^{-2n}\, dm(x)$$

defines a left Haar integral on G, and by a straightforward computation I is also right-invariant.

The group $GL(n,C)$ just considered is a Lie group with "nice" global coordinate functions, in terms of which the group translations are expressed linearly; this accounts in part for the elementary nature of the calculation of its Haar measure.   For a general Lie group the situation is more complicated.

EXERCISES

1 Show that the equation

$$J(f) = \frac{1}{2\pi} \int_0^{2\pi} f(e^{i\theta})\, d\theta$$

defines a Haar integral on the circle group, i.e., the multiplicative group of complex numbers of modulus 1 in the topology induced by the plane.   Also note that J may be described by the formula

$$J(f) = \int_0^1 f(e^{2\pi i\theta})\, d\theta.$$

2 Show that the invariant integral J in Exercise 1 may also be described by the formula

$$J(f) = \frac{1}{\pi} \int_{-\infty}^{\infty} f\left(\frac{1 + ix}{1 - ix}\right) \frac{dx}{1 + x^2}.$$

[*Hint:* If $g \in C[-\pi,\pi]$, show that

$$\int_{-\pi}^{\pi} g(\theta)\, d\theta = 2 \int_{-\infty}^{\infty} g(2 \tan^{-1} x)\, \frac{dx}{1 + x^2}.]$$

3   *a.* Show that a locally compact group G is unimodular if there exists a compact neighborhood of the identity invariant under inner automorphisms.

   b. Show that the affine group on the line has no compact neighborhood of the identity invariant under inner automorphisms.

   c. Show that a locally compact group is unimodular if it is compact, discrete, or Abelian.

   d. Construct an example of a unimodular group which is not compact, discrete, or Abelian.

4   Show that the complement of a Haar null set in a locally compact group is dense in the group.

5   Let G be a locally compact group, and I a Haar integral on $C_0(G)$.   Show that G is compact if and only if

$$\sup_{0 \le f \le 1} I(f) < \infty.$$

6   If I and J are left Haar integrals on the locally compact groups G and H, show that the direct product of I and J is a left Haar integral on $G \times H$.

7   If G is the multiplicative group of complex numbers $\ne 0$, determine the Haar measure on

$$G \times G \times \cdots \times G.$$

8   Let G be three-dimensional euclidean space.   If the product of two elements x and y is defined by

$$(x_1, x_2, x_3)(y_1, y_2, y_3) = (x_1 + y_1, x_2 + y_2 + x_3 y_1, x_3 + y_3),$$

show that G is then a locally compact non-Abelian group and that Lebesgue measure on G is a left- and right-invariant Haar measure.

9   Show that there is an essentially unique integral on the $n - 1$ sphere S in n-space which is invariant under rotations, i.e., the maps $x \to xa$, where $a \in 0^+(n,R)$. [*Hint:* Let $G = 0^+(n,R)$, and consider the transformation $f \to f'$ from $C(S)$ to $C(G)$, defined by

$$f'(a) = f(x_0 a), \qquad a \in G,$$

where x_0 is some fixed point of S.]

10   Show that there is an essentially unique integral on the hyperboloid

$$x_0{}^2 = \sum_{i=1}^{n} x_i{}^2$$

in $(n + 1)$-dimensional euclidean space which is invariant under the group of all homogeneous linear transformations, leaving the form

$$x_0{}^2 - \sum_{i=1}^{n} x_i{}^2$$

invariant, and determine its analytic form in terms of the x's.   (This group is the generalized *Lorentz* group, and when $n = 3$, the invariant measure plays a significant role in the treatment of the group representations associated with relativistic wave equations.)

11   A group of transformations on a measure space, measurable in the sense of taking measurable sets into measurable sets, is said to be *ergodic* if the only *essentially invariant* locally measurable sets are local null sets or complements thereof, where essentially invariant means that each transform of the set differs from it only by a set of measure 0.

>   *a.* Show that if the measure on the space is invariant under the group action, and if the space is a countable union of measurable sets, then the group acts ergodically if and only if every invariant measure on the same σ-ring of measurable sets is a multiple of the given one.   [*Hint:* If the group acts ergodically and m' is invariant in addition to the given measure m, then dm'/dm is an essentially invariant function, and the set where it is bounded by any given constant c is essentially invariant.   On the other hand, if every invariant measure m' is proportional to the given one m, and if E is an essentially invariant set, then setting $m'(A) = m(E \cap A)$, one gets an invariant measure whose proportionality with m implies that E or its complement is a local null set.]
>
>   *b.* Show that if a group acts ergodically and leaves the measure invariant, then the only essentially invariant locally measurable functions are essentially constants.

12   Show that rotation through an irrational angle is an ergodic transformation on the circle relative to the usual measure invariant under rotations.

13   Show that on the n-torus T_n (the direct product of n circle groups) the one-parameter group of displacements consisting of multiplication by

$$(e^{ita_1}, e^{ita_2}, \ldots, e^{ita_n}),$$

where the a_i are given real numbers which are linearly independent over the rationals, acts ergodically, relative to Haar measure on T_n.

14   Give an example to show that on a noncompact locally compact space there is in general no integral invariant under a given homeomorphism.

15   Show that the quotient of a locally compact group modulo a compact subgroup admits a unique invariant integral relative to the canonical action of the full group.   (*Hint:* Show the group acts equicontinuously on the coset space.)

16   Show that for any right-translation-invariant metric on a locally compact group satisfying the first countability axiom, right-invariant Haar measure is invariant under all isometries.

17   Let M denote the 2-torus, represented as the set of all ordered pairs of real numbers modulo 1, and let G denote the group of all transformations on M of the form $(x_1, x_2) \rightarrow (x_1, x_2) + t(y_1, y_2) \pmod 1$, where t is an arbitrary real number and (y_1, y_2) is a fixed pair of real numbers whose quotient is irrational.   Show that there exists a unique invariant measure on M relative to G.   Generalize the result to the case of the n-torus.

18   Show that the representation Δ described in Corollary 7.5.4 is the same for both left- and right-invariant Haar measures.

7.6   FUNCTION SPACES UNDER GROUP ACTION

If G is a topological transformation group on a locally compact space, it is important for analytical purposes to study the induced action of G on various function spaces, besides the continuous functions of compact support. In case there is a regular measure invariant under G, we have the useful extension of a classical result of Lebesgue:

SCHOLIUM 7.1   *If G is a topological transformation group on a regular measure space M, and the measure on M is invariant under G, then for any f in $L_p(M)$, $1 \leq p < \infty$, the map*

$$a \to L_a f,$$

where $L_a f(x) = f(a^{-1}x)$, is continuous from G into $L_p(M)$.

Proof   If f is a continuous function of compact support,

$$\sup_x |f(a^{-1}x) - f(x)| \to 0$$

as $a \to e$ by uniform continuity.   Since f has compact support, there also exists a fixed compact set supporting f and the $L_a f$ for a sufficiently near e, and it follows that $\|L_a f - f\|_p \to 0$ as $a \to e$.   Continuity as $a \to b$ follows if one replaces f by $L_b f$.

Now if f is an arbitrary element of $L_p(M)$, $1 \leq p < \infty$, and $\epsilon > 0$, there exists a continuous function g of compact support such that $\|f - g\|_p < \epsilon.$. It then results that

$$\|L_a f - f\|_p \leq \|L_a(f - g)\|_p + \|L_a g - g\|_p + \|g - f\|_p$$

and as the measure is invariant under G,

$$\|L_a f - f\|_p \leq 2\epsilon + \|L_a g - g\|_p.$$

Thus $\|L_a f - f\|_p \leq 3\epsilon$ if a is sufficiently close to e.   The continuity at points other than e follows by group translation, as earlier.

Instead of the direct action of G, it is often appropriate to consider a kind of average of the action called convolution.

DEFINITION   Suppose G is a locally compact transformation group on a regular measure space M and that the measure on M is invariant under G.   Then the *convolution* of a function f on G with a function g on M, denoted $f * g$, is said *to exist at a point x in M* if

$$y \to f(y)g(y^{-1}x)$$

is an element of $L_1(G)$, relative to left Haar measure; the value of the convolution at x is then defined by the integral

$$f * g(x) = \int_G f(y)g(y^{-1}x) \, dy,$$

where dy denotes the element of left Haar measure. The convolution is said to *exist* if it exists at almost all points x of M.

SCHOLIUM 7.2 *Let G be a locally compact transformation group on a locally compact space S. Suppose m is a regular measure on the ring $\mathfrak{B}$ of Baire subsets of S invariant under G. If M is the measure space $(S,\mathfrak{B},m)$, then for any f in $L_1(G)$ and any g in $L_p(M)$, $1 \le p < \infty$, the function*

$$f(y)g(y^{-1}x)$$

is measurable on $G \otimes M$, the convolution of f with g exists, and $f * g$ defines an element h of $L_p(M)$ such that

$$\|h\|_p \le \|f\|_1 \|g\|_p.$$

Proof The function $f(y)g(x)$ is measurable with respect to the ring of Baire sets on $G \otimes M$ by Scholiums 3.9 and 5.3, being the product of two measurable functions, and $f(y)g(y^{-1}x)$ is its composition, with the homeomorphism $(y,x) \to (y,y^{-1}x)$. Thus $f(y)g(y^{-1}x)$ is measurable on $G \otimes M$ by Scholium 5.3, and the observation made earlier that measures transform in the same way as the associated positive linear functionals on the continuous functions of compact support on a locally compact space.   It follows by the Fubini theorem that

$$k(x) = \int_G |f(y)g(y^{-1}x)| \, dy$$

is a measurable function on M.   In the case that $p = 1$, it follows by a second application of the Fubini theorem that

$$\int_M k(x) \, dm(x) = \int_G |f(y)| \left[\int_M |g(y^{-1}x)| \, dm(x) \right] dy.$$

Since $\int_M |g(y^{-1}x)| \, dm(x) = \int_M |g(x)| \, dm(x)$, it then results that

$$\int_M k(x) \, dm(x) = \|f\|_1 \|g\|_1.$$

If $p > 1$, the integral defining k may be written

$$\int |f(y)|^{1/p} |g(y^{-1}x)| \, |f(y)|^{1/q} \, dy,$$

where $1/p + 1/q = 1$.   Hölder's inequality then shows that

$$[k(x)]^p \le \left[\int |f(y)| \, |g(y^{-1}x)|^p \, dy \right] \left[\int |f(y)| \, dy \right]^{p/q}.$$

Again applying the Fubini theorem, we see that

$$\int [k(x)]^p \, dm(x) \le \|f\|_1^{p-1} \int |f(y)| \left[\int |g(y^{-1}x)|^p \, dm(x) \right] dy$$

$$= \|f\|_1^{p-1} \int |f(y)| \left[\int |g(x)|^p \, dm(x) \right] dy$$

$$= \|f\|_1 \|g\|_p^p < \infty.$$

It follows in all cases that $k(x)$ is finite a.e. Thus $f * g$ exists, and as $|f * g(x)| \le k(x)$ at all points where $k(x) < \infty$, it defines a function h in $L_p(M)$ such that $\|h\|_p \le \|f\|_1 \|g\|_p$.

> SCHOLIUM 7.3 *For an arbitrary compact neighorhood N of the unit e in the locally compact group G, let f_N denote any non-negative Baire function supported by N, with $\int f_N = 1$. Then, for any $g \in L_p(M)$, $1 \le p < \infty$, the limit in $L_p(M)$ of the net $\{f_N * g\}$ with respect to N (the neighborhoods of e being ordered by set-theoretic inclusion reversed) is g.*

Writing $g(x) - (f_N * g)(x)$ in the form

$$\int f_N(y)[g(x) - g(y^{-1}x)] \, dy,$$

which possibility derives from the circumstance that $\int f_N = 1$; writing

$$|f_N(y)[g(x) - g(y^{-1}x)]| = [|f_N(y)|^{1/p} |g(x) - g(y^{-1}x)|][|f_N(y)|^{1/q}],$$

where $p^{-1} + q^{-1} = 1$; and applying Hölder's inequality as in the foregoing, it follows that

$$\|g - f_N * g\|_p \le \left[\int f_N(y) \|g - g_y\|^p \, dm(y) \right]^{1/p} = \left[\int_N c \|g - g_y\|^p \, dm(y) \right]^{1/p}$$

$$\le cm(N) \sup_{y \in N} \|g - g_y\|_p$$

$$= \sup_{y \in N} \|g - g_y\|_p,$$

which tends to zero by Scholium 7.1

> SCHOLIUM 7.4 *If G is a locally compact unimodular group, the convolution of an element f of $L_p(G)$ with an element g of $L_q(G)$, where $1 \le p < \infty$ and $1/p + 1/q = 1$, exists everywhere and is a uniformly continuous function bounded by $\|f\|_p \|g\|_q$.*

Proof Suppose, first, that $p > 1$. Then $g(y^{-1}x)$ as a function of y is in $L_q(G)$ and has norm equal to that of g. For the modular function is identically 1 on G, so that

$$\int_G |g(y)|^q \, dy = \int_G |g(yx)|^q \, dy = \int_G |g(y^{-1}x)|^q \, dy.$$

As the product of an element of $L_p(G)$ with one in $L_q(G)$ is integrable, the convolution of f with g exists everywhere, and by Hölder's inequality,

$$|f * g(x)| \leq \left[\int |f(y)|^p \, dy\right]^{1/p}\left[\int |g(y^{-1}x)|^q \, dy\right]^{1/q} = \|f\|_p \|g\|_q.$$

Moreover, $f * g(x) - f * g(z) = \int f(y)[g(y^{-1}x) - g(y^{-1}z)] \, dy$, so that

$$|f * g(x) - f * g(z)| \leq \|f\|_p \left[\int |g(y^{-1}x) - g(y^{-1}z)|^q \, dy\right]^{1/q}$$

$$= \|f\|_p \left[\int |g(yx) - g(yz)|^q \, dy\right]^{1/q}$$

$$= \|f\|_p \|R_x g - R_z g\|_q,$$

where $R_x g(y) = g(yx)$. Now $\|R_x g - R_z g\|_q = \|R_{z^{-1}x} g - g\|_q$, and $a \to R_a g$ is continuous from G to $L_p(G)$, by the argument used in the proof of Scholium 7.1. Hence $f * g$ is uniformly continuous on G.

Now consider the case $p = 1$. Then $f \in L_1(G)$, and

$$\int |f(y)| \, dy = \int |f(xy)| \, dy = \int |f(xy^{-1})| \, dy,$$

so that $y \to f(xy^{-1})$ is also integrable for any x in G. Since $q = \infty$, that is, $g \in L_\infty(G)$, it follows that $f(xy^{-1})g(y)$ is integrable as a function of y. Moreover,

$$\int f(xy^{-1})g(y) \, dy = \int f(y^{-1})g(yx) \, dy = \int f(y)g(y^{-1}x) \, dy,$$

so that $f * g$ exists for all x. Hence

$$f * g(x) - f * g(z) = \int [f(xy^{-1}) - f(zy^{-1})]g(y) \, dy,$$

so that

$$|f * g(x) - f * g(z)| \leq \|g\|_\infty \int |f(xy^{-1}) - f(zy^{-1})| \, dy = \|g\|_\infty \|L_{z^{-1}x} f - f\|_1.$$

We mention one final result:

SCHOLIUM 7.5 *If G is a locally compact group, $L_1(G)$ relative to left Haar measure is an associative algebra with respect to the usual addition and scalar multiplication and convolution as multiplication of elements.*

With the aid of Scholium 7.2 it is straightforward to verify the conditions which the theorem asserts to be valid, the only one not immediately obvious being the associativity. But this is readily deduced from the Fubini theorem.

REMARK   This L_1-algebra of G is a generalization of the notion of a group algebra which plays an important role in the representation theory of finite groups. Of course, $C_0(G)$, $L_1(G) \cap L_p(G)$ for any value of p, etc., likewise form algebras, and in the case of a finite group these all reduce to the same algebra. These are, however, ideals in $L_1(G)$, and it is $L_1(G)$ which arises in a distinctive way in a number of analytical situations. On the other hand, from certain representation-theoretic viewpoints, the algebra of operators on $L_2(G)$ generated by the left-regular representation, or rather its closure in certain topologies, is also an important object of study, which likewise generalizes the finite-group algebras.

Still another generalization is the algebra of all differences of finite regular measures, under convolution as multiplication (see the exercises below); $L_1(G)$ is an ideal in this algebra, but it suffers from the fundamental difficulty that it is too large for group translations to act continuously on the algebra, in the Banach norm of the total variation, on the one hand, and too small to be closed in topologies which are natural in connection with Hilbert-space theory.

We mention, finally, that the restriction to locally compact groups is a natural one in connection with invariant measures. This is shown by Weil's converse to Haar's theorem, according to which a group admitting an invariant measure with certain rather minimal regularity properties may be imbedded in a locally compact group in such a way that the given invariant measure is naturally derived from the Haar measure on the locally compact group.

EXERCISES

1 Let G be a locally compact group, and for a function f on G set $f^*(x) = \overline{f(x^{-1})}\Delta(x^{-1})$. Show that $f \to f^*$ is an involution on the L_1-algebra.

2 Show that if $L(f)$ denotes the mapping $g \to f * g$ from $L_p(M)$ into $L_p(M)$, $f \in L_1(G)$, with G and M as in Scholium 7.2, then the map $f \to L(f)$ is a representation of $L_1(G)$ as an algebra under convolution and is of norm 1.

3 Show that if λ is a *character* on the reals, i.e., a map into the complex numbers of modulus 1 such that $\lambda(x + y) = \lambda(x)\lambda(y)$, and if λ is Lebesgue-measurable, then it is automatically continuous. (*Hint:* Multiply the functional equation for λ by an arbitrary function which is continuous and of compact support, integrate over y, and use the L_1-continuity of $L_x f$ as a function of x, after making a change of variable.)

4 Show that it suffices for the result of Exercise 3 that $\lambda(x + y) = \lambda(x)\lambda(y)$ a.e. on the product space, in the sense that λ may then be altered on a null set so as to be continuous.

5 Show that a continuous homomorphism of the algebra $L_1(-\infty, \infty)$ under convolution as multiplication into the complex numbers has the form

$$f(x) \to \int_{-\infty}^{\infty} e^{itx} f(x)\, dx$$

for some real number t, and that all such maps have the indicated property.   (*Hint:*
Use the representation theorem for linear functionals on L_1 and then employ the
multiplicative property to reduce the result to that of Exercise 3.)

6   Show that a representation T of a locally compact group on a finite-dimensional vector space **V** is automatically continuous if it is measurable in the sense that

$$a \to \varphi[T(a)x]$$

is measurable on G for every $x \in$ **V** and every $\varphi \in V^*$.

7   Show that in Exercise 6, the map

$$f \to \int_{-\infty}^{\infty} f(a)T(a)\, da$$

from $L_1(G)$ to the linear transformations on **V** is a representation of $L_1(G)$ as an
algebra.

8   If G is a locally compact group and f and g are functions on G in L_p and L_q,
where $p^{-1} + q^{-1} = 1 + r^{-1}$, show that $f * g$ exists and that $\|f * g\|_r \le \|f\|_p \|g\|_q$.
(*Hint:* Use Hölder's inequality.   The present inequality, named for W. H. Young,
is quite useful.)

9   If G is a compact group, show that for any f in $L_1(G)$, the map

$$g \to f * g, \qquad g \in L_p, \qquad 1 \le p < \infty,$$

is compact.

10   Show that if G is a Lie group (i.e., it has a local C^∞ structure near e such that
ab is a C^∞ function of a and b jointly), and if f is an infinitely differentiable function
with compact support, then $f * g$ is likewise infinitely differentiable for any g in
L_p.   In the case of the reals under addition, show that

$$D^n(f * g)(x) = (D^n f) * g(x),$$

where $D = d/dx$.

11   Convolutions may be extended from functions to measures.   Suppose I_S
is a complex integral on a locally compact space S and that I_G is a complex integral
on G, G being a locally compact transformation group on S.   Then the convolution

$$I = I_G * I_S$$

is defined as the integral on $C_0(S)$, given by the equation

$$I(k) = I_{G \otimes S}(k'),$$

where k' is the function on $G \otimes M$ defined by $k'(y,x) = k(yx)$, and $I_{G \otimes S}$ is the
direct product of I_G and I_S.   Show that this is indeed an extension of the earlier
notion by considering the case where I_G and I_S are indefinite integrals of L_1
functions relative to Haar measure and an invariant measure on S.

12   *a.* Show that the complex integrals on a locally compact group G form an
algebra relative to convolution.

b. Show that the map $I_G \to R(I_G)$, where $R(I_G)$ denotes the operation

$$I_S \to I_G * I_S$$

on complex integrals over S, is a representation of this algebra.

   c. Show that the absolutely continuous integrals (with respect to the Haar integral) form an ideal in this algebra.

13   Let G be a locally compact group, and I the integral defined by

$$I(f) = \int f(x)g(x)\, dx,$$

where $g \in L_1(G)$.   If J is the integral defined by a complex measure on the ring of Baire sets, show that there exists a function h in $L_1(G)$ such that

$$I * J(f) = \int f(x)h(x)\, dx$$

for all f in $C_0(G)$.   The convolution $g * J$ is then defined as h.

14   *a.* With J as in Exercise 13, show that the map $g \to g * J$ of $L_1(G)$ to itself commutes with the operation of left translation by group elements.

   b. Show that every continuous linear transformation on $L_1(G)$ commuting with all left translations is right convolution with a finite integral such as J.

VIII

ALGEBRAIC INTEGRATION THEORY

8.1 INTRODUCTION

There are a number of important analytical situations in which the approach to an integral through a measure space is unnatural or technically disadvantageous. In fact, we have developed a major part of the theory from the point of view of integration lattices, following the fundamental ideas of Daniell. However, in the examples of integration lattices considered so far, e.g., the real step functions on a basic measure space or the continuous real-valued functions of compact support on a locally compact space, there is another inherent element of structure which in many respects is more important; these particular lattices are also algebras, and there are definite advantages from a broad viewpoint to a formulation of integration theory which starts from a linear functional on an algebra rather than a lattice. This approach to the theory of integration, which might be called the algebraic approach, is not restricted to function algebras alone, arises naturally from the foundations of probability theory, and is explicitly or implicitly indicated in a variety of other situations, e.g., commutative spectral theory in Hilbert space, the theory of integration in infinite-dimensional linear spaces, harmonic analysis on Abelian and more general groups (in particular, the L_2 theory), and developments closely paralleling integration theory in the theory of rings of operators.

The intuitively simplest case is that of probability theory, and a brief account of how this leads to an algebraic viewpoint will suffice here. Probability did not become a rigorous mathematical theory until around 1930. It was then recognized that the "random variables" with which the theory dealt could be represented by measurable functions on a measure space of total measure 1; the integral represented the "expectation" of the random variable, and the mathematical model was seen to be one encompassing all the essential objects of the original heuristic theory. However, in this model the random variables are displaced as the fundamental objects by the measure space itself, in which the underlying set is the class of "events" and the measure of a given measurable set is the "probability" that the chance event is a point of the set. Now this space of events is in certain cases a relatively artificial object. For example, in order to discuss probabilities of coin tosses from this point of view, it is necessary to consider the measure space determined by the basic coin-tossing space described at the end of Chap. 2. Nevertheless, the theory of measure and measure spaces does provide a solid basis for determining what is deductive and what is inductive in a given probability situation, and it formed the basis for the great development of the mathematical theory of probability which has taken place since the early thirties.

On the other hand, from a contemporary viewpoint, the early probabilists, such as the Bernoullis some two hundred years ago, may be regarded as primitive exponents of a variety of algebraic integration theory. For they could be described as operating implicitly on an axiomatic basis. The random variables with which they worked directly were never defined or constructed mathematically, but were simply objects to be treated and manipulated algebraically according to certain rules, which were, however, not explicitly stated. In retrospect, the assumptions implicit in their work may be formulated as follows: The random variables form (or generate) a real commutative associative algebra with unit, on which a distinguished linear functional, the "expectation" functional, is defined, which is nonnegative on squares and normalized as 1 on the unit. Although these assumptions have a kind of simple general plausibility, in the early thirties they would have seemed to be far too few to provide a rigorous analytic basis for any comprehensive theory of probability.

By the early forties there were, however, mathematical developments showing that such a basis was indeed sufficient, and in fact essentially equivalent to that in terms of a measure space. Moreover, these developments did not emerge from probability theory, but from quantum mechanics, the spectral theory of operators in Hilbert space, and purely mathematical considerations; and the probabilists were by this time happy with the foundations of their subject as laid a decade earlier.   It is nevertheless of a certain foundational interest that the original ideas of the Bernoullis can provide an effective basis for probability theory, through the use of an integration theory for linear functionals on algebras which is relevant in a number of other contexts.   By this time it is also possible to cite certain probabilistic situations in which the algebraic approach may be used to advantage, such as the direct treatment of Gaussian probability distributions in function space, which can be treated in terms of measure spaces only in a circumlocutory fashion.

Instead of following the methodologically pure approach of developing the entire theory of integration in the context of abstract algebras, we shall adopt the point of view of representation theory; that is, we shall follow the procedure of showing that the various types of commutative "integration algebras" which commonly arise are isomorphic to appropriately dense subalgebras of the integration lattices already considered, thereby reducing the theory in all essential respects to that considered in the earlier chapters. For this purpose it is necessary to show that the elements of certain algebras may be represented by functions.   This is most easily done for complex algebras, and the basic result of the next section deals with such algebras, rather than the real ones, more natural as a putative model for an algebra of random variables; however, as will be shown, results for real algebras are simple consequences.   The point of treating complex algebras is that certain elementary complex variable arguments may be used.   These could be avoided, but at the cost of substantial complication, and it would be misleading to represent real analysis as a subject which cannot benefit from complex analysis.

8.2 BANACH ALGEBRAS AND THE CHARACTERIZATION
OF FUNCTION ALGEBRAS

As indicated above, it is desirable to establish conditions under which the elements of a given algebra may be represented as functions.   Virtually all known results of this general type yield functions on locally compact spaces, rather than measurable ones associated with a given σ-ring.   Although for some purposes this topological structure is irrelevant, there are others where it is technically advantageous, and some where it is involved with important features of the theory.   It suffices for the present purposes to treat algebras with a unit, and to show that under suitable conditions these algebras are essentially those consisting of some or all continuous functions on compact spaces.   The choice of conditions is largely a matter of technical convenience determined by the applications envisaged, and the preliminary theory applies to general Banach algebras.

DEFINITIONS A *Banach algebra* is a system **B** with the structure both of an algebra and a Banach space such that $\|xy\| \leq \|x\| \|y\|$ for all x and y. The *spectrum* of **B** is the collection of all nonzero continuous linear functionals ϕ on the algebra which are multiplicative in the sense $\phi(xy) = \phi(x)\phi(y)$ for arbitrary x and y in **B**.

In general, the spectrum of a Banach algebra is empty. It can, however, be shown to be nontrivial in case **B** is commutative and has a unit. When there is a unit e in **B**, it is easily seen that the condition $\phi \neq 0$ is equivalent to the condition that $\phi(e) = 1$. Not so easily seen is that in this case the continuity of ϕ is automatic; this emerges from the relation between the functional ϕ and its kernel $\phi^{-1}(0)$, which is a (proper) maximal ideal in **B**. To the study of these matters we now turn.

SCHOLIUM 8.1 *Suppose* **B** *is a Banach algebra and* $a_1, a_2, \ldots$ *is a sequence of scalars such that*

$$\sum_{n \geq 1} |a_n| r^n < \infty$$

with $r > 0$. *Then the series* $\sum_{n \geq 1} a_n x^n$ *converges and defines a continuous function in the open set* $[x: \|x\| < r]$.

Proof Suppose $\|x\| < r$. Since $\|x^n\| \leq \|x\|^n$, it then follows from the Cauchy criterion that the series $\sum_{n \geq 1} a_n x^n$ converges in **B** to an element $f(x)$. The continuity of the function f may be verified in a number of ways. It follows, for example, from the continuity properties of ordinary power series once it is shown that for any x and y in **B** and any positive integer n

$$\|(x + y)^n - x^n\| \leq (\|x\| + \|y\|)^n - \|x\|^n.$$

This offers no real difficulty, but does not follow directly from the binomial expansion because of the general lack of commutativity.

SCHOLIUM 8.2 *In a Banach algebra with unit* e, *the invertible elements form an open set on which the map* $x \to x^{-1}$ *is continuous, and includes the open sphere of radius* 1 *around* e.

Proof Suppose **B** is a Banach algebra with unit e. Then, by Scholium 8.1, the geometric series $e + \sum_{n \geq 1} x^n$ converges and defines a continuous function f in the open set $\|x\| < 1$. On the other hand, if

$$s_k(x) = e + \sum_{n=1}^{k} x^n, \qquad \|x\| < 1,$$

the standard argument for the numerical case shows that

$$s_k(x)(e - x) = (e - x)s_k(x) = e - x^{k+1}.$$

Since $s_k(x) \to f(x)$ and $x^{k+1} \to 0$, it follows by continuity that

$$f(x)(e - x) = (e - x)f(x) = e.$$

Hence $e - x$ is invertible, and $f(x) = (e - x)^{-1}$. This shows that an element y is convertible if $\|e - y\| < 1$, and since it is then true that

$$y^{-1} = e + \sum_{n \geq 1} (e - y)^n,$$

it follows that y^{-1} is in the closed subalgebra generated by y and the unit e.

Now let u be invertible and v any element of the open set

$$V = [v \colon \|u - v\| < \|u^{-1}\|^{-1}].$$

Then $e - u^{-1}v = u^{-1}(u - v)$, and

$$\|e - u^{-1}v\| \leq \|u^{-1}\| \|u - v\| < 1.$$

Hence $u^{-1}v$ is invertible, and the same is true of $v = u(u^{-1}v)$. Moreover, since $v^{-1} = (u^{-1}v)^{-1}u^{-1}$, the map $v \to v^{-1}$ is continuous on V.

SCHOLIUM 8.3   *Any maximal ideal in a Banach algebra with unit is closed.*

Proof Suppose **B** is a Banach algebra with unit e, and let **M** be a (proper) maximal ideal in **B**. Then the closure of **M** is again an ideal which, if not **M**, must be the entire algebra, and so contain e. The ideal **M** must therefore contain elements arbitrarily close to e. But by the preceding result these elements will have inverses and so can be contained in no ideal other than **B** itself.

SCHOLIUM 8.4   *If **B** is a Banach algebra with unit, its spectrum is a compact subset of **B*** in the w*-topology.*

Proof It is evident that an element of **B** having an inverse cannot map into 0 under any element ϕ of the spectrum; in particular, no element of the form $\phi(x)e - x$ has an inverse. This implies $|\phi(x)| \leq \|x\|$. For otherwise

$$\phi(x)e - x = \phi(x)\left[e - \frac{x}{\phi(x)}\right]$$

would have an inverse by the proof of Scholium 8.2. The spectrum S of **B** is therefore contained in the unit sphere of the dual. Since this is compact in the w*-topology (Alaoglu's theorem), to conclude the proof it suffices to show that the spectrum is a closed set in the w*-topology.

It is evident that the limit of any net of multiplicative linear functionals is again multiplicative: if $\phi_\lambda(xy) = \phi_\lambda(x)\phi_\lambda(y)$ for all x and y, and $\phi_\lambda \to \phi$, this means precisely that $\phi_\lambda(x) \to \phi(x)$ for all x, so that $\phi(xy) = \phi(x)\phi(y)$. Similarly, if $\phi_\lambda(e) = 1$ for all λ, then $\phi(e) = 1$. Thus the spectrum is indeed closed. (The reader can just as easily draw this conclusion by a classical approximation argument, avoiding the use of nets.)

As noted in the proof, an element having an inverse cannot be mapped into 0 by any element of the spectrum.   The converse is less obvious and more useful, and will follow from the results to be proved next.

SCHOLIUM 8.5 *If x is an element of a complex Banach algebra with unit e, the complex numbers λ such that $x - \lambda e$ fails to be invertible form a nonempty compact subset of the closed disk with center 0 and radius $\|x\|$.*

Proof The complement of the set S in question contains all complex numbers a such that $|a| > \|x\|$.   For if $|a| > \|x\|$, then $\|x/a\| < 1$, and $e - x/a$ is invertible by the proof of Scholium 8.2.   Suppose a is any complex number such that $x - ae$ is invertible.   Then by Scholium 8.2, there is a neighborhood of $x - ae$, all of whose elements are invertible.   Since $x - be$ is a continuous function of b, this implies the existence of a neighborhood of a such that $x - be$ is invertible for all b in the neighborhood.   Thus S is a closed subset of the disk $|\lambda| \leq \|x\|$, and all that remains is to prove that S is nonempty.

Suppose, on the contrary, that $x - ae$ is invertible for all complex a, and let $g(a) = (x - ae)^{-1}$.   Then it is readily verified that for any complex a and b

$$g(a + b) - g(a) = b[x - (a + b)e]^{-1}(x - ae)^{-1}.$$

Since $[x - (a + b)e]^{-1}$ is a continuous function of b, this implies the differentiability of g, in fact that

$$\lim_{b \to 0} \frac{g(a + b) - g(a)}{b} = (x - ae)^{-2}.$$

If f is a continuous linear functional on the algebra, it follows that $f \circ g$ is holomorphic in the entire complex plane.   Moreover, $|f(g(a))| \leq \|f\| \, |a|^{-1} \|e - x/a\|^{-1} \to 0$ as $a \to \infty$.   Thus $f \circ g$ is a bounded entire function vanishing at ∞, and by Liouville's theorem it vanishes identically.   Hence $f(x^{-1}) = 0$ for all f, and by the Hahn-Banach theorem this leads to the contradiction $x^{-1} = 0$.

SCHOLIUM 8.6 *A complex Banach algebra in which every nonzero element is invertible is isomorphic to the complex field.*

Proof Suppose **B** is such an algebra.   Let e be its unit and x a nonzero element.   Then by Scholium 8.5 there exists a complex number λ such that $x - \lambda e$ fails to be invertible.   By our assumptions this implies $\lambda \neq 0$ and $x - \lambda e = 0$.   Thus for each x in **B**, there exists a unique complex number λ such that $x = \lambda e$, and the map $x \to \lambda$ is the required isomorphism.

SCHOLIUM 8.7   *If* $\mathbf{I}$ *is a closed ideal in the Banach algebra* $\mathbf{B}$, *the quotient* $\mathbf{B}/\mathbf{I}$ *is a Banach algebra with respect to the quotient norm*

$$\|x + \mathbf{I}\| = \inf_{y \in \mathbf{I}} \|x + y\|$$

Proof The term ideal as used here and before means two-sided ideal, so that $\mathbf{B}/\mathbf{I}$ is indeed an algebra. The indicated norm is easily seen to be homogeneous, subadditive, and positive on nonzero elements. Moreover, since $\mathbf{I}$ is an ideal,

$$\inf_{y \in \mathbf{I}} \|x_1 x_2 + y\| \leq \inf_{y_i \in \mathbf{I}} \|(x_1 + y_1)(x_2 + y_2)\| \leq \left(\inf_{y_1 \in \mathbf{I}} \|x_1 + y_1\| \right) \left(\inf_{y_2 \in \mathbf{I}} \|x_2 + y_2\| \right)$$

so that $\mathbf{B}/\mathbf{I}$ is a normed linear algebra. If $\mathbf{B}$ has a unit e, then $e + \mathbf{I}$ is a unit for $\mathbf{B}/\mathbf{I}$. Now all that remains is to verify that $\mathbf{B}/\mathbf{I}$ is complete. For this suppose $\{x_n + \mathbf{I}\}$ is a Cauchy sequence in $\mathbf{B}/\mathbf{I}$. Then an increasing sequence n_i of positive integers and corresponding elements y_i in $\mathbf{I}$ may be chosen so that

$$\|(x_{n_i} + y_i) - (x_{n_{i+1}} + y_{i+1})\| < 2^{-i}.$$

The sequence $\{x_{n_i} + y_i\}$ then has a limit x, and it follows without difficulty that

$$\|(x_n + \mathbf{I}) - (x + \mathbf{I})\| \to 0 \quad \text{as} \quad n \to \infty.$$

SCHOLIUM 8.8   *If* $\mathbf{M}$ *is a proper maximal ideal in a complex commutative Banach algebra* $\mathbf{B}$ *with unit* e, *then there is a unique element* ϕ *in the spectrum of* $\mathbf{B}$ *whose kernel is* $\mathbf{M}$.

Proof By Scholium 8.3, $\mathbf{M}$ is closed, and therefore, by Scholium 8.7, $\mathbf{B}/\mathbf{M}$ is a Banach algebra with respect to the quotient norm. Since $\mathbf{M}$ is maximal and $\mathbf{B}$ is a commutative algebra with identity, it follows that $\mathbf{B}/\mathbf{M}$ is also a commutative algebra with identity, which moreover contains no nonzero proper ideals. Any such algebra is automatically a field. Thus $\mathbf{B}/\mathbf{M}$ is a complex Banach algebra, with the property that every nonzero element is invertible, and hence is isomorphic to the complex field, by Scholium 8.6. Specifically, this means that for each x in $\mathbf{B}$, there exists a unique complex number λ_x such that

$$x + \mathbf{M} = \lambda_x e + \mathbf{M}.$$

If ϕ is any element of the spectrum of $\mathbf{B}$ with kernel $\mathbf{M}$, it is evident that $\phi(x) = \lambda_x$ for all x; so there is at most one such ϕ. On the other hand, the map $x \to \lambda_x$ is linear and multiplicative and assigns the value 1 to e and has kernel $\mathbf{M}$.

THEOREM 8.1   *A commutative complex Banach algebra* $\mathbf{B}$ *with unit* e *has a nonempty spectrum, and if* $x \in \mathbf{B}$ *and* λ *is a complex number, there* *exists an element ϕ of the spectrum such that*

$$\phi(x) = \lambda$$

if and only if $x - \lambda e$ fails to be invertible.

Proof   Suppose ϕ is an element of the spectrum and $\phi(x) = \lambda$. Then $\phi(x - \lambda e) = 0$, and hence, as observed before, $x - \lambda e$ cannot be invertible. Conversely, suppose λ is a complex number such that $x - \lambda e$ is not invertible. Then, as **B** is commutative, the ideal **I** generated by $x - \lambda e$ consists of all multiples $y(x - \lambda e)$ of $x - \lambda e$ by elements y of **B**.   If $\mathbf{I} = \mathbf{B}$, then $y(x - \lambda e) = e$ for some y; so $x - \lambda e$ has an inverse.   Thus **I** is proper, and the collection of all proper ideals containing **I** has the property that any simply ordered subset has an upper bound in the collection, namely, the set-theoretic union, which is again an ideal, and proper since it excludes e.   Hence **I** is contained in a proper maximal ideal **M**.   By Scholium 8.8 there is an element ϕ of the spectrum whose kernel is **M**; hence $\phi(x - \lambda e) = \phi(x) - \lambda = 0$.   The fact that the spectrum of **B** is nonempty follows from the argument just given, together with the observation that there exist elements x in **B** and corresponding complex scalars λ for which $x - \lambda e$ is not invertible; for example, $x = 0$ and $\lambda = 0$. (For the existence of less trivial examples we may refer to Scholium 8.5.)

COROLLARY 8.2.1   *An element of a commutative complex Banach algebra with unit has an inverse if and only if no element of the spectrum maps it into zero.*

COROLLARY 8.2.2   *Let S be the spectrum of a complex commutative Banach algebra **B** with unit e which is generated by e and a single element x.   Then the mapping*

$$\phi \to \phi(x)$$

is a homeomorphism of S onto the set of complex numbers λ for which $x - \lambda e$ fails to be invertible.

Proof   The continuity of the mapping is implied by the definition of the weak topology in S, and its range is identified by the theorem.   Since **B** is the closure of the set of all complex polynomials in x, it is evident by the continuity and multiplicative character of an element of the spectrum that such an element is completely determined by the value it assigns to x; hence the mapping is one-to-one and therefore a homeomorphism, inasmuch as its image is Hausdorff and S is compact.

Theorem 8.1, together with the following one, may be regarded as the key foundational results in the theory of Banach algebras.

THEOREM 8.2 *For any element x of a commutative complex Banach algebra* **B** *with spectrum S and unit e,*

$$\lim_{n \to \infty} \|x^n\|^{1/n} = \sup_{\phi \in S} |\phi(x)|.$$

Proof If $\phi \in S$, then, as already noted, $\|\phi\| \leq 1$, so that

$$|\phi(x)|^n = |\phi(x^n)| \leq \|x^n\|,$$

from which it follows that

$$\sup_{\phi \in S} |\phi(x)| \leq \liminf_{n} \|x^n\|^{1/n}.$$

By Scholium 8.2, the set D of complex members μ such that $e - \mu x$ is invertible is open and evidently is nonempty. In fact, D contains the open disk $|\mu| < \|x\|^{-1}$, and for such μ

$$(e - \mu x)^{-1} = \sum_{n \geq 0} \mu^n x^n, \qquad x^0 = e.$$

On the other hand, setting

$$s = \sup_{\phi \in S} |\phi(x)|,$$

we have $\|x\|^{-1} \leq s^{-1}$, and it follows readily from Theorem 8.1 that $\mu \in D$ whenever $0 < |\mu| < s^{-1}$; thus D contains the entire disk $|\mu| < s^{-1}$. For μ in D, let $g(\mu) = (e - \mu x)^{-1}$. Then, by an argument similar to that used in the proof of Scholium 8.5,

$$\lim_{\lambda \to 0} \frac{g(\mu + \lambda) - g(\mu)}{\lambda} = x(e - \mu x)^{-2}.$$

If f is a continuous linear functional on **B**, it follows that $f \circ g$ is complex analytic in D. Moreover, for $|\mu| < \|x\|^{-1}$,

$$f \circ g(\mu) = \sum_{n \geq 0} \mu^n f(x^n),$$

and since $f \circ g$ is analytic in $|\mu| < s^{-1}$, the series must also converge and represent $f \circ g$ in this larger disk. Thus, for $0 < |\mu| = r < s^{-1}$,

$$f(x^k) = \frac{1}{2\pi i} \int_C \frac{f \circ g(\mu)}{\mu^{k+1}} \, d\mu,$$

where C is the circle of radius r with center 0. This implies

$$|f(x^k)| \leq \frac{\|f\|}{r^k} \sup_{\mu \in C} \|g(\mu)\|.$$

Setting $M_r = \sup_{\mu \in C} \|g(\mu)\|$, and using Corollary 8.2.2, we find that

$$\|x^k\| \leq \frac{M_r}{r^k},$$

and this shows that

$$\limsup_k \|x^k\|^{1/k} \le r^{-1}.$$

This being true for all positive $r < s^{-1}$, it follows that

$$\limsup_n \|x^n\|^{1/n} \le s.$$

By the inequality obtained in the first paragraph above this completes the proof.

Example 8.2.1 In the analysis of a single bounded operator on a Banach space it is helpful to investigate the (commutative) Banach algebra generated by the operator, together with the identity. Consider, for example, the Banach space $C[0,1]$ of continuous complex-valued functions on $[0,1]$ in the sup norm. Let T be the integration operator defined on $C[0,1]$ by

$$Tf(x) = \int_0^x f(y)\, dy, \qquad 0 \le x \le 1.$$

Then T is a bounded linear mapping of $C[0,1]$ into itself. In fact, since

$$T^n f(x) = \int_0^x \frac{(x-y)^{n-1}}{(n-1)!}\, f(y)\, dy,$$

it follows that $|T^n f(x)| \le \|f\|(x^n/n!)$. This of course implies that

$$\|T^n\| \le 1/n!,$$

so that $\lim_n \|T^n\|^{1/n} = 0$. Now let **B** denote the (commutative) complex Banach algebra generated by T and the identity operator I, the norm being that defined by the operator bound. Then **B** is simply the closure of all polynomials in T in the algebra of all bounded operators on $C[0,1]$. If S is the spectrum of **B**, it follows from Corollary 8.2.2 that the mapping $\phi \to \phi(T)$ is a homeomorphism of S onto the set of complex numbers λ for which $T - \lambda I$ fails to be invertible. On the other hand, by Theorem 8.2.2 and the computation above,

$$\sup_{\phi \in S} |\phi(T)| = 0.$$

Thus S consists of precisely one element ϕ for which $\phi(T) = 0$ and $\phi(I) = 1$; and $T - \lambda I$ is invertible for all complex $\lambda \ne 0$. An element x in a commutative complex Banach algebra with unit is called a *generalized nilpotent* if

$$\lim_n \|x^n\|^{1/n} = 0.$$

In this example, this property of T is a reflection of the fact that repeated applications of T produce smoother and smoother functions, i.e., functions with better differentiability properties.

It is useful to note that much of the foregoing theory can be modified so as to be applicable to Banach algebras which do not necessarily possess a unit.

The spectrum is then locally compact, rather than compact, and the Gelfand homomorphism, defined as the mapping from an element of a Banach algebra to the corresponding function on the spectrum, maps into the algebra of continuous functions which vanish at infinity. The modifications in question can be based on

> COROLLARY 8.2.3 *Any Banach algebra A over the complex field can be imbedded in a unique minimal Banach algebra A' having a unit; A is then either a maximal ideal in A' or all of A'. If A is proper in A', then the restriction mapping $\phi \to \phi \mid A$ is a homeomorphism from the spectrum of A', less the unique point whose kernel is A, onto the spectrum of A.*

The uniqueness must be understood in the sense of isomorphism: If A'' is any other extension, then there is an isomorphism of A' with a subalgebra of A'' which carries A as imbedded in A' into A as imbedded in A''. The proof of uniqueness is straightforward, once the existence is established, and we prove only the latter.

If A does not have a unit, let A' denote the set of all pairs (a,λ), with $a \in A$ and λ a complex number, as a Banach algebra relative to the definitions

$$(a,\lambda) + (a',\lambda') = (a + a',\ \lambda + \lambda'),\ \ (a,\lambda)(a',\lambda') = (aa' + \lambda a' + \lambda'a,\ \lambda\lambda'),$$

$$\lambda'(a,\lambda) = (\lambda'a,\ \lambda'\lambda),\ \ \|(a,\lambda)\| = \|a\| + |\lambda|.$$

The map $a \to (a,0)$ is then an isomorphism of A into a subalgebra of A', which subalgebra is the kernel of the homomorphism $\phi_0\colon (a,\lambda) \to \lambda$; A' contains the unit $(0,1)$. Any point ϕ of the spectrum of A' restricts on A either to a point of the spectrum of A or to the zero functional; the latter is the case, evidently, if and only if $\phi = \phi_0$. Since $\phi((a,\lambda)) = \phi(a) + \lambda$ [where $\phi(a) = \phi((a,0))$], the restriction map is one-to-one on the spectrum less the point ϕ_0. The same equation shows that any point of the spectrum of A defines a point of the spectrum of A', so that the restriction map is onto.

In order to ensure that a given complex commutative Banach algebra with unit be isomorphic to the algebra of all continuous complex functions on a compact space, it is necessary to assume some additional element of structure on the algebra. Now one obvious property of the algebra of all continuous functions on a compact space is the fact that the sup norm satisfies the condition $\|x\|^2 = \|x^2\|$. But this condition is not sufficient to characterize the algebra, inasmuch as it may also be satisfied by certain subalgebras such as that of the continuous functions on the closed unit disk which are holomorphic in the interior. This example is excluded by the assumption of an involution or adjunction operation $x \to x^*$, which, together with the assumption that $\|xx^*\| = \|x\|^2$, turns out to be generally appropriate and readily verified in a variety of cases.

DEFINITION   An *involution* on an algebra over the complex field is a *conjugate linear involutary antiautomorphism* $x \to x^*$ of the algebra. Specifically, this means that

$$(cx)^* = \bar{c}x^*$$

$$(x^*)^* = x$$

$$(x + y)^* = x^* + y^*$$

$$(xy)^* = y^*x^*$$

for any complex c and all x, y in the algebra.

A simple example is the operation of complex conjugation on the algebra of all continuous complex-valued functions on a compact space, which is in fact the involution on such an algebra relevant and assumed in the following discussion.

With the aid of the Stone-Weierstrass theorem we may now prove the following result, which characterizes the algebra of continuous complex functions on a compact space.

THEOREM 8.3   *A complex commutative Banach algebra with unit and involution $x \to x^*$ such that $\|xx^*\| = \|x\|^2$ for all elements x is isomorphic algebraically, as regards the norm and as regards the involution to the algebra of all continuous complex-valued functions on its spectrum.*

Lemma 8.2.1   *Suppose ϕ is a continuous multiplicative linear functional on a Banach algebra $\mathbf{B}$ with unit e and involution $x \to x^*$ such that $\|xx^*\| = \|x\|^2$ for all elements x. Then ϕ assumes real values on the elements x such that $x = x^*$.*

Proof   Suppose x is an element such that $x = x^*$, and let $\phi(x) = r + is$, where r and s are real.   If t is real and $y = x + ite$, then $\phi(y) = r + i(s + t)$, so that

$$|\phi(y)|^2 = r^2 + s^2 + 2st + t^2.$$

On the other hand, $|\phi(y)|^2 \leq \|y\|^2 = \|yy^*\|$, and since $y^* = x - ite$, it follows that

$$\|yy^*\| = \|x^2 + t^2e\| \leq \|x^2\| + t^2 \|e\|.$$

Thus $r^2 + s^2 + 2st \|e\| \leq \|x\|^2$, for all real t, and this implies $s = 0$.

Proof of Theorem 8.3   Let S be the spectrum of the algebra $\mathbf{B}$, and let $x(\cdot)$ denote the function on S given by the equation

$$x(\phi) = \phi(x).$$

The map $x \rightarrow x(\cdot)$ is evidently a homomorphism of **B** into the algebra **C**(S) of all continuous complex-valued functions on S. From the definition it is immediate that these functions "separate" S; that is, for any two points ϕ and ϕ' in S there is an element x in **B** such that $x(\phi) \neq x(\phi')$. Moreover, the set of all $x(\cdot)$ is closed under complex conjugation. To prove this it is sufficient to show that $x^*(\phi) = \overline{x(\phi)}$ for all x and ϕ. For this we write a general element x in the form $x = u + iv$, with $u = u^*$ and $v = v^*$, by taking $u = (x + x^*)/2$ and $v = (x - x^*)/2i$. Then, since $\phi(x) = \phi(u) + i\phi(v)$ and $\phi(x^*) = \phi(u) - i\phi(v)$, it is evidently enough to show that $\phi(x)$ is real if $x = x^*$. But since ϕ is an element of S, this follows at once from the lemma. Since $e(\phi) = 1$ for all ϕ, it follows in turn by the Stone-Weierstrass theorem that the $x(\cdot)$ form a dense subalgebra of **C**(S). Thus, to finish the proof, it suffices to show that the mapping $x \rightarrow x(\cdot)$ is an isometry of **B** into **C**(S). This is a consequence of Theorem 8.2. For if $x \in$ **B**, then $\|x\|^2 = \|xx^*\|$, and as $(xx^*)^* = xx^*$, it follows readily by induction that

$$\|xx^*\| = \|(xx^*)^{2^n}\|^{1/2^n}.$$

Hence $\|x\|^2 = \sup_{\phi} |\phi(xx^*)| = \sup_{\phi} |\phi(x)|^2$.

Example 8.2.2   If S is any compact Hausdorff space, the spectrum of **C**(S) is naturally identified with S itself. For suppose t is a fixed point of S. Then the mapping

$$\phi_t : f \rightarrow f(t), \quad f \in \mathbf{C}(S),$$

is evidently an element of the spectrum of **C**(S). Since the continuous functions separate points, the map $t \rightarrow \phi_t$ is one-to-one from S into the spectrum, and continuous by the definition of the weak topology. To prove that it is a homeomorphism, it is sufficient to show that a given element ϕ of the spectrum is "evaluation" at some point t. If this is the case, the kernel **M** of ϕ must consist of all functions f vanishing at t. Hence it is necessary and easily seen to be sufficient to prove that a given proper maximal ideal **M** in **C**(S) consists of the functions vanishing at some point t. If not, there exists for each t in S a function f in **M** such that $f(t) \neq 0$. Since $\bar{f}f \in$ **M**, we may assume in fact that f is nonnegative and that $f(t) > 0$. Then by continuity there is a neighborhood N of t in which f is bounded away from 0. A finite number of such neighborhoods cover S, and adding the corresponding functions, we obtain a positive function f in **M** which is bounded away from 0. But this implies that f is invertible, and hence that **M** = **C**(S), a contradiction. Thus there is a point t in S at which all the functions in **M** vanish, and as **M** is maximal, **M** is precisely the collection of all such functions.

Example 8.2.3   Let M be a nontrivial measure space, i.e., one in which there exist measurable sets of positive measure. Then it follows from Exercise 8, Sec. 4.1, that the algebra $\tilde{L}_\infty(M)$ of function classes determined by the essentially bounded complex-valued locally measurable functions satisfies the hypotheses of Theorem 8.3, the involution being that induced by complex conjugation.   Thus $\tilde{L}_x(M)$ is isometrically $*$-isomorphic to $C(S)$, where S is its spectrum.   In general, it is impossible to give an explicit description of S; however, it can be shown that the topology for S has a basis consisting of sets which are both open and closed, so that S is totally disconnected.   This is a reflection of the fact that the function classes defined by linear combinations of characteristic functions of locally measurable sets form a dense set in $\tilde{L}_x(M)$.   Finally, it follows from Theorem 8.2 and Exercise 9, Sec. 4.1, that the range of spectral values of an element of $\tilde{L}_\infty(M)$ is the same as its essential range, as defined earlier.

EXERCISES

1   Show that the functions in $C(S)$, S being a given compact Hausdorff space, that vanish on a given subset of S, form a closed ideal which is maximal if and only if the subset consists of a single point.

2   Show that the norm in $C(S)$, S being a compact Hausdorff space, may be defined in a purely algebraic way, so that any algebra-automorphism of $C(S)$ is necessarily an isometry.

3   Deduce that if S and S' are compact Hausdorff spaces, then $C(S)$ and $C(S')$ are algebraically isomorphic if and only if S and S' are homeomorphic.

4   *a.* Show that the algebra $C(S)$ of all bounded continuous complex-valued functions on a topological space S satisfies the conditions of Theorem 8.3 with the involution taken as complex conjugation and the norm taken as the least upper bound of the absolute value.

   b. Show that if S is completely regular, the map $x \to \phi_x$, where $\phi_x(f) = f(x)$, is a continuous one-to-one map of S into the spectrum of $C(S)$.   This spectrum is called the *Cech compactification* of S, after the man who first developed it, in a different way.

5   Show that if $\mathbf{I}$ is a closed ideal of the algebra $C(S)$, with S a compact Hausdorff space, then $\mathbf{I}$ consists of all elements of the algebra vanishing on some closed subset of S, and that the correspondence between closed ideals and closed subsets is one-to-one.

6   *a.* Let G be the additive group of integers, and $\mathbf{A}$ the algebra of all integrable functions on G relative to convolution as multiplication.   Show that the spectrum of $\mathbf{A}$ consists of all mappings of the form

$$\{a_n\} \to \sum_n a_n e^{inx}$$

   for some real x.

   b. Deduce that if a periodic function with an absolutely convergent Fourier series never vanishes, then its reciprocal likewise has an absolutely convergent Fourier series.   (This important result had been conjectured for a number of years before it was established by Wiener by classical methods, which in part implicitly recognized the algebraic context but made no deep use of this aspect.)

7   Let **A** be an associative algebra with unit, and $\mathcal{M}$ the collection of all proper maximal ideals in **A**. Let $\mathcal{M}$ be topologized by designating as a closed set the collection of all maximal ideals containing a given ideal.

 a. Show that with this topology $\mathcal{M}$ is a compact space, but give an example to show that it need not be Hausdorff.

 b. Show that in case $\mathbf{A} = C(S)$ for some compact Hausdorff space S, then the map from S into $\mathcal{M}$, sending a point into the ideal of functions vanishing at the point, is a homeomorphism.

 c. Show that the two topologies are in essential agreement also in the case of the convolution algebra of integrable functions on the positive integers. [*Hint:* Show that the function f on $(-1,1)$, given by the equation $f(x) = 1 - n|x|$ for $|x| \le n^{-1}$ and $f(x) = 0$ otherwise, has an absolutely convergent Fourier series on this interval.]

8   Show that if **E** denotes the algebra of all entire functions of a complex variable, then for any element a of a Banach algebra **B**, the map $f \to f(a)$ is a homomorphism of **E** into **B**, where for any f in **E** having the form

$$f(z) = \sum_n c_n z^n,$$

$f(a)$ is defined as $\sum_n c_n a^n$.

9   Show that if f is analytic in a region bounded by a differentiable simple closed curve C, in whose interior are contained the spectral values $\phi(a)$ of the element a of the commutative Banach algebra **B**, with unit e, then

$$\frac{1}{2\pi i} \int_C f(\lambda)(\lambda e - a)^{-1}\, d\lambda$$

exists as a vector-valued integral, and its value is an element b of **B** such that

$$\phi(b) = f(\phi(a))$$

for all ϕ in the spectrum of **B**.

10   Deduce that if f is analytic on an open set containing the range of values of a periodic function g, with absolutely convergent Fourier series, then $f \circ g$ also has an absolutely convergent Fourier series.

11   Let **A** be a real commutative Banach algebra with unit e and the properties that $\|a\|^2 = \|a^2\|$, $(e + a^2)^{-1}$ exists for all a. Show that **A** is isomorphic, algebraically and regards the norm, to the algebra of all continuous real-valued functions on a compact Hausdorff space.

12   a. Repeat Exercise 11, with the condition that $(e + a^2)^{-1}$ exists replaced by the condition that $\|a^2 - b^2\| \le \max(\|a^2\|, \|b^2\|)$ for all a and b.

 b. Repeat (a) with the altered condition $\|a^2 + b^2\| \ge \|a^2\|$.

13   Show that a Banach space **B** which is a topological algebra such that $\|xy\| \le c\|x\|\,\|y\|$, where c is a constant, may be renormed with an equivalent norm $\|\cdot\|'$ (i.e., one defining the same topology) such that $\|xy\|' \le \|x\|'\|y\|'$.

8.3 INTRODUCTORY FEATURES OF HILBERT SPACES

The development of the theory of commutative Banach algebras was to a large extent motivated and influenced by properties of commutative self-adjoint algebras of operators on Hilbert space, and one of the most important applications of the material in the preceding section is to the "spectral theory" of such algebras. Operator algebras are also relevant in the present context since they arise naturally at an intermediate stage in the process of representing integration algebras as algebras of functions. It is therefore appropriate to develop the elementary geometrical theory of Hilbert spaces, which depends fundamentally on simple properties of nonnegative hermitian forms.

DEFINITIONS A *hermitian form* on a real or complex vector space $\mathbf{V}$ is a function h on $\mathbf{V} \times \mathbf{V}$ which assigns to each pair of vectors (x,y) in $\mathbf{V}$ a scalar $h(x,y)$ in the relevant field in such a way that

(1) $h(cx + y, z) = ch(x,z) + h(y,z)$

(2) $h(x,y) = \bar{h}(y,x)$

for all x, y, and z in $\mathbf{V}$ and any scalar c. The *quadratic form* associated with h is the function q on $\mathbf{V}$, defined by

$$q(x) = h(x,x).$$

We turn now to some general observations about hermitian forms. For this, suppose that h is a given hermitian form on a real or complex vector space and that q is its associated quadratic form. Then it follows from (1) and (2) that

$$h(x, cy + z) = \bar{c}h(x,y) + h(x,z)$$

for any vectors x, y, and z and any scalar c, the complex conjugates appearing here and in (2) being superfluous in the real case. From the above properties of h it follows that q is real-valued and "homogeneous" in the sense that, for any scalar c and vector x,

$$q(cx) = |c|^2 q(x).$$

Moreover, for any vectors x and y,

$$q(x + y) = q(x) + 2 \operatorname{Re} h(x,y) + q(y),$$

where $\operatorname{Re} h(x,y)$ is the real part of $h(x,y)$. If we replace y by $-y$ and add the result to the original equation, we obtain the so-called *parallelogram law*,

$$q(x + y) + q(x - y) = 2q(x) + 2q(y).$$

On the other hand, by subtracting, we find that

$$\operatorname{Re} h(x,y) = \tfrac{1}{4}q(x + y) - \tfrac{1}{4}q(x - y).$$

In the complex case, the imaginary part of $h(x,y)$ is evidently related to its real part by the equation

$$\operatorname{Im} h(x,y) = \operatorname{Re} h(x,iy)$$

so that

$$h(x,y) = \tfrac{1}{4}q(x + y) - \tfrac{1}{4}q(x - y) + \tfrac{1}{4}q(x + iy) - \tfrac{1}{4}q(x - iy).$$

Thus h is always completely determined by q. The equations which express h in terms of q are called *polarization identities*.

We turn now to the consideration of nonnegative forms. A hermitian form h and its associated quadratic form q are said to be *nonnegative* if $q(x) \geq 0$ for all vectors x. As one might suspect, nonnegative hermitian forms have a number of special properties. The most distinctive property and the one we shall derive next is the *Cauchy-Schwarz inequality*.

SCHOLIUM 8.9   *Let h be a nonnegative hermitian form on a real or complex vector space, and q its associated quadratic form. Then, for all vectors x and y,*

$$|h(x,y)|^2 \leq q(x)q(y).$$

Proof The standard proof actually gives a somewhat stronger inequality, which is occasionally useful. Let x and y be fixed vectors and b any real number such that

$$b \leq \inf_{c} q(x - cy),$$

where the infimum is taken over all scalars c. We shall show that

$$|h(x,y)|^2 \leq [q(x) - b]q(y).$$

For this observe that

$$b \leq q(x - cy) = q(x) - 2 \operatorname{Re} \bar{c}h(x,y) + |c|^2 q(y).$$

Then, with $c = rh(x,y)$,

$$b \leq q(x) - 2r\,|h(x,y)|^2 + r^2\,|h(x,y)|^2\,q(y)$$

for all real r. If $q(x)$ and $q(y)$ are both 0, it follows that $h(x,y) = 0$ as well, and the desired inequality is valid trivially. Since $|h(x,y)|$ is symmetric in x and y, it suffices now to consider the case where $q(y) \neq 0$. Then, taking $r = 1/q(y)$, we find that

$$b \leq q(x) - \frac{|h(x,y)|^2}{q(y)},$$

and this implies

$$|h(x,y)|^2 \leq [q(x) - b]q(y).$$

Now let q be the quadratic form associated with a given nonnegative hermitian form h, and for any vector x, let $\|x\|$ be defined as the nonnegative square root of $q(x)$.   Then the Cauchy-Schwarz inequality may be written in the form

$$|h(x,y)| \leq \|x\| \|y\|.$$

Thus, for any vectors x and y,

$$\|x + y\|^2 = \|x\|^2 + 2 \operatorname{Re} h(x,y) + \|y\|^2$$

$$\leq \|x\|^2 + 2 \|x\| \|y\| + \|y\|^2$$

$$= (\|x\| + \|y\|)^2.$$

Taking square roots, we obtain the *Minkowski inequality*

$$\|x + y\| \leq \|x\| + \|y\|.$$

Since $\|cx\| = |c| \|x\|$, it follows that $x \to \|x\|$ is a pseudonorm on the underlying vector space.   The most important situation is that in which $x \to \|x\|$ is actually a norm, i.e., in which $\|x\| > 0$ when $x \neq 0$.

DEFINITIONS   A hermitian form h is called an *inner product* if its associated quadratic form q is *strictly positive* in the sense that $q(x) > 0$ when $x \neq 0$.   An *inner-product space* is a real or complex vector space equipped with a particular inner product h.   If x and y are vectors in such a space, the scalar $h(x,y)$ is called the *inner product* of x and y and is usually denoted simply by (x,y).

Inner-product spaces are associated with arbitrary nonnegative hermitian forms in the following natural fashion.

SCHOLIUM 8.10   *Let h be a nonnegative hermitian form on a real or complex vector space* $\mathbf{V}$, *and q its associated quadratic form.   Then the vectors w, such that $q(w) = 0$ form a subspace* $\mathbf{W}$ *of* $\mathbf{V}$, *and the equation*

$$(x + \mathbf{W}, y + \mathbf{W}) = h(x,y)$$

defines an inner product on the quotient space $\mathbf{V}/\mathbf{W}$.

Proof   The fact that $\mathbf{W}$ is a subspace of $\mathbf{V}$ is an immediate consequence of the Minkowski inequality and the homogeneity of q.   The Schwarz inequality shows that if $w \in \mathbf{W}$, then $h(v,w) = 0$ for all v in $\mathbf{V}$.   Thus, for any x and y in $\mathbf{V}$,

$$h(x + w, y + w') = h(x,y)$$

for all w and w' in $\mathbf{W}$.   This shows that the putative inner product is well defined.   The fact that it is indeed an inner product is essentially obvious.

In the context of inner-product spaces, the Cauchy-Schwarz inequality may be regarded as a special case of another important inequality concerning orthonormal sets.

DEFINITION   A subset S of an inner-product space **V** is said to be *orthonormal* in case $(x,y) = \delta_{xy}$ for all x and y in S, where $\delta_{xy} = 1$ if $x = y$ and $\delta_{xy} = 0$ if $x \neq y$.

SCHOLIUM 8.11   (*Bessel's inequality*)   *For any vector x and any orthonormal set S in an inner-product space* **V**,

$$\sum_{y \in S} |(x,y)|^2 \leq \|x\|^2.$$

Proof   Since the sum over S is the least upper bound of the corresponding sums over the finite subsets of S, it is enough to consider the case in which S is finite and consists, say, of $y_1, y_2, \ldots, y_n$.   Then the vector $u = \sum_i (x,y_i)y_i$ has the property that $(x - u, y_j) = 0$ for all j.   From this it follows that $(x - u, u) = 0$, so that

$$\|x - u\|^2 = \|x\|^2 - (u,x) = \|x\|^2 - \sum_i |(x,y_i)|^2 \geq 0.$$

The Schwarz inequality follows by homogeneity from the case in which S consists of a single element.

As one might suspect, no analytic results concerning inner-product spaces, of any significant depth, are obtainable without further restrictions of an analytic nature on the space itself.   The most natural additional requirement to impose is that of completeness relative to the metric determined by the norm.

DEFINITION   A Hilbert space is an inner-product space which is complete in the metric induced by the norm.

The first axiomatic treatment of Hilbert space was given in the late twenties by von Neumann, Hilbert himself having considered only L_2-space realizations.   In von Neumann's original definition, Hilbert spaces were required to be infinite-dimensional and separable, which fixed the space within isomorphism, but it is now generally regarded as more convenient to drop these restrictions.   It is also possible to consider Hilbert spaces over the quaternions, and even the Cayley numbers, but the theory is technically more complicated, and really novel results do not emerge; these cases will not be considered here.   A different kind of theory, employing other locally compact fields, such as the p-adic numbers, has been investigated recently for number-theoretic purposes, but the spectral theory is limited to the analysis of special completely continuous operators, and is developed in a rather different way, so that it also will not be considered here.

It should be observed that there is no essential loss of generality in restricting one's attention to Hilbert spaces.   For a given inner-product space, **V** is always isometrically isomorphic to a dense linear subset of a Hilbert space **H**, and the space **H** is essentially unique in the sense that any two such imbeddings of **V** are connected by a unitary equivalence between the corresponding Hilbert spaces. A *unitary equivalence* between two Hilbert spaces is a vector-space isomorphism preserving inner products. The space **H**, or more precisely the equivalence class of such spaces, is naturally called the *completion* of **V**.

The existence of the completion **H** may be proved by obvious modifications of the usual argument for completing any metric space. But the existence is most economically seen with the aid of results from Chap. 6. Recall that the dual of a normed linear space **V** is automatically complete relative to the metric induced by the norm. Hence any closed subset of the second dual **V**** is complete. Thus the completion of **V** as a normed linear space may be described simply as the closure **H** of the image of **V** under the isometric imbedding $x \rightarrow x$** of **V** into **V****. If **V** is an inner-product space,

$$x,y \rightarrow (x,y)$$

is a uniformly continuous function on **V** $\times$ **V**, and so extends uniquely to a continuous function on **H** $\times$ **H**. The fact that the extended function is an inner product follows easily by continuity.

If M is a measure space, the Riesz-Fischer theorem and other results of Chap. 4 show that the space $\tilde{L}_2(M)$ of square-integrable function classes on M is a Hilbert space. This space, obtained from $L_2(M)$ by forming a quotient, as in Scholium 8.10, is perhaps the most typical and frequently encountered example of a Hilbert space. On the other hand, other concrete realizations of Hilbert spaces besides the familiar L_2-spaces also arise quite naturally. We shall give a number of examples; however, the proofs that the indicated spaces are indeed Hilbert spaces involve in most instances a fair amount of analysis, which we shall omit.

Example 8.3.1   The Paley-Wiener class **H**$_2$ *in the right half plane.* Here the Hilbert space **H**$_2$ consists of all functions $f(z)$ holomorphic in Re $z > 0$ and such that

$$\int |f(x + iy)|^2 \, dy$$

is finite and uniformly bounded for $x > 0$. The norm of f may be defined as having as its square the greatest lower bound of the indicated integrals, and the inner product may be deduced from the square of the norm by polarization; alternatively, it may be defined by the equation

$$(f,g) = \lim_{\epsilon \rightarrow 0+} \int_{-\infty}^{\infty} f(\epsilon + iy)\bar{g}(\epsilon + iy) \, dy.$$

The basic theory shows that this limit actually exists. The completeness of **H**$_2$ depends upon the fact that, for the particular holomorphic functions under consideration, norm convergence implies uniform convergence over compact sets. Finally, it should be mentioned that the right half plane is a special case of a so-called "tube domain" over a convex cone, and the $\mathbf{H_2}$ theory generalizes naturally to this setting.

Example 8.3.2 The normalizable relativistic positive frequency fields. Here the Hilbert space $\mathbf{H}$ consists of certain complex-valued solutions φ of the relativistic equation

$$\square \varphi = m^2 \varphi,$$

where φ is a function of the real variables t, x, y, z, and $\square$ denotes the differential operator

$$\frac{-\partial^2}{\partial t^2} + \frac{\partial^2}{\partial x^2} + \frac{\partial^2}{\partial y^2} + \frac{\partial^2}{\partial z^2}.$$

The inner product is defined by

$$(\varphi,\psi) = \int \left(\varphi \frac{\overline{\partial \psi}}{\partial t} - \bar{\psi} \frac{\partial \varphi}{\partial t} \right) dx\, dy\, dz,$$

and one takes only those solutions φ for which (φ,φ) is finite and which satisfy the positive-frequency condition, to the effect that the Fourier transform of φ vanish for negative values of the dual to the time.

Example 8.3.3 The set $\mathbf{H}$ of all entire functions $f(z)$ in the complex plane for which

$$\int |f(z)|^2 e^{-|z|^2/4}\, dx\, dy$$

is finite, with the corresponding inner product

$$(f,g) = \int f(z)\bar{g}(z) e^{-|z|^2/4}\, dx\, dy.$$

This space and its generalization to n dimensions is useful in elementary quantum mechanics; its extension to the case of infinitely many complex variables is useful in quantum field theory.

Example 8.3.4 The set of all infinite complex matrices $A = (a_{jk})$, for which $\sum_{jk} |a_{jk}|^2$ is finite, with the inner product

$$(A,B) = \sum_{jk} a_{jk}\overline{b_{jk}}.$$

There is, incidentally, no natural and useful way in which this Hilbert space may be represented as a function space (that is, L_2-space) by means of a suitable transform.

The classification of Hilbert spaces within unitary equivalence is very simple. If $\mathbf{H}$ is a given Hilbert space, the collection of all orthonormal subsets of $\mathbf{H}$ is partially ordered by set inclusion, and has the property that any simply ordered subset has an upper bound in the collection, namely, its set-theoretic union. It follows that any orthonormal subset of **H** is contained in some maximal orthonormal set. The cardinal number of a maximal orthonormal set is called the *dimension* of **H**. This terminology is justified by the following theorem.

THEOREM 8.4   *The dimension of a Hilbert space is unique, and any two Hilbert spaces of the same dimension are isomorphic.*

Lemma 8.3.1   *If S is an orthonormal set and x a vector in the Hilbert space* **H**, *the sum* $\sum_{y \in S} (x,y)y$ *is convergent.*

Proof   This means, to be quite specific, that the corresponding sums over finite subsets of S have a limit with respect to the net of all finite subsets. By the Cauchy criterion, it suffices for the proof to show that if P and Q denote finite subsets of S, and u_P and u_Q the corresponding partial sums $u_P = \sum_{y \in P} (x,y)y$, then $\|u_P - u_Q\| \to 0$ as $P, Q \to S$. It is readily verified that

$$\|u_P - u_Q\|^2 = \sum_{y \in R} |(x,y)|^2,$$

where $R = (P - Q) \cup (Q - P)$, and these sums converge to 0 by Bessel's inequality.

If a vector x is a finite linear combination $x = \sum_i c_i y_i$ of distinct vectors y_i chosen from an orthonormal set, it is evident that $c_i = (x,y_i)$. Thus an orthonormal set is necessarily an independent set. Moreover, a maximal orthonormal set in a Hilbert space is always a *spanning set*, in the following sense:

SCHOLIUM 8.12   *If S is a maximal orthonormal set and x a vector in the Hilbert space* **H**, *then* $x = \sum_{y \in S} (x,y)y.$

Proof   We must prove that x is the limit of the partial sums $\sum_{y \in F} (x,y)y$, taken over all finite subsets F of S. By Lemma 8.3.1, these sums converge to some element x' in **H**, and it must be shown that $x' = x$. For this suppose that z is a fixed element of S. If F is a finite subset of S which contains z, it then follows that

$$\left(\sum_{y \in F} (x,y)y, \, z \right) = (x,z).$$

Thus $(x',z) = (x,z)$ by the continuity properties of the inner product. This implies $(x' - x, z) = 0$ for all z in S, and hence $x' - x = 0$. For otherwise we could construct a larger orthonormal set by adjoining the vector

$$\frac{x' - x}{\|x' - x\|}$$

to S, contradicting the maximality of S.

SCHOLIUM 8.13   *(Parseval)   For any vector x and any maximal orthonormal set S in* **H**,

$$\|x\|^2 = \sum_{y \in S} |(x,y)|^2.$$

Proof   This follows easily by continuity from Lemma 8.3.1.

SCHOLIUM 8.14   *Suppose S is a maximal orthonormal set in a Hilbert space* **H**, *and c is a scalar-valued function on S such that*

$$\sum_{y \in S} |c(y)|^2 < \infty.$$

Then the series $\sum_{y \in S} c(y)y$ converges to an element x in **H**, *and x has the property that $(x,y) = c(y)$ for all y in S.*

Proof   If F is a finite subset of S,

$$\| \sum_{y \in F} c(y)y \|^2 = \sum_{y \in F} |c(y)|^2.$$

Thus the series in question converges to an element x in **H**, by the argument based on the Cauchy criterion employed earlier in the proof of Lemma 8.3.1. Next observe that for a given z in S

$$(\sum_{y \in F} c(y)y, z) = c(z)$$

for any finite subset F of S containing z.   Thus $(x,z) = c(z)$ by continuity.

Proof of Theorem 8.4   Let S and T be maximal orthonormal sets in the Hilbert space **H**.   Then, for a given element x of **H**, (x,y) can be nonzero for at most countably many y in T, by virtue of Bessel's inequality

$$\sum_{y \in T} |(x,y)|^2 \leq \|x\|^2.$$

Thus each vector in S can have nonvanishing inner products with at most countably many elements of T.   But every element of T must have a non-vanishing inner product with some element of S, since otherwise the element could be adjoined to S to obtain a properly larger orthonormal set.   It follows that

$$n(T) \leq \aleph_0 \, n(S),$$

where $n(S)$ and $n(T)$ are the cardinal numbers of S and T.   Since the situation is symmetric between S and T, it follows that if both cardinals are infinite, then they are equal.   On the other hand, if one set, say, S, is finite, it is necessarily a basis for **H** in the algebraic sense, by Scholium 8.12.   Since T is an independent set, this implies $n(T) \leq n(S)$, and by symmetry $n(S) \leq n(T)$; so $n(S) = n(T)$ in all circumstances.

To complete the proof it is therefore enough to show that if S is any maximal orthonormal set in **H**, then **H** is unitarily equivalent to $L_2(S)$, via the map $x \to (x,\cdot)$, where S is given the discrete measure in which any one-point set has measure 1. The map is clearly linear. It is isometric by Scholium 8.13, and maps **H** onto $L_2(S)$ by Scholium 8.14.

The geometrical term orthogonal is quite suggestive, and as in the case of a finite-dimensional space, two vectors in **H**, or two sets of vectors, are called *orthogonal* in case $(x,y) = 0$ whenever x is in the first set, and y in the second.

COROLLARY 8.3.1 *For any subspace* **K** *of a Hilbert space* **H** *and element* x *of* **H**, *there is a unique decomposition* $x = y + z$, *with* y *in* K *and* z *orthogonal to* K.

Proof A subspace, i.e., a closed linear subset, of a Hilbert space is a Hilbert space in its own right, and therefore has a maximal orthonormal set. If S is a maximal orthonormal set in **K**, it is evidently contained in a maximal orthonormal set T in **H**. By Scholium 8.13,

$$x = \sum_{y \in T} (x,y)y = \sum_{y \in S} (x,y)y + \sum_{y \in T-S} (x,y)y.$$

The sum over S may be taken as y, and the sum over $T - S$ as z. If $x = y' + z'$ is another such decomposition, then $y - y' = z' - z$; the vector $y - y'$ is thus both in **K** and orthogonal to **K**, hence 0.

DEFINITION The mapping $x \to y$ is called the (orthogonal) *projection* of **H** onto **K**, or the *projection with range* **K**. A linear operator P on **H** is called a *projection* if it has this form.

COROLLARY 8.3.2 (F. RIESZ) *For every continuous linear functional f on a Hilbert space* **H**, *there exists a unique vector y such that*

$$f(x) = (x,y)$$

for all elements x of H, and y has the further property that $\|y\| = \|f\|$.

Proof We may assume that $f \neq 0$. Then the null space of f is a proper closed subspace **K** of **H**, so that by Corollary 8.3.1 there exists a nonzero vector z orthogonal to **K**. After multiplying z by a suitable constant, we may assume that $f(z) = 1$. Then $x - f(x)z$ is an element of **K** for any vector x. Since z is orthogonal to **K**, it follows that $(x,z) = f(x)(z,z)$. Setting $y = z/\|z\|^2$, we obtain an element of **H**, with the property that $f(x) = (x,y)$ for all x. If y' were another such vector, then $y - y'$ would be orthogonal to **H**, and hence 0. The Cauchy-Schwarz inequality shows that $\|f\| \leq \|y\|$, and taking $x = y/\|y\|$, we obtain the reverse inequality, $\|y\| \leq \|f\|$.

The overriding distinction between Hilbert spaces and other Banach spaces is that, in the former, there is a natural adjunction operation, or involution, on the continuous linear operators on the space.   In the case of a Banach space, the adjoint of an operator may be defined as an operator on the dual space, and as such it has many useful, although purely linear, properties.   On the other hand, in the Hilbert-space context, it is also meaningful to multiply an operator and its adjoint, and this is essential for the development of an effective spectral theory.

SCHOLIUM 8.15   *For any continuous linear operator T on a Hilbert space* **H**, *there exists a unique linear operator T^*, the adjoint of T, defined on* **H**, *such that*

$$(Tx,y) = (x,T^*y)$$

for all x and y in **H**.   *Moreover,* $\|T^*\| = \|T\|$,

$$\|T^*T\| = \|T\|^2,$$

and the map $T \to T^*$ is an involution on the algebra of all continuous linear continuous operators on **H**.

Proof   Suppose T is a continuous linear operator on **H**.   Then for any fixed vector y, $x \to (Tx,y)$ is a continuous linear functional on **H**, and hence has the form $(Tx,y) = (x,y')$.   Since y' is unique, T^* is necessarily the (well-defined) map $y \to y'$, which is easily verified to be linear.   Moreover,

$$\|T^*y\| = \sup_{\|x\|=1} (Tx,y)$$

so that $\|T^*y\| \leq \|Tx\|\|y\| \leq \|T\|\|y\|$.   This shows that $\|T^*\| \leq \|T\|$, and therefore T^* is continuous.   On the other hand, it follows from the defining equation for T^* that

$$(T^*y,x) = (y,Tx)$$

for all x and y in H.   Thus $T = (T^*)^*$, so that $\|T\| \leq \|T^*\|$, by the foregoing inequality applied to T^*.

Next observe that

$$\|T^*T\| \leq \|T^*\|\|T\| = \|T\|^2$$

and that for any vector x

$$\|Tx\|^2 = (T^*Tx,x) \leq \|T^*Tx\| \cdot \|x\|;$$

so $\|T\|^2 \leq \|T^*T\|$.

The fact that $T \to T^*$ is an involution on the algebra of all continuous linear operators on **H** now follows easily from the equation defining T^*.

DEFINITIONS   An algebra **B** of continuous linear operators on a Hilbert space **H** is said to be *uniformly closed* if it is closed as a subset of the algebra of all bounded operators on **H** in the *uniform topology*.   This is the topology defined by the metric in which the distance between any two operators is the bound of their difference. A *self-adjoint* algebra is one that is closed under adjunction, i.e., the formation of adjoints. A *C*-algebra* is a uniformly closed self-adjoint algebra of continuous linear operators on a Hilbert space.

COROLLARY 8.3.3   *A complex commutative C*-algebra with identity is isomorphic algebraically and as regards the norm and involution to the algebra of all continuous complex-valued functions on its spectrum.*

Proof It follows from the assumptions and Scholium 8.15 that any such algebra is a Banach algebra satisfying the hypotheses of Theorem 8.3.

REMARK   Much information can be derived from this result. On the other hand, it is a purely algebraic theorem, inasmuch as it says nothing about the way in which the operators in the algebra act on the underlying Hilbert space. For analytical purposes, what is needed is a spatial version of the spectral theorem, characterizing the algebra within unitary equivalence. The distinction is that two operator algebras are *algebraically equivalent* if they are related by an algebraic isomorphism preserving the norm and involution, while a *spatial isomorphism*, or *unitary equivalence*, between two operator algebras is one which can be implemented by an isomorphism between the underlying Hilbert spaces. Spatial isomorphism in this sense of two operator algebras implies their algebraic isomorphism; so Corollary 8.3.3 carries implications for the problem of determining the structure of commutative self-adjoint operator algebras within this more exacting type of isomorphism. We shall consider this question in the next section.

EXERCISES

1   Let $\mathbf{H}$ be a Hilbert space, and $\mathbf{K}$ a closed subspace of $\mathbf{H}$. If x is a vector in $\mathbf{H}$, and y is the orthogonal projection of x on $\mathbf{K}$, show that

$$\|x - y\| = \inf_{z\in\mathbf{K}} \|x - z\|.$$

2   Let $\mathbf{V}$ be an inner-product space, and $\mathbf{W}$ a linear subspace of $\mathbf{V}$. Let $x \in \mathbf{V}$, and suppose y_n in $\mathbf{W}$ are chosen so that

$$\inf_{z\in\mathbf{W}} \|x - z\| = \lim_{n} \|x - y_n\|.$$

Show that $\|y_n - y_m\| \to 0$ as $m, n \to \infty$ and that $\lim_{n} (z, x - y_n) = 0$ for every z in $\mathbf{W}$.

3   Let $\mathbf{V}$ be an inner-product space, and $\mathbf{V}^*$ its dual. Use Exercise 2 to show that the mapping $y \to f_y$ is a conjugate linear isometry of $\mathbf{V}$ onto a dense subset of $\mathbf{V}^*$, where $f_y(x) = (x,y)$, for x in $\mathbf{V}$.

4   Let **H** be a Hilbert space, and S a subset of **H**. Then the orthogonal complement of **S**, denoted by $\mathbf{S}^\perp$, is the set of all x in **H** such that $(x,y) = 0$ for all y in **S**. Show that

   a. $\mathbf{S}^\perp$ is a closed subspace of **H**.
   b. $(\mathbf{S}^\perp)^\perp$ is the closed subspace of **H** generated by **S**.

5   An orthonormal set **S** in an inner-product space **V** is *complete* if the finite linear combinations of elements of **S** are dense in **V**. Show that in a Hilbert space **H** the following statements concerning an orthonormal set $\mathbf{S} = \{e_\alpha\}$ are all equivalent.

   a. **S** is a maximal orthonormal set.
   b. **S** is complete.
   c. $x = \sum_\alpha (x,e_\alpha)e_\alpha$ for every x in **H**.
   d. $\|x\|^2 = \sum_\alpha |(x,e_\alpha)|^2$ for every x in **H**.
   e. $(x,y) = \sum_\alpha (x,e_\alpha)\overline{(y,e_\alpha)}$ for all x,y in **H**.

6   Let **V** be the space of continuous functions on [0,1], with the inner product

$$(f,g) = \int_0^1 f(x)\overline{g(x)}\, dx.$$

Show that the functions $e^{2\pi i n x}$, $n = 0, \pm1, \pm2, \ldots$, form a complete orthonormal set in **V**.

7   Construct an inner-product space **V** and a maximal orthonormal set in **V** which is not complete.

8   Suppose $f \in L_2(0,1)$ and

$$\sum_n |c_n| < \infty,$$

where $c_n = \int_0^1 f(x)e^{-2\pi i n x}\, dx$.   Show that the partial sums

$$S_N(x) = \sum_{n=-N}^{N} c_n e^{2\pi i n x}$$

converge uniformly to a continuous function g and that $f = g$ a.e.

9   If T is a bounded linear operator on an inner-product space **V**, show that

$$\|T\| = \sup_{\|x\|=\|y\|=1} |(Tx,y)|.$$

10   If T is a self-adjoint linear operator on a Hilbert space **H**, show that

$$\|T\| = \sup_{\|x\|=1} |(Tx,x)|.$$

11   Show that a projection on a Hilbert space may be characterized as a linear operator P, with the property that $P^2 = P$ and $P^* = P$.

12   Show that if W is an isometric linear operator on a Hilbert space, then WW^* is a projection, and that $WW^* = W^*W = I$ if and only if W is unitary, i.e., an isomorphism.

13   If $\mathbf{H} = L_2(0,1)$ and

$$Tf(x) = \int_0^x f(y)\,dy, \qquad 0 \le x \le 1,$$

show that $\|T\| \le 1$ and that $T^* = -T + P$, where P is the orthogonal projection on the subspace of constant functions.

14   Let $\mathbf{H}$ be a Hilbert space with a countable maximal orthonormal set $\{e_n\}$, and suppose $\{z_n\}$ is an orthonormal set such that

$$\sum_n \|e_n - z_n\|^2 < \infty.$$

Show that $\{z_n\}$ is also a maximal orthonormal set in $\mathbf{H}$.

15   Let $\mathbf{B}$ be an algebra of linear operators on a Hilbert space $\mathbf{H}$ satisfying the hypotheses, and hence the conclusion of Corollary 8.3.1.   If ϕ is an isolated point of the spectrum of S, show that there exists a nonzero vector x in $\mathbf{H}$ such that $Tx = \phi(T)x$ for all T in $\mathbf{B}$.

8.4   INTEGRATION ALGEBRAS

In this section we shall study the elementary properties of integration algebras and determine the structure of commutative integration algebras with units. In so doing we establish the connection indicated at the beginning of this chapter between the "Bernouilli" approach to probability theory, in which the random variables are dealt with directly as primary objects, and the conventional approach, which takes a space of elementary events as primary. A further reason for considering integration algebras is their utility in the treatment of the spectral analysis of commuting operators on Hilbert space.

Although the theory in its simplest, most intuitive form deals with a positive linear functional on a real commutative algebra with unit, it is technically advantageous to treat complex algebras, and for some applications it is desirable to consider algebras without units.   It is also desirable for logical clarity and general mathematical reasons to formulate the definitions in such a way that they apply to noncommutative algebras as well.

DEFINITIONS   An *integration algebra* is a system $(\mathbf{A},E)$, more precisely a system $(\mathbf{A},E,{}^*)$, in which $\mathbf{A}$ is a real or complex associative algebra, $*$ is an involution on $\mathbf{A}$, and E is a linear functional on $\mathbf{A}$ such that

(1) $E(f^*) = \overline{E(f)}$
(2) $E(f^*f) \ge 0$
(3) $E(fg) = E(gf)$
(4) $|E(g^*fg)| \le c(f)E(g^*g)$

for all f and g in $\mathbf{A}$, where $c(f)$ is positive and depends only on f.   If $(\mathbf{A},E)$ and $(\mathbf{B},F)$ are integration algebras, a *homomorphism* of $(\mathbf{A},E)$

into $(\mathbf{B},F)$ is an algebraic homomorphism, say, λ of A into B, such that $\lambda(f^*) = [\lambda(f)]^*$, and

$$E(f) = F[\lambda(f)]$$

for all f in $\mathbf{A}$.

Example 8.4.1   The most obvious example is any self-conjugate sub-algebra of the bounded integrable functions on a conventional measure space, with E as the integral and $*$ as complex conjugation.

Example 8.4.2   Let $\mathbf{A}$ be any commutative self-adjoint algebra of operators on a Hilbert space $\mathbf{H}$, and x a vector in $\mathbf{H}$.   Then the equation

$$E(T) = (Tx,x)$$

defines a linear functional E on $\mathbf{A}$, and $(\mathbf{A},E)$ is an integration algebra, the involution on $\mathbf{A}$ being the map which sends an element of $\mathbf{A}$ into its adjoint. The verification of conditions *(1)* and *(2)* is immediate; *(3)* follows by commutativity; and for *(4)* observe that

$$E(S^*TS) = (S^*TSx,x) = (TSx,Sx).$$

Then, by the Cauchy-Schwarz inequality,

$$|E(S^*TS)| \leq \|TSx\| \, \|Sx\| \leq \|T\| \, \|Sx\|^2 = \|T\| \, E(S^*S),$$

where $\|T\|$ is the bound of T.

Example 8.4.3   The most elementary noncommutative example is the algebra of all real or complex $n \times n$ matrices, with E defined as the trace and $*$ as conjugate transpose.   In this context conditions *(1)* to *(3)* are simply well-known properties of the trace function.   On the other hand, *(4)* is less well known, and its proof requires some work.

The following general result is suggested by the negligibility of null sets in conventional measure theory.

SCHOLIUM 8.16   *If* $(\mathbf{A},E)$ *is an integration algebra, there exists a homomorphism* $f \to \tilde{f}$ *of* $(\mathbf{A},E)$ *onto another such algebra* $(\tilde{\mathbf{A}},\tilde{E})$ *whose kernel consists of the elements* f *such that* $E(ff^*) = 0$.

Proof   We shall show first that the set $\mathbf{N}$ of all elements f in $\mathbf{A}$ such that $E(ff^*) = 0$ is a self-adjoint two-sided ideal in $\mathbf{A}$.   The fact that $\mathbf{N}$ is self-adjoint is an immediate consequence of condition *(3)*.   On the other hand, it is evident from *(1)* and *(2)* that the equation

$$h(f,g) = E(fg^*)$$

defines a nonnegative hermitian form h on $\mathbf{A}$.   Thus $\mathbf{N}$ is a linear subspace of $\mathbf{A}$, by Scholium 8.10.   Now observe that the Cauchy-Schwarz inequality for h, applied to the pair (f,g^*), may be written in the form

$$|E(fg)|^2 \leq E(ff^*)E(g^*g).$$

If $f \in \mathbf{N}$, it follows that $E(fg) = 0$ for any element g of $\mathbf{A}$, and in particular, $E(fgg^*f^*) = 0$. Thus $\mathbf{N}$ is a right ideal and, being self-adjoint, is automatically a left ideal as well.

It now follows by standard algebra that we obtain an integration algebra $(\tilde{\mathbf{A}}, \tilde{E})$ and a homomorphism $f \to \tilde{f}$, with the required properties, by setting $\tilde{\mathbf{A}} = \mathbf{A}/\mathbf{N}$, $\tilde{f} = f + \mathbf{N}$, $(\tilde{f})^* = (f^*)^\sim$, and $\tilde{E}(\tilde{f}) = E(f)$.

The *quotient algebra* $(\tilde{\mathbf{A}}, \tilde{E})$ has the property that $E(\tilde{f}\tilde{f}^*) > 0$ when $\tilde{f} \neq 0$, and for most integration-theoretic purposes, it is technically convenient, and involves no essential loss of generality, to replace $(\mathbf{A}, E)$ by $(\tilde{\mathbf{A}}, \tilde{E})$. In other words, we may ordinarily confine our attention to algebras $(\mathbf{A}, E)$, in which $E(ff^*) > 0$ when $f \neq 0$. We shall call such algebras *nonsingular* integration algebras.

SCHOLIUM 8.17 *If $(\mathbf{A}, E)$ is a nonsingular integration algebra, the equation*

$$(f,g) = E(fg^*)$$

defines an inner product on $\mathbf{A}$ such that

(1) $(f,g) = (g^*, f^*)$

(2) $(fg,h) = (g, f^*h)$

(3) $\|fg\|^2 \leq c(f^*f) \|g\|^2$

for all f, g, and h in $\mathbf{A}$.

Proof The fact that $(f,g) = E(fg^*)$ is an inner product satisfying *(1)* follows immediately from the nonsingularity of $\mathbf{A}$ and the definition of an integration algebra. To prove *(2)* note that

$$(fg,h) = E(fgh^*) = E[g(f^*h)^*] = (g, f^*h).$$

For *(3)* observe that

$$\|fg\|^2 = E(fgg^*f^*) = E(g^*f^*fg) \leq c(f^*f) \|g\|^2.$$

It is technically useful to know that the study of real integration algebras may always be reduced to the study of complex algebras by complexification. If $(\mathbf{A}, E, *)$ is a real integration algebra, its *complexification* is the complex integration algebra, in which the algebra is the set $\mathbf{A} \times \mathbf{A}$, with the algebraic operations

$$(f,g) + (h,k) = (f + h, g + k)$$
$$(f,g)(h,k) = (fh - gk, fk + gh)$$
$$(a + ib)(f,g) = (af - bg, bf + ag),$$

the linear functional is the map

$$(f,g) \to E(f) + iE(g),$$

and the involution sends (f,g) into $(f^*,-g^*)$. It is a matter of straightforward verification to check that the complexification of the real algebra $(\mathbf{A},E)$ is indeed a complex integration algebra in which $(\mathbf{A},E)$ is imbedded as a real subalgebra.

We turn now to some definitions that will be useful in formulating the basic results of this section.

DEFINITIONS   A *standard measure on a locally compact space* S is a regular measure m such that

$$\int_S f(x)\, dm(x) > 0$$

whenever $f \in \mathbf{C}_0(S)$, $f \geq 0$, and $f \neq 0$. A regular measure space is called *standard* if its measure is such. A *standard integration algebra* on S is an integration algebra $(\mathbf{A},E)$ in which $\mathbf{A}$ is a uniformly dense self-conjugate subalgebra of $\mathbf{C}_0(S)$, and E is the restriction to $\mathbf{A}$ of the integral defined by a standard measure on S.

Lebesgue measure is of course standard, as is Haar measure on any locally compact group.

THEOREM 8.5   *Any complex commutative nonsingular integration algebra* $(\mathbf{A},E)$ *with unit is isomorphic to a standard algebra on a compact space.*

Proof   With the inner product $(f,g) = E(fg^*)$, the algebra $\mathbf{A}$ acquires the structure of an incomplete Hilbert space whose completion will be denoted by $\mathbf{H}$. If f is a fixed element of $\mathbf{A}$, it follows from Scholium 8.17 that $g \to fg$ is a bounded linear operator on $\mathbf{A}$ which extends uniquely by continuity to a bounded linear operator L_f on $\mathbf{H}$, with the property that $(L_f)^* = L_{f^*}$. Thus the map $f \to L_f$ is an involution-preserving homomorphism of $\mathbf{A}$ onto a commutative self-adjoint algebra $\mathbf{B}$ of continuous linear operators on $\mathbf{H}$, and since $\mathbf{A}$ has a unit e, it is actually an isomorphism. Moreover, the linear functional E may be expressed in the form

$$E(f) = (L_f e, e).$$

Let $\mathbf{C}$ be the uniform closure of $\mathbf{B}$, and F the linear functional on $\mathbf{C}$ defined by $F(T) = (Te,e)$. Then $(\mathbf{A},E)$ is isomorphic, via the map $f \to L_f$, to the subalgebra $(\mathbf{B},F)$ of the integration algebra $(\mathbf{C},F)$.

Next note that for any T in $\mathbf{C}$, $F(TT^*) = F(T^*T) = \|Te\|^2$. Thus $F(TT^*) = 0$ if and only if $Te = 0$. If $Te = 0$ and T' is any element of $\mathbf{C}$, it follows by commutativity that

$$TT'e = T'Te = 0.$$

Since e is a *cyclic vector for* $\mathbf{C}$, that is, since the vectors of the form $T'e$ are dense in $\mathbf{H}$, we see that $Te = 0$ if and only if $T = 0$. Thus $(\mathbf{C},F)$ is a nonsingular integration algebra.

It should also be noted that the inequalities

$$|F(T)| \leq \|Te\|\|e\| \leq \|T\|\|e\|^2 = E(e)\,\|T\|$$

give a direct proof of the fact that F is bounded on **C**.

Now let S be the spectrum of **C**, and for T in **C**, let $\hat{T}$ be the function on S defined by

$$\hat{T}(\phi) = \phi(T).$$

Then S is a compact Hausdorff space, and by Theorem 8.3 and Corollary 8.3.3, $T \to \hat{T}$ is a norm and involution-preserving isomorphism of **C** onto $\mathbf{C}(S)$, the algebra of all continuous complex-valued functions on S. Thus if J is the linear functional defined on $\mathbf{C}(S)$ by

$$J(\hat{T}) = F(T),$$

it follows that $(\mathbf{C}(S),J)$ is an integration algebra isomorphic to $(\mathbf{C},F)$. Moreover, J is a nonnegative linear functional which is strictly positive on nonnegative functions different from 0. For suppose g is a nonnegative element of $\mathbf{C}(S)$ not identically 0. Then g has a square root h with the same properties, and $h = \hat{T}$, with T a nonzero self-adjoint operator in **C**. From this it follows that

$$J(g) = J(h^2) = F(T^*T) > 0.$$

It follows in turn that J is the integral relative to a unique finite standard measure on the Baire subsets of S. Hence $(\mathbf{C}(S),J)$ is a standard algebra on S, and setting $\hat{f} = (L_f)^{\wedge}$ for f in **A**, we obtain an isomorphism $f \to \hat{f}$ of $(\mathbf{A},E)$ onto a standard subalgebra of $(\mathbf{C}(S),J)$.

Although reasonably simple, the theorem just proved has a number of consequences of substance. Using it and some elementary ideas and methods of operator theory, one may be led to derive many of the most important results of the spectral theory of operators in Hilbert space. It enables such questions to be reduced to pure measure or integration theory, in terms of which they may be dealt with concretely by relatively elementary means. Indeed, commutative spectral theory as developed by von Neumann and others showed many parallels with integration theory—at least if they were looked for. Thus in retrospect it is not surprising that there is no need to repeat in different settings arguments reminiscent of integration theory, inasmuch as the subject largely reduces to such questions.

EXERCISES

1 Show that if **A** is the algebra of all polynomials on a bounded interval on the real line, and if $E(f)$ is the integral of f with respect to any bounded Lebesgue-Stieltjes measure, then $(\mathbf{A},E)$ is an integration algebra; but that if the interval is infinite, this is in general not the case.

2   Let **A** denote the algebra of all continuous complex-valued functions on a compact group, relative to convolution as multiplication. For any element $f \in$ **A**, let $E(f) = f(e)$, where e is the group unit. Show that $(\mathbf{A}, E)$ is an integration algebra and that it is nonsingular. Does this remain the case if the group G is noncompact and **A** is replaced by $C_0(G)$?

3*   Show that a real subalgebra of the real functions in L_∞ over a finite measure space containing 1 is dense in L_2 if and only if it separates measure-theoretically in the following sense: Every measurable set differs by a null set from a set in the minimal σ-ring with respect to which all elements of **A** are measurable.

4   If for each index j in the index set J, $\mathbf{H}_j$ is a Hilbert space, the (external) direct sum $\mathbf{K} = \overline{Z}_{j \in J} \oplus \mathbf{H}_j$ is defined as the set of all functions $x(j)$ from J to $\bigcup_j \mathbf{H}_j$ such that $x(j) \in \mathbf{H}_j$, and for which $\overline{Z}_{j \in J} \|x(j)\|^2 < \infty$, with the usual linear operations and the inner product $(x, y) = \sum_j (x(j), y(j))$ between any two elements x and y of **K**.

Show that the direct sum as thus defined is again a Hilbert space.

Now suppose that for each $j \in J$, $(\mathbf{A}_j, E_j)$ is a nonsingular integration algebra. Let **A** denote the algebra which as a linear space is the algebraic direct sum of the A_j, and whose multiplication is defined by the equation $aa' = 0$ if $a \in \mathbf{A}_j$ and $a' \in \mathbf{A}_{j'}$ with $j \neq j'$, and as in $\mathbf{A}_j$ if $j = j'$. Let E be defined on **A** by the equation: $E(\sum_j a_j) = \sum_j E_j(a_j)$ if $a_j \in \mathbf{A}_j$. Show that $(\mathbf{A}, E)$ is again an integration algebra; and that the Hilbert space **H** associated with $(\mathbf{A}, E)$ as in the proof of Theorem 8.5 is the Hilbert space direct sum of the respective Hilbert spaces $\mathbf{H}_j$ similarly associated with the $(\mathbf{A}_j, E_j)$, in a canonical way.

IX

SPECTRAL ANALYSIS
IN HILBERT
SPACE

9.1 INTRODUCTION

From the last result of the preceding chapter, a good deal of nontrivial information can be obtained about self-adjoint operators on Hilbert space, incomparably more than is available for operators in Banach spaces, other than those of special compactness properties, or for arbitrary (non-self-adjoint) operators on Hilbert space. This result fails, however, to give a structure theorem for a self-adjoint operator on a Hilbert space, i.e., a result giving a simple explicit form for the operator, within the isomorphism appropriate to Hilbert space (unitary equivalence). What one would like, of course, is an analog in Hilbert space to the diagonalization of a self-adjoint operator in a finite-dimensional unitary space. Simple examples such as the operation of multiplication by x, acting on $L_2(0,1)$, show that a direct analog does not exist: there need be no eigenvalues or eigenspaces whatever. However, this example points the way to an appropriate analog, which is that in which invariant subspaces are decomposed, not into a discrete direct sum of eigenspaces, but into a continuous direct integral of infinitesimal eigenspaces, any one of which corresponds to a point of measure zero in the formation of the direct integral.

The indicated operation of multiplication by x is already fully diagonalized from this standpoint, as well as from the standpoint of the applications of spectral theory to the theory of differential equations, quantum mechanics, etc.   More generally, if M is any measure space on which k is a real bounded locally measurable function, the operation

$$T: f \to kf \qquad f \in L_2(M)$$

is easily seen to be a bounded self-adjoint operator on $L_2(M)$; it is likewise fully diagonalized.   The spectral values or eigenvalues of this operator are the same as the essential values of the function k (i.e., the values λ such that $k - \lambda$ fails to have an essentially bounded inverse).   For any bounded Baire function F on the real line, $F(T)$ may be defined as the operation on $L_2(M)$ of multiplication by $F \circ k$, making the "operational calculus" highly explicit.   The analogs to the eigenspaces are similarly easy to describe.

To know that every self-adjoint operator on Hilbert space is unitarily equivalent to one of the form just indicated therefore renders them quite transparent.   This is true not only of one operator, but of any finite or infinite set of commuting self-adjoint operators; the same unitary transformation will simultaneously transform all the operators in any given such set into multiplication operators of the type just described.   The so-called spectral theorem, for one or for a finite number of commuting self-adjoint operators, can be read off from this result, which serves to reduce to integration theory most general questions concerning self-adjoint operators on Hilbert space.

Roughly speaking, the result just cited asserts that any set of commuting self-adjoint operators may be simultaneously diagonalized.   This may be subsumed under a more precise result giving the structure of the general maximal Abelian self-adjoint algebras of operators on a Hilbert space.   For it is not difficult to see that any such set of operators may be imbedded in some such maximal Abelian algebra.   Now if one knows that such an algebra is, within unitary equivalence, precisely the algebra of all multiplications of elements of L_2 by elements of L_∞ on some measure space, the earlier indicated result follows.   This structure theorem is indeed valid, as is essentially the converse; the algebra of all multiplications by elements of L_∞, acting on L_2, is maximal Abelian and self-adjoint provided the measure space in question is localizable (in particular, whenever it is σ-finite).   Knowing the structure of maximal Abelian self-adjoint algebras, one may then proceed to the determination of the structure of the most general Abelian self-adjoint operator algebra having certain natural closure properties, but this subject (the so-called multiplicity theory for Abelian rings of operators) is beyond the scope of this volume, as are related results on the structure of noncommutative rings of operators.

We turn now to the precise and detailed treatment of the theory indicated, and to some further developments.

9.2   THE STRUCTURE OF MAXIMAL ABELIAN SELF-ADJOINT ALGEBRAS

The main result of this section is the structure theorem indicated in the introduction.   Before stating the result, it will be illuminating to give an example of a maximal Abelian self-adjoint algebra of operators on a Hilbert space.

DEFINITION: THE MULTIPLICATION ALGEBRA OF A MEASURE SPACE   Let M be a measure space, m the measure on the σ-ring of measurable sets, and k an element of $L_\infty(M)$, that is, a complex-valued locally measurable and essentially bounded function on M.   Then, for any f in $L_2(M)$,

$$\|kf\|_2^2 = \int |k(x)f(x)|^2 \, dm(x) \leq \|k\|_\infty^2 \, \|f\|_2^2.$$

From this it follows easily that the equation

$$M_k \tilde{f} = (kf)^\sim$$

defines a bounded linear operator on the Hilbert space $\tilde{L}_2(M)$.   Moreover,

$$\|M_k\| = \|k\|_\infty.$$

To prove this it suffices to show that $\|M_k\| \geq \|k\|_\infty$, as we have already shown that $\|M_k\| \leq \|k\|_\infty$.   For this suppose that $\epsilon > 0$, and let E be a measurable set such that $0 < m(E) < \infty$ and $|k(x)| \geq \|k\|_\infty - \epsilon$ for $x \in E$.   If $f = [m(E)]^{-\frac{1}{2}}\varphi_E$, it then follows that $\|f\|_2 = 1$ and $|kf| \geq (\|k\|_\infty - \epsilon)\,|f|$.   Hence

$$\|kf\|_2 \geq (\|k\|_\infty - \epsilon)\,\|f\|_2 = \|k\|_\infty - \epsilon,$$

and since this is true for all $\epsilon > 0$, we see that $\|M_k\| \geq \|k\|_\infty$.   Next observe that if f and g are elements of $L_2(M)$, then

$$(kf,g) = \int k(x)f(x)\bar{g}(x) \, dm(x) = (f,\bar{k}g).$$

This implies the fact that $(M_k)^* = M_{\bar{k}}$.   Since $M_{kk'} = M_k M_{k'}$ and $M_{ck+k'} = cM_k + M_{k'}$ for any scalar c, it is evident that $k \to M_k$ is a homomorphism of $L_\infty(M)$ onto a self-adjoint algebra of continuous linear operators on $\tilde{L}_2(M)$.   This algebra is called the *multiplication algebra of M*, and its elements, *multiplication operators*.   Since $M_k = 0$ if and only if $\|M_k\| = \|k\|_\infty = 0$, it follows that M_k depends only upon $\tilde{k}$, and we may write $M_k = M_{\tilde{k}}$.   The above homomorphism then becomes a norm and involution-preserving isomorphism $\tilde{k} \to M_{\tilde{k}}$ of $\tilde{L}_\infty(M)$ onto the multiplication algebra.

DEFINITION   A *maximal Abelian* algebra **A** of bounded linear operators on a Hilbert space **H**—more precisely, maximal Abelian in the algebra of all bounded linear operators on **H**—is one which is Abelian (i.e., commutative) and such that any bounded linear operator commuting with all members of **A** is again an element of **A**.

SCHOLIUM 9.1   *The multiplication algebra of a finite measure space M is maximal Abelian in the algebra of all bounded linear operators on $L_2(M)$.*

Proof Let T denote a bounded linear operator on $L_2(M)$ which commutes with all multiplication operators, and set $g = T1$, where 1 stands for the function identically 1 on M. Then, since $TM_k = M_kT$ for all $k \in L_\infty(M)$, in particular, $TM_k1 = M_kT1$, that is, $Tk = gk$ for all $k \in L_\infty$. It does not immediately follow that $T = M_g$, for at this point it is not known that g is essentially bounded, but only that it is in $L_2(M)$. However, the argument given in the discussion of the multiplication algebra shows that g must be essentially bounded by $\|T\|$. Thus T and the bounded operator agree on the dense set $\tilde{L}_\infty(M)$, and hence are identical.

At this point we could go on and show that, more generally, the multiplication algebra of a measure space is maximal Abelian if and only if the space is localizable. However, the foregoing discussion will suffice to indicate the strength of the analogy between the multiplication algebras in relation to infinite-dimensional Hilbert spaces and the algebra of all diagonal matrices in relation to finite-dimensional spaces (see, however, the exercises). In its simplest form the result we are presently concerned with is

THEOREM 9.1   *Any maximal Abelian and self-adjoint algebra of bounded linear operators on a complex Hilbert space is unitarily equivalent to the multiplication algebra of some measure space.*

The proof is basically by reduction to the structure theorem for integration algebras. In the course of it, however, some results and concepts of independent interest will be found helpful.

DEFINITION   The *uniform algebra* of a regular measure space is the subalgebra of the multiplication algebra, consisting of all multiplications by bounded continuous functions on the underlying space.

If k is a bounded continuous function on a standard measure space, it is easily verified that the sup norm of k is the same as its $\tilde{L}_\infty$ norm. Thus, if **C** denotes the Banach algebra of all bounded continuous functions on the space, it follows that $k \to M_k$ is a norm and involution-preserving isomorphism of **C** onto the uniform algebra.

DEFINITION   A *cyclic vector* for a set **A** of linear operators on a topological linear space **L** is a vector z such that the finite linear combinations of the vectors of the form Az, $A \in \mathbf{A}$, form a dense subset of **L**. In case **A** is an algebra, it is the same to say that $\mathbf{A}z$ is dense in **L**.

SCHOLIUM 9.2   *A complex commutative C*-algebra* **A** *with identity and cyclic vector z is spatially isomorphic to the uniform algebra of a standard measure space on its spectrum.*

Proof   Let $\mathbf{H}_0$ be the subset of the Hilbert space **H** consisting of all vectors of the form $x = Tz$ with T in **A**.   Then $\mathbf{H}_0$ is a dense linear subset of **H**, and since **A** is commutative, $Tz = T'z$ if and only if $T = T'$.   If $E(T) = (Tz, z)$, it follows that $(\mathbf{A}, E)$ is an integration algebra satisfying the conditions of Theorem 8.5.   Hence, if S is the spectrum of **A**, there exists, by the proof of Theorem 8.5, a unique standard measure m on the σ-algebra $\mathcal{B}$ of Baire subsets of S such that

$$E(T) = \int_S \hat{T}(\phi)\, dm(\phi)$$

for all T in **A**.   Let M be the measure space $(S, \mathcal{B}, m)$.   We shall show that **H** is isomorphic to $\tilde{L}_2(M)$.   For this, observe that with each vector $x = Tz$ in $\mathbf{H}_0$, we may associate a function f_x in $C(S)$ by setting $f_x = \hat{T}$.   Then $x \to f_x$ is a well-defined linear mapping of $\mathbf{H}_0$ onto $C(S)$.   Moreover,

$$\|x\|^2 = \|Tz\|^2 = E(T^*T) = \int_S |f_x(\phi)|^2\, dm(\phi).$$

Thus the mapping $U_0 : x \to \tilde{f}_x$ is an isometry of $\mathbf{H}_0$ onto a dense subset of $\tilde{L}_2(M)$.   Because $\mathbf{H}_0$ is dense in **H**, U_0 extends uniquely by continuity to a unitary transformation U of **H** onto $\tilde{L}_2(M)$.

Now suppose $K \in \mathbf{A}$ and set $\hat{K} = k$.   Then, for $x = Tz$, we have

$$UKU^{-1}\tilde{f}_x = UKTz = (kf_x)^{\sim} = M_k\tilde{f}_x.$$

Thus the bounded operators UKU^{-1} and M_k agree on a dense subset of $\tilde{L}_2(M)$ and so are equal.   This shows that U provides a unitary equivalence

$$K \to UKU^{-1}$$

between **A** and the uniform algebra of the standard measure space M.

SCHOLIUM 9.3   *If* **B** *is any self-adjoint algebra of bounded linear operators on a Hilbert space* **H** *containing the identity operator, then* **H** *is a direct sum of subspaces left invariant by all elements of* **B**, *each subspace having a cyclic vector for the relative action of* **B** *restricted to the subspace.*

Proof   Let $\mathcal{C}$ be any collection of mutually orthogonal subspaces of **H**, each of which is invariant under **B** and has a relative cyclic vector z under **B**, that is $\mathbf{B}z$ is dense in the subspace, and which moreover is maximal with respect to this property.   The existence of such a collection follows readily by transfinite induction.   If **K** is the subspace of **H** generated by the subspaces in $\mathcal{C}$, it suffices to show that $\mathbf{K} = \mathbf{H}$.   If this is not the case, the orthogonal complement of **K**, say, **L**, is nonzero and invariant under **B**.   The invariance of **L** under **B** results from the self-adjointness of **B**.   For

$$(Tx,y) = (x,T^*y) = 0$$

whenever $T \in \mathbf{B}$, $x \in \mathbf{L}$, and $y \in \mathbf{K}$. If z is any nonzero vector in **L**, the closure of $\mathbf{B}z$ is a nonzero subspace of **L** orthogonal to every element of $\mathcal{C}$, and so if adjoined to $\mathcal{C}$ would yield a properly larger collection with the same general properties as $\mathcal{C}$, contradicting its assumed maximality.

We turn now to the problem of generalizing the result of Scholium 9.2 to algebras without cyclic vectors.

SCHOLIUM 9.4 *Any commutative self-adjoint set of bounded linear operators on a complex Hilbert space is unitarily equivalent to a subset of the uniform algebra of a standard measure space.*

Proof Let **C** be a commutative self-adjoint set of bounded linear operators on a complex Hilbert space **H**, and **A** the complex uniformly closed algebra generated by **C** and the identity operator.   It follows by continuity that **A** is both commutative and self-adjoint.   By Scholium 9.3 **H** is a direct sum of subspaces $\mathbf{H}_i$ $(i \in I)$ invariant under **A** with relative cyclic vectors z_i. Each of the uniformly closed restriction algebras

$$\mathbf{A}_i = \overline{\mathbf{A} \,|\, \mathbf{H}_i}$$

is a complex commutative C^*-algebra with identity and cyclic vector z_i.

Let S_i be the spectrum of $\mathbf{A}_i$, and set $R = \bigcup_i S_i$.   Then, since the S_i are mutually disjoint, there is a unique well-defined topology on R in which a subset V is open if and only if $V \cap S_i$ is an open subset of S_i for all i.   In this topology R is a locally compact space in which each set S_i is both open and compact.   A unique function f is defined on R by any selection of functions f_i on S_i $(i \in I)$.   The function f is continuous if and only if the same is true of each f_i, and the statement that f has compact support is equivalent to the assertion that $f_i \neq 0$ for at most finitely many i.

By Scholium 9.2 $\mathbf{A}_i$ is unitarily equivalent to the uniform algebra of a standard measure space M_i whose underlying space is S_i and in which the measure m_i has the property that

$$\| T_i z_i \|^2 = \int_{S_i} |\hat{T}_i(r)|^2 \, dm_i(r)$$

for all T_i in $\mathbf{A}_i$.   Thus the equation

$$J(f) = \sum_i \int_{S_i} f(r) \, dm_i(r)$$

defines an integral on $\mathbf{C}_0(R)$ such that $J(f) > 0$ when $f \geq 0$ and $f \neq 0$. Hence there is a unique standard measure m on the ring of Baire sets and corresponding measure space M on R such that J is the integral relative to m.

We shall show next that $\mathbf{H}$ is isomorphic to $\tilde{L}_2(M)$. For this, let $\mathbf{H}_0$ denote the linear subset of $\mathbf{H}$ consisting of all vectors of the form $x = \sum_i T_i z_i$, where $T_i \in A_i$ and $T_i \neq 0$ for at most finitely many i. If

$$x = \sum_i T_i z_i = \sum_i T_i' z_i,$$

it follows that $T_i z_i = T_i' z_i$, and hence that $T_i = T_i'$. Thus, with each such vector $x = \sum_i T_i z_i$, we may associate a function f_x in $\mathbf{C}_0(R)$ by setting

$$f_x(r) = \hat{T}_i(r)$$

for $r \in S_i$. Now $x \to f_x$ is evidently a linear mapping U_0 of $\mathbf{H}_0$ onto $\mathbf{C}_0(R)$, and moreover,

$$\|x\|^2 = \sum_i \|T_i z_i\|^2 = \sum_i \int_{S_i} |f_x(r)|^2 \, dm_i(r) = \int_R |f_x(r)|^2 \, dm(r).$$

Since $\mathbf{H}_0$ is dense in $\mathbf{H}$, and $\mathbf{C}_0(R)$ dense in $L_2(M)$, it follows that U_0 defines a unitary map U of $\mathbf{H}$ onto $\tilde{L}_2(M)$.

To complete the proof it suffices to show that UAU^{-1} is a subset of the uniform algebra of M. For this purpose, let T be an element of $\mathbf{A}$, and suppose that $k_i = (T \mid H_i)^\wedge$. Then k_i is a continuous function on S_i, and there is a unique continuous function k on R which coincides with k_i on S_i. Since $\|T\| \geq \|T \mid H_i\|$, it follows that k is a bounded function on R. Now suppose that $x = \sum_i T_i z_i$ is a vector in H_0. Then

$$UTU^{-1}(\tilde{f}_x) = UTx = U \sum_i TT_i z_i = U \sum_i (T \mid H_i) T_i z_i = (kf_x)^\sim = M_k(\tilde{f}_x),$$

and therefore $UTU^{-1} = M_k$. This shows that UAU^{-1} is a subset of the uniform algebra, and completes the proof.

We may observe that the measure space M above can be constructed quite explicitly as the direct sum of the measure spaces M_i without the use of the representation theory of integrals on locally compact spaces.

> DEFINITION A measure space $M = (R,\mathfrak{R},m)$ is said to be the *direct sum* of an indexed collection $M_i = (S_i,\mathcal{S}_i,m_i)$ $(i \in I)$ of *finite* measure spaces if there exists an isomorphism θ_i of M_i into M such that the sets $\theta_i(S_i)$ are mutually disjoint and have union R, and if in addition every element of $\mathfrak{R}$ meets at most countably many of the $\theta_i(S_i)$.

If (M',θ_i') defines another such direct sum, then there is an isomorphism between M and M', exchanging θ_i and θ_i' in an obvious fashion. In fact, the direct sum may be constructed in the following way when the S_i are mutually disjoint. The set R is the union of the S_i; the σ-ring $\mathfrak{R}$ consists of all sets expressible as countable unions of elements of the rings S_i; and

$$m(E) = \sum_i m(E \cap S_i)$$

for any $E \in \mathcal{R}$.

The direct sum of any collection of finite regular measure spaces is again regular and standard if each summand has this property.

Proof of Theorem 9.1   By Scholium 9.4, the given self-adjoint and maximal Abelian algebra **A** is unitarily equivalent to a subalgebra of the multiplication algebra of some measure space M, via a unitary transformation U.   Now since **A** is maximal Abelian, so also is UAU^{-1}, which means that it contains every bounded linear operator on $\tilde{L}_2(M)$ which commutes with all its elements. Since any two multiplications by bounded measurable functions commute, it follows that UAU^{-1} contains all such multiplications, and hence is precisely the multiplication algebra of M.

As a direct corollary of Theorem 9.1 one obtains the more conventional and materially weaker form of the spectral theorem, involving integrals relative to spectral measures.   In developing such applications it is useful to bear in mind

SCHOLIUM 9.5   *Any self-adjoint set* **S** *of commuting bounded linear operators on a complex Hilbert space* **H** *is contained in some maximal Abelian and self-adjoint algebra of such operators.*

Proof   Let **B** denote the algebra generated by **S**; then by transfinite induction, **B** is contained in an algebra **A** of bounded linear operators on **H** which is maximal with respect to being Abelian and self-adjoint.   However, this implies that it is also maximal Abelian, for if T commutes with every element of **A**, so also does T^*, from which it follows that the self-adjoint operators $(T + T^*)/2$ and $(T - T^*)/2i$, at least one of which is nonzero if T is nonzero, commute with every element of **A**, in contradiction with its maximality.

The presentation of the conventional form of the spectral theorem requires some development of the integral calculus relative to a projection-valued measure.

DEFINITIONS   Suppose S is a set and **B** is a self-conjugate complex algebra of bounded functions on S which is closed under pointwise convergence of sequences and contains the unit function $e(x) = 1$. We shall then define a *spectral integral on* **B** as a linear map P of **B** to the bounded linear operators on a complex Hilbert space **H** such that

(1) $P(fg) = P(f)P(g)$.

(2) $P(f)^* = P(\bar{f})$.

(3) $P(e)$ is the identity operator I.

(4) If $f_n \to f$ pointwise and $\sup_n \|f_n\| < \infty$, then

$$\|P(f)x - P(f_n)x\| \to 0$$

for all x in **H**, where the norm of a function f in **B** is the sup norm

$$\|f\| = \sup_{x \in S} |f(x)|.$$

We also assume—although it can be shown to hold automatically, for suitable algebras **B**—

(5) $\|P(f)\| \leq \|f\|$.

We point out that condition *(4)* is independent of the other conditions and may be formulated as the statement that P is continuous relative to appropriate topologies on **B** and the bounded linear operators on **H**.   The topology on **B** is that of *bounded sequential convergence* in which a subset **F** is closed if **F** contains f whenever $f_n \in \mathbf{F}$, $\sup_n \|f_n\| < \infty$, and $f_n \to f$ pointwise. All operations of **B**, including the involution, are continuous in this topology. The appropriate topology on the bounded operators is the *strong operator topology*, defined as the weakest one, in which all the maps $T \to Tx$, $x \in \mathbf{H}$, are continuous.   With this topology, the operators form a locally convex topological vector space.   But in general, $T \to T^*$ is not continuous, and the product TT' is only separately continuous in its factors, although jointly continuous on any norm-bounded subset.

Now let $\mathcal{S}$ denote the collection of all subsets E of S whose characteristic functions φ_E are elements of **B**.   It then follows readily from the methods and results of Chap. 3 that $\mathcal{S}$ is a σ-algebra of sets and that **B** is precisely the class of bounded functions measurable with respect to $\mathcal{S}$.   If P is a spectral integral on **B**, the map Q on $\mathcal{S}$ taking a set E into $P(\varphi_E)$ then has the following properties:

(6) $Q(E \cap F) = Q(E)Q(F)$

and $Q(E)Q(F) = 0$ if E and F are disjoint.

(7) $Q(E)^* = Q(E) = Q(E)^2$.

(8) $Q(S) = I$.

(9) If $E = \bigcup_n E_n$, where the E_n are mutually disjoint, then

$$Q(E) = \sum_n Q(E_n)$$

in the sense that

$$Q(E)x = \sum_n Q(E_n)x$$

for all x in **H**.

Thus the operators $Q(E)$ form a commutative family of projections on $\mathbf{H}$ which contains the identity operator and has the property that $Q(E)$ and $Q(F)$ project onto orthogonal subspaces of $\mathbf{H}$ whenever E and F are disjoint elements of S.   Moreover, if $E = \bigcup_n E_n$, with the E_n mutually disjoint, it follows from condition (9) that $Q(E)$ is the projection on the closed subspace of $\mathbf{H}$ generated by the ranges of the $Q(E_n)$.

Any map Q from a σ-algebra of sets to the projections on a complex Hilbert space which satisfies conditions (6) to (9) is called a *projection-valued measure*, or simply a *spectral measure*.   If S is a locally compact space, we shall use the term *spectral integral on S* for a spectral integral on the algebra of bounded unrestricted Baire functions on S, and the term *spectral measure on S* for a spectral measure on the σ-algebra of unrestricted Baire sets.

If Q is the spectral measure on S arising as above from a spectral integral P, and f is any element of $\mathbf{B}$, the operator $P(f)$ is called the *integral of f relative to Q* and is denoted by

$$\int_S f(x)\, dQ(x).$$

The basic calculus, or set of rules, for manipulating such integrals is for the most part simply a translation of properties (*1*) to (*5*) of P.

> COROLLARY 9.2.1   *Let $\mathbf{A}$ be a complex commutative C^*-algebra with identity, and S the spectrum of $\mathbf{A}$.   Then there exists a unique spectral integral P on the algebra $\mathbf{B}$ of bounded Baire functions on S such that*
>
> $$T = P(\hat{T}), \qquad T \in \mathbf{A},$$
>
> *where $T \to \hat{T}$ is the canonical isomorphism of $\mathbf{A}$ onto $\mathbf{C}(S)$; and such that for all vectors x, the measure $A \to (P(c_A)x,x)$ is regular.   Moreover, the operators $P(f)$, $f \in \mathbf{B}$, commute with all bounded linear operators commuting with every member of $\mathbf{A}$.*

Proof   By Scholium 9.4 we may assume that $\mathbf{A}$ is a subalgebra of the uniform algebra of a standard measure space M on a locally compact space R.   Then each point r of R determines an element

$$\phi_r : M_k \to k(r)$$

of S.   Let λ be the mapping $r \to \phi_r$.   Then since the topology on S is the weak topology defined by the functions

$$\hat{M}_k : \phi \to \phi(M_k), \qquad M_k \in \mathbf{A},$$

and since $\hat{M}_k \circ \lambda = k$, a (bounded) continuous function on R, we see that λ is a continuous mapping of R into S.   If f is a bounded Baire function on S, it follows that $f \circ \lambda$ is a bounded unrestricted Baire function on R, and setting $P(f) = M_{f \circ \lambda}$, we obtain a spectral integral P with the required property. To verify that P is continuous in the appropriate sense, suppose that the sequence of Baire functions f_n converges pointwise and boundedly to f. Then for any g in $L_2(M)$,

$$\|P(f)\tilde{g} - P(f_n)\tilde{g}\|^2 = \int_R |f(\phi_r) - f_n(\phi_r)|^2 \, |g(r)|^2 \, dm(r),$$

which tends to 0 as n goes to infinity, by the Lebesgue dominated convergence theorem.

To show the uniqueness of P, suppose that P' is another such spectral integral. Then $P'(f) = P(f)$ for all f in $\mathbf{C}(S)$; so to show that $P' = P$, it suffices by continuity to prove that $\mathbf{C}(S)$ is dense in $\mathbf{B}$ relative to the given topology. The crucial observation for this is in fact that on bounded subsets of $\mathbf{B}$ the topology of bounded sequential convergence coincides with the topology of sequential convergence used in the definition of the class of Baire functions. This reduces the problem to the elementary question of showing that the class of continuous functions on S, bounded by a given constant, is dense in the corresponding class of Baire functions.

Suppose now that T is a bounded linear operator commuting with all members of $\mathbf{A}$, and let $\mathbf{B}_0$ be the set of all f in $\mathbf{B}$ such that $TP(f) = P(f)T$. Then $\mathbf{B}_0$ contains $\mathbf{C}(S)$ and is evidently closed in the topology of bounded sequential convergence. This shows that $\mathbf{B}_0 = \mathbf{B}$, and completes the proof.

A bounded linear operator on a Hilbert space is said to be *normal* if it commutes with its adjoint. The conventional form of the spectral theorem for a normal operator is given as follows:

COROLLARY 9.2.2   *For any bounded normal operator T on a complex Hilbert space* $\mathbf{H}$, *there exists a unique spectral integral*

$$f \to \int f(\lambda) \, dE(\lambda)$$

from the algebra of bounded Baire functions on the disk $|\lambda| \leq \|T\|$ to the bounded operators on $\mathbf{H}$ *such that*

$$T = \int \lambda \, dE(\lambda).$$

Moreover, the operators $\int f(\lambda) \, dE(\lambda)$ commute with all bounded linear operators commuting with T and T^*.

Proof Let $\mathbf{A}$ be the uniformly closed algebra generated by T, T^*, and the identity operator. Then $\mathbf{A}$ is both commutative and self-adjoint. Thus if S is the spectrum of $\mathbf{A}$, there exists a spectral integral P on S with the properties indicated in Corollary 9.2.1. Since $\hat{T}$ is a continuous function on S with values in the disk $D:|\lambda| \leq \|T\|$, it follows immediately that the equation $F(f) = P(f \circ \hat{T})$ defines a spectral integral F on D. Let E be the spectral measure associated with F. Then

$$F(f) = \int f(\lambda)\, dE(\lambda),$$

and in particular,

$$T = \int \lambda\, dE(\lambda).$$

If L is a bounded linear operator commuting with T and T^*, it necessarily commutes with all members of $\mathbf{A}$, and by Scholium 9.4, L commutes with the operators $\int f(\lambda)\, dE(\lambda)$.

Now suppose F' is another spectral integral with the same properties as F. Then as F and F' agree on the identity function and its complex conjugate, it follows that they agree on all complex polynomial functions in λ and $\bar{\lambda}$. Since these functions form a self-conjugate algebra of continuous functions on D, which separates points and includes the constant functions, it follows by the Stone-Weierstrass theorem and continuity that F and F' agree on $\mathbf{C}(D)$. Since $\mathbf{C}(D)$ is dense in the algebra of bounded Baire functions, relative to the topology of bounded sequential convergence, we see by continuity that $F' = F$.

REMARKS   (*1*) With regard to the result just proved, if T is self-adjoint, then $\hat{T}$ has values in the real interval $[-\|T\|, \|T\|]$. Thus, by an obvious modification, the spectral integral may be defined on $[-\|T\|, \|T\|]$ rather than on the disk $|\lambda| \le \|T\|$, and this change involves no loss of generality.

(2) If T is a unitary operator on the space $\mathbf{H}$, then $TT^* = T^*T = I$, so that $\hat{T}$ has its values in the unit circle $|\lambda| = 1$. Hence the spectral integral associated with T is naturally defined on the unit circle.

For the sake of completeness, we now prove the rather elementary result that any spectral measure gives rise to a spectral integral.

SCHOLIUM 9.6   *If Q is a projection-valued measure on a σ-algebra $\mathfrak{R}$ of subsets of a set S, there is a unique spectral integral P on the bounded functions measurable with respect to $\mathfrak{R}$ such that $P(\varphi_E) = Q(E)$ for all E in $\mathfrak{R}$.*

Proof   We shall first define P on the algebra $\mathbf{A}$ of complex-valued step functions based on $\mathfrak{R}$. If $f \in \mathbf{A}$ and the distinct values a_i of f are assumed on the sets E_i in $\mathfrak{R}$, let $P(f) = \sum_i a_i Q(E_i)$. Then $P(f)^* = \sum_i \bar{a}_i Q(E_i) = P(\hat{f})$, and for any x in the Hilbert space $\mathbf{H}$,

$$\|P(f)x\|^2 = \sum_i |a_i|^2\, \|Q(E_i)x\|^2 \le \|f\|^2 \sum_i \|Q(E_i)x\|^2 \le \|f\|^2\, \|x\|^2.$$

Thus $\|P(f)\| \leq \|f\|$, and by the argument used in the proof of Scholium 2.1, it follows that P is a linear map from **A** to the bounded linear operators on **H** such that $P(\varphi_E) = Q(E)$ for all E in $\mathfrak{R}$. With f as above and E in $\mathfrak{R}$,

$$f\varphi_E = \sum_i a_i \varphi_{E_i \cap E}$$

so that $\quad P(f\varphi_E) = \sum_i a_i Q(E_i \cap E) = \sum_i a_i Q(E_i)Q(E) = P(f)P(\varphi_E).$

From this it follows that $P(fg) = P(f)P(g)$ for all f and g in **A**.

The family **B** of bounded functions on S measurable with respect to $\mathfrak{R}$ is a Banach algebra relative to the sup norm, and by Example 3.2.2, **A** is dense in **B**. Since P is uniformly continuous on **A**, it follows that P has a unique continuous extension to all of **B**. By an elementary continuity argument, the extended P is a *-homomorphism of **B** to the bounded linear operators on **H** such that $\|P(f)\| \leq \|f\|$ for all f in **B**.

To show that P is continuous relative to the topology of bounded sequential convergence, observe that for each x in **H**, the equation

$$m_x(E) = (Q(E)x,x)$$

defines a finite countably additive measure on $\mathfrak{R}$ such that

$$(P(f)x,x) = \int_S f(t)\,dm_x(t)$$

for all f in **A**. But since **A** is dense in **B**, it follows immediately that this holds for all f in **B** as well. If $f_n \in$ **B** and $f_n \to f$ pointwise and boundedly, it follows from the *-homomorphism property of P that

$$\|P(f)x - P(f_n)x\|^2 = (P(|f - f_n|^2)x,\, x) = \int |f(t) - f_n(t)|^2\, dm_x(t),$$

and the integral tends to 0 as n goes to infinity.

One useful observation about operator theory is that a given operator T on a Hilbert space **H** is determined, at least implicitly, by the form $x,y \to (Tx,y)$. Thus, if the bounded linear operators T and T' have the property that $(Tx,y) = (T'x,y)$ for all x,y in **H**, or simply for all x,y in some subset generating **H**, then $T = T'$. Conversely, we have the following result:

SCHOLIUM 9.7   *If $L(x,y)$ is a form defined for the vectors x and y in a dense linear subset **V** of a complex Hilbert space **H**, which is linear in x, conjugate linear in y, and bounded in the sense that*

$$|L(x,y)| \leq \text{const } \|x\|\,\|y\|$$

*for all x and y, then there exists a unique bounded operator T on **H** such that*

$$L(x,y) = (Tx,y).$$

Proof   For the proof, observe that $x \to L(x,y)$ is for any fixed y a bounded linear functional on **V** which extends to a continuous linear functional on **H**, and hence has the form $x \to (x,y')$ for some unique y' in **H**. It is easily seen that the map $y \to y'$ is linear and bounded on **V**, and the adjoint of its extension to **H** is the desired operator T.

As an application we shall obtain the following result concerning spectral integrals.

> SCHOLIUM 9.8   *If S is a locally compact space, a norm-decreasing *-homomorphism P' from $C_0(S)$ to the bounded linear operators on a complex Hilbert space H, such that $P'[C_0(S)]H$ generates H, extends uniquely to a spectral integral P on the one-point compactification of S.*

Proof   First observe that by uniform continuity P' extends uniquely to a norm-decreasing *-homomorphism on the Banach algebra $C_\infty(S)$ of all continuous functions vanishing at ∞ on S. If S' is the one-point compactification of S, every function f in $C(S')$ decomposes uniquely as a sum $f = g + \lambda e$, where $g \in C_\alpha(S)$, and e is the unit function on S'. Setting

$$P'(f) = P'(g) + \lambda I,$$

we then obtain a *-homomorphism from $C(S')$ to the bounded linear operators on **H**. Moreover, $\|P'(f)\| \leq \|f\|$ for all f in $C(S')$, as is easily seen by representing the operators $P'(f)$ as continuous functions on a compact space.

Next, note that any pair of vectors x,y in **H** defines a complex integral

$$J'_{xy} : f \to (P'(f)x,y)$$

such that

$$|J'_{xy}(f)| \leq \|f\| \, \|x\| \, \|y\|$$

for all f in $C(S')$. We wish to show now that J'_{xy} extends uniquely to the integral J_{xy}, defined by a complex measure on the Baire subsets of S'. For this we could refer to the general facts about complex integrals stated in Exercises 2 and 3, Sec. 5.2. But in the present situation one can proceed more directly. For J'_{xx} is positive, and so is the restriction to $C(S')$ of the integral defined by a unique regular measure m_x on the Baire sets. The measure m_x is finite, and in fact

$$\left| \int_{S'} f(t) \, dm_x(t) \right| \leq \|f\| \, m_x(S') = \|f\| \, \|x\|^2$$

for any bounded Baire function f. If f is a real-valued function in $C(S')$, the form $x,y \to (P'(f)x,y)$ is hermitian, and hence determined by the polarization identities. This suggests that J_{xy} be defined as the integral relative to the complex measure m_{xy}, specified by the equation

$$4m_{xy} = m_{x+y} - m_{x-y} + im_{x+iy} - im_{x-iy}.$$

On the algebra **B** of bounded Baire functions, J_{xy} is then continuous relative to the topology of bounded sequential convergence and, as readily verified, such that $J_{xy} = J'_{xy}$ on $C(S')$.   Since $C(S')$ is dense in **B** with respect to the indicated topology, it follows that J_{xy} is the unique continuous functional extending J'_{xy} to **B**.   Writing $J(x,y)$ for J_{xy}, we now see that

$$J(cx + y, z) = cJ(x,z) + J(y,z)$$

on $C(S')$ and by continuity that this equation also holds on **B**.   By a similar argument J_{xy} is conjugate linear in y.   Moreover, for f in $C(S')$,

$$|J_{xy}(f)| \leq \|f\| \, \|x\| \, \|y\|,$$

and this relation also extends by continuity to **B**.   For a fixed f in **B**, it results that $x,y \to J_{xy}(f)$ is a bounded form on **H**, linear in x, and conjugate linear in y.   By Scholium 9.7, there exists a unique bounded linear operator, which we shall denote by $P(f)$, such that

$$J_{xy}(f) = (P(f)x,y)$$

for all x,y in **H**.   The map $P: f \to P(f)$ is then obviously linear, and $P(f) = P'(f)$ for f in $C(S')$.   Again utilizing appropriate continuity arguments involving the topology of bounded sequential convergence on **B**, one shows that $P(f)^* = P(\bar{f})$ and that $P(fg) = P(f)P(g)$ on **B**, starting from the fact that these relations hold on $C(S')$.   The continuity of P now follows, as in the proof of Scholium 9.6.   It remains to prove that if e is the unit function, then $P(e) = I$.   For this observe that $P(e)P(f) = P(f)$ for all f in **B**.   Since $P[C_0(S)]$**H** generates **H**, it follows that necessarily $P(e) = I$.

EXERCISES

1   If T is a normal operator with spectral resolution $T = \int \lambda \, dE(\lambda)$ and f is a bounded Baire function defined on the support of E, the operator $\int f(\lambda) \, dE(\lambda)$ is frequently denoted by $f(T)$.

   a. Show that if T is a multiplication operator M_k on L_2 over a measure space, then $f(T)$ is $M_{f \circ k}$.

   b. Show that $f(T)$ is always normal, and that if $|f(\lambda)| = 1$ for all λ, then $f(T)$ is unitary.

   c. Show that every unitary operator U may be expressed as e^{iH}, with H a bounded self-adjoint operator, and that this may be done in infinitely many ways.

2   The *spectrum* of a bounded linear operator T is sometimes defined as the set of all values λ for which $\lambda I - T$ fails to have a bounded (two-sided) inverse in the algebra of all bounded operators on the space.

   a. If T is normal, show that its spectrum as just defined is the same as the essential range of any bounded measurable function representing T in any diagonalization, i.e., representation of T as a multiplication operator on an L_2-space.

    b. Show that a point λ is in the spectrum of a normal operator T if and only if for every $\epsilon > 0$ there exists a unit vector x such that $\|Tx - \lambda x\| < \epsilon$.

3   Let $\mathbf{H} = L_2(C)$, where C is the unit circle group under Haar measure. Show that the transformation $f(x) \to f(e^{i\theta}x)$ is unitary for real θ, and determine its spectrum.   (*Hint:* Use Fourier series.)

4   If A and B are bounded self-adjoint operators on a Hilbert space $\mathbf{H}$, $B \leq A$ means that $(Bx,x) \leq (Ax,x)$ for all $x \in H$.

    a. Show that $A \geq 0$ if and only if $A = B^2$ for some self-adjoint operator B.

    b. If A is self-adjoint and $A \geq 0$, show that there is a unique self-adjoint $B \geq 0$ such that $A = B^2$.

    c. If T is a bounded linear operator on a Hilbert space, show that T has a *polar decomposition* $T = UR$, where R is self-adjoint, $R \geq 0$, and U is a partial isometry.   [*Hint:* Let $R = (T^*T)^{\frac{1}{2}}$.]

5   Let $\mathbf{A}$ be a complex commutative Banach algebra with unit and spectrum S. The *joint spectrum* of a *finite sequence of elements* $x_1, x_2, \ldots, x_n$ *in* $\mathbf{A}$ is then defined as the image S' of S under the mapping

$$\phi \to (\phi(x_1),\phi(x_2), \ldots ,\phi(x_n)).$$

    a. Show that the joint spectrum of $x_1, x_2, \ldots, x_n$ is a compact subset of C^n.

    b. If $\mathbf{A}$ is generated by $x_1, x_2, \ldots, x_n$ and the unit, show that S is homeomorphic to S' via the given map.

    c. If $\mathbf{A}$ is a complex commutative C^*-algebra generated by the normal operators $T_1,T_1^*,\ T_2,T_2^*,\ \ldots,\ T_n,T_n^*$ and the identity, show that S is also homeomorphic to its image S'' under the mapping

$$\phi \to [\phi(T_1),\phi(T_2), \ldots ,\phi(T_n)].$$

    d. With $\mathbf{A}$ as in (*c*), let $\mathbf{B}$ denote the subalgebra of $\mathbf{A}$ generated algebraically by $T_1,T_1^*,\ T_2,T_2^*,\ \ldots,\ T_n,T_n^*$. If a general point of S'' is denoted by $\lambda = (\lambda_1, \lambda_2, \ldots, \lambda_n)$, show that $\mathbf{B}$ is isometrically *-isomorphic to the algebra of all complex polynomial functions in $\lambda_1,\bar{\lambda}_1,\ \lambda_2,\bar{\lambda}_2, \ldots,\ \lambda_n,\bar{\lambda}_n$ on S'', with 0 constant term, and that this is a self-conjugate algebra separating the points of S''. If $\bar{\mathbf{B}}$ is the uniform closure of $\mathbf{B}$, deduce that $\bar{\mathbf{B}} = \mathbf{A}$ or that S'' contains the point 0, and that $\bar{\mathbf{B}}$ is isomorphic to the algebra of all continuous functions on S'' vanishing at 0.

    e. With $\mathbf{A}$ as in (*c*), show that there is a unique spectral measure Q on S'' such that $T_i = \displaystyle\int_{S''} \lambda_i\, dQ(\lambda)$ for $i = 1, 2, \ldots, n$.

    f. With Q as in *e*, show that the operators

$$L_n = \int_{S''} \min\,(1,n\,|\lambda|)\, dQ(\lambda)$$

form a monotone-increasing sequence $L_1 \leq L_2 \leq \cdots$ of self-adjoint elements of $\bar{\mathbf{B}}$ which converges strongly to the projection P on the closed subspace of the Hilbert space **H**, generated by the set **B(H)**.   Construct an example in which $\overline{\mathbf{B}} \neq \mathbf{A}$ and $P = I$.

6   Let E be a spectral measure on the reals, and for any real t, let $U(t) = \int e^{it\lambda} \, dE(\lambda)$.   Show that $U(t + t') = U(t)U(t')$ and that $U(t)$ is a strongly continuous function of t (such a function is called a *continuous one-parameter unitary group*).

7   The *weak-operator topology* on the bounded linear operators on a Hilbert space **H** is the weakest topology in which all the maps $T \rightarrow (Tx,y)$ are continuous.

> *a.* Show that the strong- and weak-operator topologies on the group of all unitaries on a Hilbert space are the same.
>
> *b.* Give an example to show that the uniform topology on the unitaries is distinct from these.
>
> *c.* Show that in either topology the group of unitaries is topological [i.e., the map $(U,V) \rightarrow UV^{-1}$ is continuous].

8   If **A** is a maximal Abelian self-adjoint algebra of bounded linear operators on a complex Hilbert space **H**, show that **A** contains the projection on any subspace of **H** invariant under **A**.

9   Suppose **A** and **A'** are maximal Abelian self-adjoint algebras (m.a.s.a.) on **H** and **H'** and that $T \rightarrow T'$ is an algebraic isomorphism of **A** onto **A'**.   If U is an isometry of **H** into **H'** such that $UT = T'U$ for all T in **A**, show that U maps **H** onto **H'**, that is, that U is unitary.

10   Show that the multiplication algebra of a direct sum of finite measure spaces is maximal Abelian.

11   Show that in general the multiplication algebra of a measure space is maximal Abelian if and only if the dual of L_1 is canonically isomorphic with L_∞.

12   Show that two m.a.s.a. algebras on Hilbert spaces are unitarily equivalent if and only if they are algebraically isomorphic.   (*Hint:* First consider the case in which one of the measure spaces is finite.)

13   Show that any m.a.s.a. algebra is closed in the weak-operator topology.

14   Show that a weakly closed self-adjoint algebra with a cyclic vector is m.a.s.a.

15   Show that any m.a.s.a. algebra on a separable Hilbert space has a cyclic vector.   More generally, if the m.a.s.a. algebra **A** is countably decomposable in the sense that any collection of mutually orthogonal projections in **A** is at most countable, show that **A** has a cyclic vector.

16   A standard regular measure space is called *perfect* in case, for every bounded measurable function k, there exists a bounded continuous function k' such that k and k' differ only on a local null set.

> *a.* Show that in Scholium 9.2, the measure space in question is perfect if and only if **A** is maximal Abelian.
>
> *b.* Show that in Theorem 9.1, the measure space in question may be taken to be the direct sum of finite perfect spaces.

   c. Let M be a finite perfect measure space.   Show that
     (i) The measure of every open set is positive.
     (ii) The closure of any open set is open.
     (iii) Any set of measure zero is of first category.

17   The *point spectrum* of an operator T on a Hilbert space **H** is the name sometimes given to the set of all values λ for which there exists a vector $x \neq 0$ in **H** such that $Tx = \lambda x$.   The vector x is then called a proper (or eigen-) vector for T belonging to the proper (or eigen-) value λ.   The operator T is said to have *pure point spectrum* if its proper vectors span the space.

Now let **H** be the Hilbert space of all square-integrable functions on the reals modulo 1, and let K denote the operator

$$f(x) \rightarrow \int k(x - y) f(y)\, dy,$$

where k is a given element of **H**.   Show that K has pure point spectrum and that the spectral (eigen-) values of K are simply the Fourier coefficients of k, and determine the corresponding proper functions.

18   An operator T is said to be *compact* if it maps bounded sets onto sets whose closures are compact.   Show that a compact self-adjoint operator T on a Hilbert space has pure point spectrum, and in fact that

    a. There exists a maximal orthonormal set consisting of proper vectors for T.
    b. For any $\epsilon > 0$ there exist only finitely many elements of such a maximal orthonormal system for which the corresponding proper values exceed ϵ in absolute value.

19   If T is an operator on a Hilbert space, show that T is compact if and only if T^*T is compact.

20   Show that the uniform limit of a sequence of compact operators is compact.

21   Let $K(x,y)$ be a square integrable function on $M \otimes M$, M being a given measure space.   Let T be the linear transformation defined on $L_2(M)$ by

$$Tf(x) = \int_M K(x,y) f(y)\, dy.$$

    a. Show that T maps L_2 into L_2 and that $\|T\| \leq \|K\|_2$.
    b. Show that

$$T^*f(x) = \int_M \overline{K(y,x)} f(y)\, dy$$

and that $T = T^*$ if and only if

$$K(x,y) = \bar{K}(y,x) \qquad \text{a.e. on } M \otimes M.$$

    c. Show that T is compact.   [*Hint:* Consider the expansion of K relative to an appropriate orthonormal base for $L_2(M \otimes M)$.]

d. If T is self-adjoint, show that there exist scalars $\lambda_n \to 0$ and a countable orthonormal system $\{g_n\}$ in $L_2(M)$ such that

$$K(x,y) = L_2 - \lim_{i < n} \sum \lambda_i g_i(x)\overline{g_i(y)}.$$

22   Prove the following form of the *Peter-Weyl* theorem (see Exercises 3 and 17). For any compact group G, $L_2(G)$ is the direct sum of finite-dimensional subspaces of continuous functions, each of which is invariant and irreducible under the operations

$$L(a): f(x) \to f(a^{-1}x)$$

for all a in G.   [*Hint:* Let $\mathbf{H}$ be the direct sum of a maximal collection of mutually orthogonal such subspaces.   If $\mathbf{H}$ is not all of $L_2(G)$, choose an element $f \neq 0$ which is orthogonal to $\mathbf{H}$, set $g = f^* * f$, where $f^*(x) = \overline{f(x^{-1})}$, and consider the operator

$$T: h \to h * g, \qquad h \in L_2.$$

Then observe that (i) $T \neq 0$; (ii) T is compact and self-adjoint; (iii) T commutes with all the $L(a)$; infer that the $L(a)$ leave invariant the eigenspace belonging to any eigenvalue of T.]

23   A self-adjoint operator is said to have *absolutely continuous spectrum* in case the corresponding spectral measure E is such that the measures $m(B) = (E(B)x,x)$ are absolutely continuous with respect to Lebesgue measure for all vectors x.

    a. Give an example of an operator whose point spectrum is empty which does not have absolutely continuous spectrum.

    b. Show that the operation of multiplication by a bounded continuously differentiable function on $L_2(-\infty, \infty)$ has absolutely continuous spectrum, provided the derivative of the function is nowhere vanishing.

24   Show that the operation $f \to k * f$ on $L_2(-\infty, \infty)$, where k is an integrable function, is compact only if $k = 0$, and determine its spectrum.   Show, in particular, that if k has compact support, then the point spectrum is empty.   Construct an example in which the point spectrum is not empty.

25   For any set S of bounded linear operators on a Hilbert space $\mathbf{H}$, the *commutor* of S, denoted S', is defined as the set of all bounded operators on $\mathbf{H}$ which commute with each element of S.

    a. Show that S' is always weakly and strongly closed, and self-adjoint provided S is such.

    b. Show S is Abelian if and only if $S \subseteq S'$, and maximal Abelian if and only if $S = S'$.

    c. Show that $S \subseteq S''$.

26   Show that the algebra of finite linear combinations of the translation operators

$$U(t): f(x) \to f(x + t)$$

on $L_2(-\infty, \infty)$ is dense in the strong-operator topology in a maximal Abelian self-adjoint algebra.

X

GROUP REPRESENTATIONS AND UNBOUNDED OPERATORS

10.1 REPRESENTATIONS OF LOCALLY COMPACT GROUPS

Among a number of important developments emerging from the foregoing discussions none is more logical and significant than the extension of spectral theory to unbounded operators, and related developments in group-representation theory. The theorem of Stone, giving the structure of the general one-parameter continuous unitary group on a Hilbert space, connects with both these matters and is intrinsically important. Its analogs for groups which are either compact or locally compact Abelian are central for harmonic analysis on these groups.

This chapter is devoted to the foundations of the theory of unitary representations of locally compact groups including such results as those just indicated, and to the treatment of unbounded operators. This important material serves also to indicate some of the fruitful connections between the results of Chaps. 7 and 9 and to show how these results are typically applied to more structured situations.

In this first section some theorems on group representations will be proved. In order to put these and other results in an appropriate general setting, it is helpful to have

DEFINITIONS   A *representation* of a topological group G on a topological linear space $\mathbf{V}$ is a map $a \to T(a)$ from G to the continuous linear operators on $\mathbf{V}$ such that

$$T(ab) = T(a)T(b), \qquad T(e) = I,$$

for all a and b in G, where e denotes the group unit, and I the identity operator.   If $\mathbf{V}$ is a normed linear space, such a representation is called *continuous* in the *uniform, strong,* or *weak* topology according as the map is continuous from G to the operators on $\mathbf{V}$ in the relevant topology.

For infinite-dimensional spaces $\mathbf{V}$, the uniformly continuous representations are, however, far too special to be of much interest; for example, the analytically fairly protypical, although very simple, representation described in Exercise 26, Sec. 9.2, is not uniformly continuous.   For unitary representations on a Hilbert space $\mathbf{H}$, that is, those for which $T(a)$ is unitary for every a, the notions of strong and weak continuity agree, and the convention will be adopted that, without further qualification, *continuous representation* means *strongly continuous representation.*   If T and T' are representations of G on the Hilbert spaces $\mathbf{H}$ and $\mathbf{H}'$, then T' is *unitarily equivalent* to T if there exists a unitary transformation U of $\mathbf{H}$ onto $\mathbf{H}'$ such that

$$UT(a)U^{-1} = T'(a)$$

for all a in G.   A continuous unitary representation T is *irreducible* if there are no nonzero proper subspaces of the Hilbert space invariant under each $T(a)$.

One important method for constructing unitary representations is given in the following example.

Example 10.1.1   Suppose G is a topological transformation group acting on the left on the underlying topological space of a regular measure space M which leaves the measure invariant.   Then Scholium 7.1 may be reformulated as the statement that the translations

$$L_a : f(x) \to f(a^{-1}x)$$

define a continuous isometric representation of G on $L_p(M)$, or more precisely, on $\dot{L}_p(M)$.

Thus every locally compact group G has at least one strongly continuous unitary representation.   For G acts on itself by left translation, and we may take M as the measure space determined by a left Haar measure on the ring of Baire sets.   The unitary representation corresponding to the case $p = 2$ is called the *left-regular representation.*

In the study of representations of locally compact groups, as in a variety of other analytic situations, it is frequently useful to employ a smoothing method, often referred to in other contexts as *regularization*.   One of its main applications with regard to representation theory may be formulated as follows:

THEOREM 10.1   *For any continuous unitary representation T of a locally compact group G, the map*

$$T: f \rightarrow \int_G f(b)T(b)\, db$$

*is a norm-decreasing *-homomorphism of $L_1(G)$, as an algebra under convolution with respect to left Haar measure, onto an algebra of operators whose strong closure contains $T(G)$.   Moreover, for any f in $L_1(G)$,*

$$T(a)T(f) = T(L_a f)$$

for all a in G.

Proof   From the strong continuity of T it is readily verified that for any x, y in the Hilbert space $\mathbf{H}$,

$$a \rightarrow (T(a)x, y)$$

is a continuous function on G.   Since $|(T(a)x, y)| \leq \|x\| \, \|y\|$, it follows from Scholium 9.7 that the operator-valued integral

$$T(f) = \int_G f(b)T(b)\, db$$

may be defined weakly for any f in $L_1(G)$ by the condition that

$$(T(f)x, y) = \int_G f(b)(T(b)x, y)\, db$$

for all x, y in $\mathbf{H}$.   It follows then that $\|T(f)\| \leq \|f\|_1$.   Moreover, T is obviously linear on $L_1(G)$, and

$$(T(L_a f)x, y) = \int f(a^{-1}b)(T(b)x, y)\, db$$

$$= \int f(b)(T(a)T(b)x, y)\, db$$

$$= (T(f)x, T(a)^* y) = (T(a)T(f)x, y)$$

so that $T(a)T(f) = T(L_a f)$.   With $f^*(x) = \bar{f}(x^{-1})\, \Delta(x^{-1})$, similar elementary manipulations with the inner product justify the computation

$$\left[\int f(b)T(b)\, db \right]^* = \int \overline{f(b)}T(b)^*\, db = \int \bar{f}(b^{-1})\, \Delta(b^{-1})T(b)\, db,$$

showing that $[T(f)]^* = T(f^*)$.   Analogously, one shows by means of the

Fubini theorem that

$$\int f * g(b)T(b)\,db = \int\left[\int f(a)g(a^{-1}b)\,da\right]T(b)\,db$$

$$= \int f(a)\left[\int g(a^{-1}b)T(b)\,db\right]da$$

$$= \int f(a)\left[\int g(b)T(a)T(b)\,db\right]da$$

$$= \left[\int f(a)T(a)\,da\right]\left[\int g(b)T(b)\,db\right]$$

i.e., that $T(f * g) = T(f)T(g)$.

Next let $\epsilon > 0$, and N be the strong basic neighborhood of the identity operator I determined by ϵ and the vectors $x_1, x_2, \ldots, x_n$ in **H**.   Then there exists a neighborhood V of the identity e in G such that

$$\|T(b)x_i - x_i\| \leq \epsilon$$

for all b in V and any i.   Now choose a nonnegative function f in $C_0(G)$ whose support is contained in V such that

$$\int f(b)\,db = 1.$$

Then for any vectors x, y in **H**,

$$(T(f)x - x, y) = \int f(b)(T(b)x - x, y)\,db,$$

and hence

$$|(T(f)x - x, y)| \leq \sup_{b \in V} \|T(b)x - x\|\, \|y\|.$$

For $y = T(f)x - x$, we have

$$\|y\| \leq \|T(f)x\| + \|x\| \leq \|f\|_1\|x\| + \|x\| = 2\|x\|.$$

Thus, in particular,

$$\|T(f)x_i - x_i\|^2 \leq 2\epsilon\|x_i\|$$

for all i.

Since $T(a)$ is unitary and $T(a)T(f) = T(L_a f)$, it follows that

$$\|T(L_a f)x_i - T(a)x_i\|^2 \leq 2\epsilon\|x_i\|$$

for all a in G.   In view of the arbitrary nature of ϵ, this shows that $T(G)$ is contained in the strong closure of $T[L_1(G)]$ and completes the proof.

Example 10.1.2   Consider the one-parameter unitary group U on L_2 of the reals, with the property that

$$U(t):g(s) \to e^{its}g(s).$$

Then, for any f in L_1 and any g,h in L_2,

$$(U(f)g,h) = \int_{-\infty}^{\infty} f(t)\left[\int_{-\infty}^{\infty} e^{its}g(s)\bar{h}(s)\,ds\right]dt$$

$$= \int_{-\infty}^{\infty} g(s)\bar{h}(s)\left[\int_{-\infty}^{\infty} e^{its}f(t)\,dt\right]ds.$$

Hence $U(f)$ is the multiplication operator $M_{\hat{f}}$, where $\hat{f}$ is the Fourier transform of f defined by

$$\hat{f}(s) = \int_{-\infty}^{\infty} e^{its}f(t)\,dt.$$

SCHOLIUM 10.1   *If L is the left-regular representation of the locally compact group G, then for any f in $L_1(G)$, the operator*

$$L(f) = \int_G f(a)L_a\,da$$

is left convolution by f, and $L(f) = 0$ if and only if $f = 0$ a.e.

Proof   For f in $L_1(G)$ and g,h in $L_2(G)$,

$$(L(f)g,h) = \int_G f(a)\left[\int_G g(a^{-1}b)\bar{h}(b)\,db\right]da.$$

Since $f(a)g(a^{-1}b)\bar{h}(b)$ is measurable as a function on $G \otimes G$ and

$$\int |f(a)|\left[\int |g(a^{-1}b)\bar{h}(b)|\,db\right]da < \infty,$$

it follows from the Fubini theorem that

$$(L(f)g,h) = \int \bar{h}(b)\left[\int f(a)g(a^{-1}b)\,da\right]db.$$

By Scholium 7.2 the convolution

$$k(b) = \int f(a)g(a^{-1}b)\,da$$

exists almost everywhere and defines an element of $L_2(G)$.   Since

$$(L(f)g,h) = (k,h)$$

for all h in $L_2(G)$, it results that $L(f)g = k$ a.e.   If $L(f) = 0$ and g is any bounded measurable function vanishing outside a measurable set of finite measure, the function

$$k(b) = \int f(a)g(a^{-1}b)\,da$$

is continuous and 0 a.e.   As the complement of a Haar null set is dense, this implies the fact that k is identically 0.   Hence

$$\int f(a)g(a)\, da = 0$$

for every measurable g supported by a set of finite measure, and therefore $f = 0$ a.e.   Since $L(f)$ is evidently 0 when $f = 0$ a.e., this completes the proof.

It should be noted that the first part of the proof applies equally well to the more general situation considered in Example 10.1.1.

The particular convenience of $L_1(G)$ as a generalization of the notion of the "group ring" of a finite group, as treated in algebra, is indicated by the validity of a converse to Theorem 10.1.

THEOREM 10.1′   *Let ϕ be any continuous *-representation of $L_1(G)$, as a convolution algebra, by operators on a Hilbert space $\mathbf{H}$, having the (nontriviality) property that if $\phi(f)x = 0$ for all $f \in L_1(G)$, where x is a vector in $\mathbf{H}$, then $x = 0$.   Then there exists a unique continuous unitary representation U of G on $\mathbf{H}$ such that $\phi(f) = \int U(a)f(a)\, da$.*

Proof   Observe first that if $f \in L_1(G)$, $z \in \mathbf{H}$, and $a \in G$, then with the notation $f_a(x) = f(a^{-1}x)$, the following equality is valid: $\|\phi(f)z\| = \|\phi(f_a)z\|$. For $\|\phi(f_a)z\|^2 = (\phi(f_a)z, \phi(f_a)z) = (\phi(f_a)^*\phi(f_a)z, z)$;   now $(f_a)^*(x) = \bar{f}(a^{-1}x^{-1})\Delta(x^{-1})$, so that

$$((f_a)^* * f_a)(x) = \int \bar{f}(a^{-1}y^{-1})\, \Delta(y^{-1}) f(a^{-1}y^{-1}x)\, dy$$

$$= \int \bar{f}(a^{-1}y) f(a^{-1}yx)\, dy = \int \bar{f}(y) f(yx)\, dy,$$

which is independent of the value of a.

Now suppose that z is a cyclic vector for $\phi[L_1(G)]$, and let $\mathbf{H}_0 = \phi(L_1(G))z$; $\mathbf{H}_0$ is then a linear submanifold which is dense in $\mathbf{H}$.   The mapping

$$U_0(a): \phi(f)z \to \phi(f_a)z$$

is, by the preceding paragraph, single-valued from $\mathbf{H}_0$ into $\mathbf{H}_0$ and is, in fact, an isometry.   It therefore extends uniquely to an isometry $U(a)$ of all of $\mathbf{H}$ into $\mathbf{H}$.   Since $U_0(a)$ has the inverse $U_0(a^{-1})$, its range is dense, and $U(a)$ is therefore unitary.   By virtue of the continuity of ϕ and the continuity of f_a as a function of a, the mapping $a \to U(a)\phi(f)z$ is continuous from G into $\mathbf{H}$.   Thus $U(a)w$ is a continuous function of a for each vector w in a dense class, and it follows readily from the unitarity of the $U(a)$ that the continuity is valid for all vectors w; this means that U is continuous in the strong operator topology.

From the definition of $U_0(a)$ it follows that $U_0(ab) = U_0(a)U_0(b)$; it follows in turn that $U(ab) = U(a)U(b)$; thus U is a continuous unitary representation of G.   Now if $\phi'(f)$ denotes the integral $\int U(a)f(a)\,da$, then

$$\phi'(f)\phi(g)z = \int \phi(g_a)zf(a)\,da = \phi(f * g)z = \phi(f)\phi(g)z;$$

thus the continuous linear operators $\phi'(f)$ and $\phi(f)$ agree on the dense set $\mathbf{H}_0$.   They must therefore be identical, which means that $\phi(f) = \int U(a)f(a)\,da$.

The proof is therefore complete for the case in which a cyclic vector exists. In the general case, $\mathbf{H}$ is a direct sum of cyclic invariant subspaces under $\phi(L_1(G))$, by a simple argument employed earlier; relative to each such subspace, there is a corresponding continuous unitary representation, and the direct sum of these representations is the required one, as is easily seen; its uniqueness follows in the same fashion as in the cyclic case.

> **REMARK**   It follows that ϕ is necessarily of bound not greater than 1. Actually, the assumption that ϕ is a self-adjoint homomorphism of $L_1(G)$ is sufficient to establish this fact, for this assumption automatically implies that ϕ is continuous (see Exercise 24 below).

Theorem 10.1$'$ enables the determination of the spectrum of the L_1-algebra of an arbitrary locally compact Abelian group in relatively explicit terms. A *character* of a topological Abelian group is defined as a continuous homomorphism of G into the multiplicative group of all complex numbers of absolute value 1.

> **COROLLARY** 10.1.1   *Every nonzero multiplicative linear functional ϕ on the convolution algebra $L_1(G)$ of a locally compact Abelian group G has the form*
>
> $$\phi(f) = \int \lambda(a)f(a)\,da$$
>
> *for some character λ of G.*

Proof Let f be any element of $L_1(G)$ such that $\phi(f) \neq 0$.   Defining $\lambda(a) = \phi(f_a)/\phi(f)$, it follows readily that $\lambda(ab) = \lambda(a)\lambda(b)$; that $\lambda(a)$ is a continuous function of a (noting the automatic continuity of ϕ, compare Corollary 8.2.1); and that λ is bounded on G, which implies, by virtue of the multiplicativity of λ, that $\lambda(a)$ is of absolute value 1 for all $a \in G$.   If $g \in L_1(G)$, then $f_a * g = g_a * f$, from which it follows that if $\phi(g) \neq 0$, then $\lambda(a) = \phi(g_a)/\phi(g)$ also; it follows in turn, by a simpler version of an argument in the foregoing proof, that $\phi(f) = \int \lambda(a)f(a)\,da$ for all $f \in L_1(G)$.

The other major class of locally compact groups, besides the Abelian ones, for which representation theory is particularly simple, are the compact groups. The general features of the representation theory for compact groups are maximally parallel, among topological groups, to the representation theory for finite groups; in particular, we have

THEOREM 10.2 *Every continuous unitary representation of a compact group is a direct sum of irreducible finite-dimensional such representations.*

Proof Let G be a compact group, and T a continuous unitary representation of G on a complex Hilbert space **H**. Let **K** be the direct sum of a maximal collection of mutually orthogonal finite-dimensional subspaces irreducible and invariant under T. Such a collection is easily seen to exist by transfinite induction. Now let x be any vector orthogonal to **K**, and **W** any finite-dimensional subspace of $L_2(G)$ irreducible and invariant under the left-regular representation. Since G is compact, its measure is finite and may be normalized as 1. Then, by the Cauchy-Schwarz inequality,

$$\|f\|_1 \leq \|f\|_2 < \infty$$

for any $f \in L_2(G)$. Thus the set **W**$'$ of all vectors of the form $T(f)x$ with f in **W** is a finite-dimensional subspace of **H**, invariant under T by virtue of the relation $T(a)T(f) = T(L_a f)$. Since **W** is irreducible under the left-regular representation, it is readily seen that **W**$'$ is irreducible under T, and hence that **W**$'$ consists only of 0.

According to the form of the Peter-Weyl theorem outlined in Exercise 22, Sec. 9.2, any f in $L_2(G)$ may be decomposed as a sum $f = \sum_n f_n$ of pairwise orthogonal functions f_n from subspaces such as **W**. If $g_n = \sum_{i \leq n} f_i$, it follows that $T(g_n)x = 0$. Since $\|f - g_n\|_2 \to 0$, $\|T(f) - T(g_n)\| \to 0$ as well. Hence $T(f)x = 0$ for all f in $L_2(G)$. On the other hand, for any $\epsilon > 0$, there exists a continuous function f such that $\|T(f)x - x\| < \epsilon$; so $x = 0$.

10.2 REPRESENTATIONS OF ABELIAN GROUPS

In the case of a unitary representation of a noncompact Abelian group, such as the additive group of the reals, the new phenomenon arises, essentially that of continuous spectrum, that there may be no irreducible invariant subspaces, except in a highly modified sense. Certain "infinitesimal" irreducibly invariant submanifolds may be identified, and the entire Hilbert space becomes the *direct integral* rather than the direct sum of such subspaces. In general, a direct integral splits into two parts, one involving a *continuous direct sum*, and the other a discrete sum. Without making these notions precise, we mention that an L_2-space in which the measure has no discrete part may be regarded as a fairly typical example of a continuous direct sum of one-dimensional spaces. The reduction into "irreducible parts," or diagonalization, of a continuous unitary representation of the additive group of the reals may now be formulated precisely, as follows:

THEOREM 10.3   (STONE)   *Any continuous one-parameter unitary group on a complex Hilbert space is unitarily equivalent to a group of multiplication operators of the form*

$$U(a): g(x) \to e^{iah(x)}g(x)$$

on $L_2(M)$ for some standard measure space M and real measurable function h thereon.

It should be mentioned that the algebraic outlook adopted in the discussion above is really ex post facto; Stone's theorem originates in concrete analytical and applied problems. However, the additive group of the reals is the simplest and most important continuous group, whose theory is basic to the study of more complicated groups, and the theorem of Stone has many abstract mathematical ramifications.

Proof of Theorem 10.3.   By Theorem 10.1, the set of all operators of the form $U(a)$ for some a and

$$U(f) = \int_{-\infty}^{\infty} f(a)U(a)\, da$$

for some integrable function f on the reals is a self-adjoint commutative set, and so is simultaneously diagonalizable.   It is no essential loss of generality to assume that the Hilbert space $\mathbf{H}$ in question is $L_2(M)$, where M is a standard measure space on a locally compact space R, and that $U(a)$ is the multiplication operator defined by a bounded continuous function u_a on R.   From the group property

$$U(a + b) = U(a)U(b)$$

and the unitarity of $U(a)$ it results that $u_{a+b}(x) = u_a(x)u_b(x)$ and that $|u_a(x)| = 1$ for all x in R.   Thus the function λ_x on the reals, defined by the equation

$$\lambda_x(a) = u_a(x),$$

is a unitary character of the additive group of the reals; if this character is continuous, it is of the form

$$\lambda_x(a) = e^{iah(x)}$$

for a uniquely determined real number $h(x)$.

We shall use Theorem 10.1 to investigate the continuity of the characters λ_x.   For this purpose let f be any Lebesgue-integrable function on the reals. Then, since $U(b)U(f) = U(L_b f)$, it results that

$$u_b(x)v_0(x) = v_b(x)$$

for all x in R, where v_b is the bounded continuous function defining the multiplication operator $U(L_b f)$.   Now

$$|v_b(x) - v_0(x)| \leq \|U(L_b f) - U(f)\| \leq \|L_b f - f\|_1 \to 0$$

as $b \to 0$.   Thus $\lambda_x(b)v_0(x)$ is a continuous function of b at $b = 0$, and hence λ_x is similarly continuous for all x such that $v_0(x) \neq 0$; being a character, λ_x is then continuous everywhere.

In general, R may contain points x such that $v_0(x) = 0$ for all f.   However, if F is the set of all such x, its complement is an open subset S, and the functions g in $\mathbf{C}_0(R)$ which vanish outside S are dense in $L_2(M)$.   For if g is any function in $\mathbf{C}_0(R)$ and $\epsilon > 0$, there exists an integrable f such that $\| U(f)g - g \|_2 < \epsilon$.   From this it follows that the points of F (if any) may be "removed" from R, that is, that we may assume F is empty.   For let $\mathcal{S}$ be the subring of measurable sets E on $M = (R,\mathfrak{R},m)$ such that $E \subset S$, set $M' = (S,\mathcal{S},m)$, and give S the induced topology.   Then M' is a standard measure space, and the mapping $W:g(x) \to g(x) \,|\, S$ is a unitary transformation from $L_2(M)$ onto $L_2(M')$, which carries $U(a)$ into the operator:

$$U'(a):g(x) \to u'_a(x)g(x),$$

where $u'_a = u_a \,|\, S$; and $U'(f)$ into multiplication by $v_0 \,|\, S$ when $U(f)$ is multiplication by v_0.

Thus we may assume that F is empty, and hence that the continuous characters λ_x have the form

$$\lambda_x(a) = e^{iah(x)},$$

where h is a real-valued function on R.   Since $h(x) = \lim_{n \to \infty} -i\, n[\lambda_x(n^{-1}) - 1]$, it follows that h is measurable.

From its present formulation one may readily derive the conventional form of Stone's theorem.

COROLLARY 10.2.1   *For any continuous one-parameter unitary group U there exists a spectral measure E on the reals such that*

$$U(a) = \int_{-\infty}^{\infty} e^{ia\lambda}\, dE(\lambda).$$

Proof   There is a unitary transformation V such that $VU(a)V^{-1}$ has the form given in the statement of the theorem.   Then simply define E by the condition that

$$\int_{-\infty}^{\infty} g(\lambda)\, dE(\lambda) = V^{-1}M_{g \circ h}V$$

for any bounded Baire function g.

It is perhaps not clear that the real line plays two distinct roles in the above result, one as the additive group of the reals and the other as its dual, or character, group.   The distinction is brought out when one generalizes the corollary to the situation in which the additive group of the reals is replaced by an arbitrary locally compact Abelian group.

DEFINITIONS   The *character group* G^* of a locally compact Abelian group G is defined as the set of all continuous characters of G, as a group relative to the usual pointwise product of functions, and equipped with the *compact-open* topology: a basic neighborhood $N(\lambda_0, C, \epsilon)$ of a point λ_0 in G^* depends on an arbitrary compact set C in G and positive number ϵ, and consists of all points $\lambda \in G^*$ such that $|\lambda(a) - \lambda_0(a)| < \epsilon$ for all $a \in C$. It is straightforward to verify that G^* forms a topological group, whose further properties are explicated by

Lemma 10.2.1 G^* *is a locally compact Abelian group.   The mapping from G^* to the spectrum of the convolution algebra $L_1(G)$, which is defined by the equation*

$$\lambda \to \phi_\lambda, \qquad \phi_\lambda(f) = \int \lambda(a) f(a) \, da,$$

is a homeomorphism.

Proof   The continuity of the mapping $\lambda \to \phi_\lambda$ is a consequence of the regularity of Haar measure on G and a simple estimate of the integral defining ϕ_λ. The inverse map is well defined by virtue of Corollary 10.1.1. To show the continuity of this inverse map, note that, by an elementary compactness argument, for any Banach space **B** with dual **B***, the set of all functions of $f \in \mathbf{B}^*$ of the form $f \to f(x)$, where x ranges over a compact subset of **B**, is equicontinuous on bounded sets. In particular, for any element $f \in L_1(G)$ and compact subset C of G, the functions $\phi(f_a)$ of the point ϕ in the spectrum of $L_1(G)$, where f is fixed and a ranges over the compact subset C of G, are equicontinuous. Now the inverse map carries a given point ϕ into the character λ, given by the equation $\lambda(a) = \phi(f_a)/\phi(f)$, f being so chosen that $\phi(f) \neq 0$. It follows that for any character $\lambda_0 \in G^*$, the difference $\lambda(a) - \lambda_0(a)$ may be made uniformly small throughout C by choosing the corresponding ϕ_λ in a sufficiently small w^*-neighborhood of ϕ_{λ_0}, which means that the inverse map is continuous. Since the spectrum is locally compact, so also is G^*.

COROLLARY 10.2.2   *Any continuous unitary representation U of a locally compact Abelian group G on a complex Hilbert space* **H** *has the form*

$$U(a) = \int_{G^*} \lambda(a) \, dE(\lambda)$$

for some spectral measure E on the character group G^*.

Proof   The proof of Corollary 10.2.2 applies essentially without change except for the part which depends on the specific form of the continuous characters on the reals. As before, we may assume that the Hilbert space **H** is $L_2(M)$ for some standard measure space M, and that $U(a)$ is multiplication by a bounded continuous function u_a. Defining the character λ_x by $\lambda_x(a) = u_a(x)$, one may prove by a similar argument that it may be assumed that

$\lambda_x(a)$ is jointly continuous in a and x. Specifically, observe that with the notation used above,

$$|v_a(x) - v_b(x)| \le \|L_a f - L_b f\|_1.$$

It follows that

$$|v_a(x) - v_b(y)| \le |v_a(x) - v_b(x)| + |v_b(x) - v_b(y)|$$

$$\le \|L_a f - L_b f\|_1 + |v_b(x) - v_b(y)|,$$

from which it results that $v_a(x)$ is jointly continuous in a and x. Now, for a given x_0 outside the negligible set F indicated earlier, there exists an f such that $v_0(x_0) \ne 0$. Then $v_0(x) \ne 0$ for all x sufficiently close to x_0, and since

$$\lambda_x(a)v_0(x) = v_a(x),$$

the joint continuity of $v_a(x)$ in a and x implies that of $\lambda_x(a)$.

It now follows by a compactness argument that for any point x_0, any compact set K in G, and any $\epsilon > 0$, there exists a neighborhood N of x_0 such that $|\lambda_x(a) - \lambda_{x_0}(a)| < \epsilon$ for all x in N and all a in K. Indeed, for any element a of K, there exists a neighborhood V_a of a and a neighborhood N_a of x_0 such that $|\lambda_x(a') - \lambda_{x_0}(a)| < \epsilon/2$ for $a' \in V_a$ and $x \in N_a$. A finite number V_{a_i} of such neighborhoods cover K. Let N be the intersection of the N_{a_i}. Then, for $a \in K$, say, $a \in V_{a_i}$, and $x \in N$,

$$|\lambda_x(a) - \lambda_{x_0}(a)| \le |\lambda_x(a) - \lambda_{x_0}(a_i)| + |\lambda_{x_0}(a_i) - \lambda_{x_0}(a)| < \epsilon.$$

This means that the mapping $x \to \lambda_x$ is continuous from M into G^*.

If B is an unrestricted Baire subset of G^*, $E(B)$ is defined as multiplication by the inverse image of B under this map, and E is then a spectral measure with the indicated property.

The uniqueness of the spectral measure in the generalized Stone theorem is implied by a result of some general utility.

Lemma 10.2.2 *If m is a (bounded) complex Baire measure on G^* such that $\int \lambda(a)\, dm(\lambda) = 0$ for all $a \in G$, then $m = 0$.*

Proof If $f \in L_1(G)$, then by Fubini's theorem,

$$\int f(a)\left[\int \lambda(a)\, dm(\lambda)\right] da = \int \hat{f}(\lambda)\, dm(\lambda),$$

where

$$\hat{f}(\lambda) = \int \lambda(a)f(a)\, da;$$

thus $\int \hat{f}(\lambda)\, dm(\lambda) = 0$ for all $f \in L_1(G)$. Now the mapping $f \to \hat{f}$ is the Gelfand homomorphism of an element of a Banach algebra into the corresponding function on the spectrum of the algebra, within the identification of G^*, with this spectrum provided by Lemma 10.2.1. It follows from the Stone-Weierstrass theorem, in view of the evident self-conjugacy of the set of all $\hat{f}$, that the latter set is uniformly dense in the algebra $C(G^*)$ of all continuous functions vanishing at infinity on G^*. Hence $\int g(\lambda)\,dm(\lambda) = 0$ for all $g \in C(G^*)$, implying that $m = 0$.

COROLLARY 10.2.3 *The spectral measure E described in Corollary 10.2.2 is unique.*

Proof If $U(a) = \int \lambda(a)\,dE(\lambda) = \int \lambda(a)\,dE'(\lambda)$ for two spectral measures E and E', then for any vectors x and y in $\mathbf{H}$,

$$(U(a)x, y) = \int \lambda(a)\,d(E(\lambda)x, y) = \int \lambda(a)\,d(E'(\lambda)x, y).$$

Lemma 10.2.2 implies that $(E(B^*)x, y) = (E'(B^*)x, y)$ for any Baire subset B^* in G^*, for all x and y, which means that $E = E'$.

DEFINITIONS If m is any (bounded) complex Baire measure on the locally compact Abelian group G, its *Fourier-Stieltjes transform $\hat{m}$* is the function on G^* defined by the equation

$$\hat{m}(\lambda) = \int \lambda(a)\,dm(a);$$

if m is absolutely continuous with derivative f, in which case $\hat{m} = \hat{f}$, $\hat{m}$ is the *Fourier transform* of f.

The univalency (i.e., one-to-one character) of the two transforms just defined follows from Lemma 10.2.2, together with the Pontrjagin duality theorem, readily deduced as a corollary to the Plancherel theorem proved later, to the effect that every character of G^* has the form $\lambda \to \lambda(a)$ for some element $a \in G$. Although we shall have no occasion to use this univalency before this point is reached, it may be illuminating to show how it may be proved at the present stage.

COROLLARY 10.2.4 *If G is a locally compact Abelian group and f a function in $L_1(G)$, then*

$$\int \lambda(a)f(a)\,da = 0$$

for all λ in G^* if and only if $f = 0$ a.e.

Proof Let L be the (left) regular representation of G on $L_2(G)$ and E the spectral measure on G^* such that

$$L_a = \int_{G^*} \lambda(a)\,dE(\lambda).$$

Then for any f in $L_1(G)$ and g in $L_2(G)$,

$$(L(f)g,g) = \int_G f(a)(L_a g, g)\, da = \int_G f(a)\left[\int_{G^*} \lambda(a)\, dm_g(\lambda)\right] da,$$

where m_g is the finite measure on the unrestricted Baire subsets of G^*, with the property that

$$m_g(B) = (E(B)g, g)$$

for any unrestricted Baire set B. By Fubini's theorem

$$(L(f)g,g) = \int_{G^*}\left[\int_G \lambda(a)f(a)\, da\right] dm_g(\lambda) = \int_{G^*} \hat{f}(\lambda)\, dm_g(\lambda).$$

Now let $E(\hat{f}) = \int_{G^*} \hat{f}(\lambda)\, dE(\lambda)$. It then results that $(L(f)g, g) = (E(\hat{f})g, g)$ for all g in $L_2(G)$. From this it follows readily that $(L(f)g, h) = (E(\hat{f})g, h)$ for all g,h in $L_2(G)$, and hence that $L(f) = E(\hat{f})$. If $\hat{f} = 0$, then $L(f) = 0$, and by Scholium 10.1, $f = 0$ a.e. Conversely, if $f = 0$ a.e., it is evident that $\hat{f} = 0$.

On any locally compact group, the set **B** of all complex Baire measures forms an algebra relative to convolution as multiplication, in which the algebra $L_1(G)$ is imbedded via the correspondence just indicated. Specifically, then, $m * m'$ is the measure m'' such that for any element $f \in C_0(G)$,

$$\int f(x)\, dm''(x) = \iint f(xy)\, dm(x)\, dm'(y);$$

the integral on the right is easily seen to exist and to define a complex integral on G, by representation of m and m' as linear combinations of positive measures. As in the case of convolution of functions, associativity of the convolution operation is a simple consequence of the Fubini theorem, and it follows that **B** is an algebra, as indicated. It is noteworthy that $L_1(G)$ is imbedded not only as a subalgebra, but as an ideal in **B**: if m' is absolutely continuous, then so also is $m * m'$. Indeed, if $m'(E) = \int_E h(x)\, dx$, where it is no essential loss of generality to take h to be a Baire function whose support is σ-compact (see Sec. 5.2), then for $f \in C_0(G)$,

$$\int f(x)\, dm''(x) = \iint f(xy)\, dm(x)h(y)\, dy = \iint f(y)\, dm(x)h(x^{-1}y)\, dy$$

$$= \int f(y)k(y)\, dy, \qquad k(y) = \int h(x^{-1}y)\, dm(x),$$

showing that m'' is absolutely continuous (with density function k). In the Abelian case one has, further,

COROLLARY 10.2.5   *The Fourier-Stieltjes transform is an isomorphism of the algebra of all complex Baire measures on a locally compact Abelian group G into the algebra of all continuous functions on the dual group G^*.*

The continuity of a Fourier-Stieltjes transform and the multiplicativity of the transform on convolutions follow in the same fashion as in the case of the Fourier transform. The only nontrivial point is the unicity of the transform. Now if $\hat{m} = 0$, then the same is true of the convolution of m with any complex Baire measure; choosing an absolutely continuous measure and using the observation that these form an ideal in **B** and the unicity for Fourier transforms on $L_1(G)$, it follows that $m * m' = 0$ whenever m' is absolutely continuous. If m' has the density function g in $C_0(G)$, this means that $\int f(xy)\,dm(x)g(y)\,dy = 0$ for all f in $C_0(G)$, which implies that $\int k(x)\,dm(x) = 0$ for all functions k on G of the form $k(x) = \int f(xy)g(y)\,dy$. It is easily seen that any element of $C_0(G)$ is the uniform limit of a net of functions having the form of k for suitable elements f and g in $C_0(G)$, from which it follows that $\int k(x)\,dm(x) = 0$ for all k in $C_0(G)$, so that $m = 0$.

REMARK   If the function h in Stone's theorem were bounded, multiplication by it would be a bounded operator, say, H, and it would result that $U(t) = e^{itH}$, in the sense of the "operational calculus" given in the exercises of Sec. 9.2. It can be shown that this is the case if and only if the given one-parameter group $U(\cdot)$ is uniformly continuous, that is, $\|U(t) - U(t')\| \to 0$ as $t \to t'$. This is rather rarely the case in concrete analytical situations. The simplest and most familiar one-parameter group arising in analysis is that given by the regular representation L of the reals in which

$$L_t: f(x) \to f(x - t).$$

Now L is strongly but not uniformly continuous, and it may be shown that in a formal way

$$L_t = e^{itH}, \qquad \text{where } H = \frac{1}{i}\frac{d}{dx}.$$

Here H is an operator of obvious analytical importance, but it is not at all continuous, and so is outside the domain of the applicability of the spectral theory developed up to this point. On the other hand, Stone's theorem suggests that H should be regarded as unitarily equivalent to a multiplication operator on an L_2-space, by a necessarily unbounded function h. It also suggests that other "unbounded" operators besides H may be effectively diagonalizable as well. The next section deals with an extension of the theory to a natural limit, which provides, as one of many applications, a precise way to deal with the situation just described.

EXERCISES

1 Show that if H is a continuous self-adjoint operator on a Hilbert space, then $U(t) = e^{itH}$ defines a uniformly continuous one-parameter unitary group.

2 Show that any uniformly continuous one-parameter unitary group has the form given in Exercise 1. (*Hint:* Otherwise the function h in Stone's theorem is essentially unbounded. There is then a sequence of disjoint sets E_n of finite positive measure such that $|h|$ is bounded from below by n on E_n. Now note that if s_n is any sequence of numbers such that $|s_n| \to \infty$, then the supremum of $|e^{is_n t} - 1|$ over n and t, for $|t| < \epsilon$, does not go to zero with ϵ.)

3 Let U be a one-parameter unitary group on a Hilbert space $\mathbf{H}$, let $x \in \mathbf{H}$, and let

$$f(t) = (U(t)x, x).$$

Show that f has the property that, for any finite sequence $t_1, t_2, \ldots, t_n$ of real numbers and corresponding ordered set of complex numbers $c_1, c_2, \ldots, c_n$,

$$\sum_{i,j} c_i \bar{c}_j f(t_i - t_j) \geq 0$$

and $f(-t) = \overline{f(t)}$ for all t. (Such a function is called *positive definite*.)

4 Extend Exercise 4 to the case of a unitary representation of an arbitrary group, where the positive-definiteness condition is replaced by the conditions

$$\sum_{i,j} c_i \bar{c}_j f(a_i a_j^{-1}) \geq 0$$

and $f(a^{-1}) = \overline{f(a)}$, where $a_1, \ldots, a_n$ are elements of the group.

5 Show that any positive definition function ϕ on a group G can be obtained in the form

$$\phi(a) = (U(a)x, x),$$

where U is a unitary representation. (Take $\mathbf{H}$ as the Hilbert space of all functions f on G which vanish except at countably many points and for which

$$\sum_{a,b} \phi(ab^{-1}) f(a)\overline{f(b)} < \infty.$$

Define the inner product (f, g) as

$$\sum_{a,b} \phi(ab^{-1}) f(a)\overline{g(b)},$$

and take $U(a)$ as the map $f(x) \to f(a^{-1}x)$ for f in $\mathbf{H}$.)

6 *a.* Show that a representation U of a topological group G by unitary operators on a Hilbert space $\mathbf{H}$ is continuous provided $a \to (U(a)x, x)$ is continuous for a dense set of vectors in $\mathbf{H}$.

   b. Conclude that if the positive definite function ϕ in Exercise 5 is continuous, then so is the corresponding representation.

7 Show that if ϕ is any continuous positive definite function on the reals, then there exists a regular measure m on the reals such that

$$\phi(t) = \int e^{itx}\, dm(x).$$

(*Hint:* Apply Stone's theorem to the one-parameter group associated with ϕ as in Exercise 5.)

8 Extend Exercise 7 to the case of a continuous positive definite function on an arbitrary locally compact Abelian group. (This result is due originally to Herglotz for the case of an infinite cyclic group, Bochner for the reals under addition, and Weil for the general case, by methods not involving unitary groups.)

9 Show that if U is a unitary representation of a locally compact group G on a Hilbert space **H** which is weakly measurable in the sense that $a \to (U(a)x,y)$ is measurable for all x,y in **H**, then **H** is the direct sum of two invariant subspaces, on one of which the restriction of U is continuous, and on the other of which it is singular. Here singular means that $a \to (U(a)x,x)$ is equivalent to 0 for all vectors x in the subspace. [*Hint:* Take as the singular subspace the set of all x such that

$$\int_G f(a)(U(a)x,y)\, da = 0$$

for all f in $L_1(G)$.]

    10 *a.* Show in Exercise 9 that if **H** is separable, then the singular subspace must be absent.

        b. Show that the representation U associated with the positive definite function ϕ on the reals, which is 1 for $t = 0$ and 0 otherwise, as in Exercise 5, is weakly measurable but completely singular.

11 Show that for any locally measurable positive definite function ϕ on a locally compact group, there is a unique continuous one equal to it almost everywhere. [*Hint:* Employ as the Hilbert space for the construction of the associated representation the completion of the set of all continuous functions f on G, relative to the inner product

$$(f,g) = \iint \phi(ab^{-1})f(a)\overline{g(b)}\, da\, db$$

after deflation relative to this inner product; this involves showing that a positive definite function is always bounded, for which the case $n = 2$ of the defining inequality should be examined.]

12 A measure m is said to be *quasi-invariant* under an invertible bimeasurable transformation T in case m and the transformed measure m_T given by $m_T(E) = m[T^{-1}(E)]$ are mutually absolutely continuous. Show that if the measure space is a countable union of measurable sets, then the mapping $U(T)$,

$$f(x) \to k(T,x)f(T^{-1}x), \qquad f \in L_2,$$

where

$$k(T,x) = \left(\frac{dm_T}{dm}\right)^{\frac{1}{2}}$$

is a unitary transformation of L_2 onto itself (see Exercise 3b, Sec. 7.2).

13   Show that U in Exercise 12 gives a unitary representation of the group of all invertible bimeasurable transformations relative to which the given measure is quasi-invariant.

14   *a.* Show that $U(T)$ of Exercise 12 has the property that

$$U(T)M_kU(T)^{-1} = M_{k\circ T}$$

for all bounded locally measurable k.

   b. Show that any unitary representation satisfying the condition of (*a*) has the form given in Exercises 12 and 13, with some function $k(T,x)$.

   c. Determine a necessary and sufficient condition that a given function $k(T,x)$ have the property that the mapping $U(T)$ given in Exercise 12 be a unitary operator and that $T \to U(T)$ be a representation.   Show that the function given there is the only nonnegative such function.   (Such functions are called multipliers, and are important in the theory of group representations.)

15   Show that any connected locally compact group has a finite measure, quasi-invariant under left and right translations.   (*Hint:* There is a symmetric compact neighborhood N of e, and the union of the sets N_k, $k = 1, 2, \ldots$, is the entire group.   Start with Haar measure on N, and build up a measure by induction, by appropriately altering Haar measure on $N^k - N^{k-1}$.)

16   Show that any regular measure on the Baire subsets of a locally compact group which is quasi-invariant under left translation is equivalent to left Haar measure.   (*Hint:* Employ the Fubini theorem in a fashion similar to the proof of the uniqueness of Haar measure.)

17   Let G be a locally compact group, and H a closed subgroup.   The coset space G/H is then a locally compact space in the canonical topology in which a set is open if and only if it is the image of an open set in G.

   a. Show that if G is a countable union of compact sets, then there exists a finite regular measure m on G/H quasi-invariant under the natural action of G.

   b. If m' is any other such measure, show that m and m' are mutually absolutely continuous.

   c. Construct a strongly continuous unitary representation of G on $L_2(G/H,m)$ where m satisfies the conditions of *a*.

18   *a.* Show that, on a perfect measure space, any real measurable function is continuous relative to the complement of a null set.

   b. Deduce that the function h in Stone's theorem may be assumed to be continuous.

19   Show that the (bounded) complex Baire measures on any locally compact group form a Banach algebra relative to convolution as multiplication and the total variation as norm.

20*   Show that if $f \in L_1(G)$, where G is a locally compact Abelian group, and if ϕ is a complex analytic function on an open set containing the closure of the range of values of f, then there exists an element $g \in L_1(G)$ such that $\hat{g}(\lambda) = \phi(\hat{f}(\lambda))$

for all $\lambda \in G^*$.   (*Hint:* Use the general theorem on Banach algebras on the existence of inverses in combination with the Cauchy-integral formula.)

21* Show that the assumption in Theorem 10.1′ that ϕ be continuous is super-fluous. [*Hint:* Use a Schwarz-inequality type of argument, coupled with the use of the analytic functions $(1 \pm z)^{\frac{1}{2}}$ applied to suitable elements of $L_1(G)$ of norm less than 1.]

10.3   UNBOUNDED DIAGONALIZABLE OPERATORS

From an a priori algebraic standpoint a theory of partially defined operators in a Banach space appears to be an unpromising thing.   Yet it turns out that in the case of Hilbert space a clean and simple theory is possible which is applicable to many important specific analytical situations.   In fact, it is from the complex of such considerations, rather than algebraic ones, that the theory emerges, although ex post facto it is simpler to express the theory, and it may even be well motivated, in algebraic terms.

It is natural to take the multiplication operator by an unbounded locally measurable function on an L_2-space as the prototype of a tractable unbounded operator; a *diagonalizable* operator in Hilbert space is naturally definable as one which is similar, via an invertible transformation, to such a multiplication operator.   In a similar vein it is natural to define a generalized normal or self-adjoint operator as one which is similar, via a unitary transformation, to such a multiplication operator, with the added restriction of reality in the self-adjoint case.   In a way, the virtue of Hilbert space in this connection is that it is possible to give algebraically simple and practically verifiable conditions on a given operator that it be normal or self-adjoint.   The crucial success in this direction was that due to von Neumann and Stone, independently, for the self-adjoint case.   The operation of multiplication by an unbounded locally measurable function k is not definable on all of L_2 to L_2; so one defines the operator M_k to have as domain all functions f in L_2 such that $k(x)f(x)$ is a function of x which is again in L_2, and to operate on this domain in the obvious fashion.   An intrinsic condition on an operator that it be unitarily equivalent to an M_k with k real, but expressed in terms of spectral measures, was proposed originally by E. Schmidt, according to a footnote in von Neumann's article, and independently arrived at by Stone.   It is simply that the operator be identical with its suitably defined adjoint.   With appropriate special devices, many differential and integral operators may be shown to satisfy this condition, and therefore be diagonalizable by a unitary transformation.   In a few situations, the diago-nalization may be carried out in such a closed form that it may be deduced without the use of the cited theorem, but in general this is not at all the case.   The normal unbounded operators, i.e., those which are unitarily equivalent to a multiplication by a complex-valued function, may also be characterized similarly, although somewhat less effectively.

On the other hand, there are many serious complications in dealing with unbounded self-adjoint operators, even though these are the best behaved of the unbounded ones.   Such an operator, for example, is not determined by its action on a dense domain.   Two distinct self-adjoint operators with distinct spectral values (the essential values of the corresponding locally measurable functions) may agree on a dense domain.   In general, two self-adjoint operators have no common vectors in their domains except 0, and there is no effective or reasonable nontrivial way to define their sum. Even when they have a common dense domain, their sum need not be diagonalizable in any sense; this is the case even if it is assumed that they leave a common dense domain invariant and commute on it.   In fact, they need not then be simultaneously diagonalizable.   The operators that provide examples of such behavior, which is pathological from the algebraic finite-dimensional standpoint, are not at all pathological in themselves.   It is fair to say that, generally, almost all kinds of pathology not excluded by general theorems can be shown to exist by fairly simple explicit examples involving linear differential operators.

DEFINITIONS   An *operator in* (as opposed to *on*) a Hilbert space **H** is a map T from subset **D**, called the *domain* of T, to **H**.   If D is a linear set and T is linear on **D**, T is called a *linear operator*.   The operator T is called *bounded* in case $\|Tx\| \le c\,\|x\|$ for some constant c, for all x in **D**.   By a familiar theorem on the extension of uniformly continuous functions on dense subsets of a metric space, a bounded linear operator whose domain is dense extends uniquely to a continuous linear operator on the entire space.   This extension is, strictly speaking, a bounded everywhere-defined operator, but sometimes "bounded" is used loosely to include the additional implications of being everywhere defined; it is generally either clear from the context or else immaterial which usage is intended.   An operator T is said to be *closed* if whenever $x_n \in$ **D** and both $x_n \to x$ and $Tx_n \to y$ for some y in **H**, then x is in the domain of T and $Tx = y$.   The *graph* of an operator T consists of all pairs $[x,Tx]$ regarded as a subset of the Hilbert space **H** $\oplus$ **H**. It is straightforward to verify that an operator is closed if and only if its graph is a closed subset of **H** $\oplus$ **H**.   An operator T' is called an *extension* of T, symbolically, $T \subset T'$, in case the domain of T is contained in that of T' and the two operators agree on their common domain.   Some care is needed in forming algebraic combinations of partially defined operators.   Suppose S and T are linear operators in H with domains D_S and D_T, and let c be a scalar.   Then

(*1*) $S + T$ is defined as the operator whose domain is $D_S \cap D_T$ and which has the value $Sx + Tx$ for any vector x in this domain.

(*2*) ST is defined as the operator whose domain consists of all vectors

x in D_T such that $Tx \in D_S$ and which has the value $S(Tx)$ on any such vector.

(3) cT is defined as the operator with domain D_T and having the value $c(Tx)$ for any vector in this domain.

These rules for algebraic operations are similar to, but more complicated and inhibited than, the usual ones for everywhere-defined operators. Note, for example, that $0 \cdot T \neq 0$ if T is not everywhere defined and all that can be said is that $0 \cdot T \subset 0$. This may seem pedantic, but there is sufficient pathology with unbounded operators so that one cannot afford to be lax about such little points. The familiar associative laws

$$(S + T) + U = S + (T + U), \qquad (ST)U = S(TU),$$

are, however, easily seen to be valid.

If T is an operator in H with dense domain D, the *adjoint* T^* of T is defined as the linear operator whose domain consists of all vectors y in $\mathbf{H}$ for which there exists another vector y' in $\mathbf{H}$ such that

$$(Tx,y) = (x,y')$$

for all x in $\mathbf{D}$, and which has the value y' on such a vector y (the value y' is unique in view of the density of $\mathbf{D}$).

Special care must be observed in forming adjoints of sums and products. If S and T are densely defined operators, then $S^* + T^*$ is well defined, but the domain of $S + T$ need not be dense (indeed, it might consist of 0 alone), so that $(S + T)^*$ would not be defined. If, however, S, T, and $S + T$ are all densely defined, then it is straightforward to verify that

$$S^* + T^* \subset (S + T)^*.$$

Similarly, if S, T, and ST are densely defined, then

$$T^*S^* \subset (ST)^*.$$

Finally, if S is densely defined and $S \subset T$, then $T^* \subset S^*$.

The principal, and almost only, nontrivial result at this structural level explains the relation between an operator T and its second adjoint T^{**}.

SCHOLIUM 10.2 *The adjoint T^* of a given densely defined operator T in a Hilbert space $\mathbf{H}$ is itself densely defined, so that T^{**} also exists, if and only if T has a closed linear extension, in which event T^{**} is the minimal such extension.*

Proof Like many other questions concerning unbounded operators, this can be treated effectively through consideration of the graphs of T and T^*. Let J denote the linear operator $[x,y] \rightarrow [y,-x]$ on $\mathbf{H} \oplus \mathbf{H}$. Then J is unitary, and therefore commutes, in its induced action on all subsets, with

the operation of forming the orthocomplement. The basic point is this:
If the graph of an operator T is denoted as $\mathbf{G}_T$, and the orthocomplement of
a subset $\mathbf{S}$ as $\mathbf{S}^\perp$, then for any densely defined operator T,

$$\mathbf{G}_{T*} = (J\mathbf{G}_T)^\perp = J[(\mathbf{G}_T)^\perp].$$

The first equality follows immediately from the definition of T^*, and the
second from the unitary character of J. This shows, incidentally, that T^*
is always closed, inasmuch as the orthocomplement of any set is closed.

Now suppose T has a closed linear extension T' and that x is a vector
orthogonal to $\mathbf{D}_{T*}$. Then $[0,x]$ is orthogonal to all $[T^*y,-y]$ for y in $\mathbf{D}_{T*}$.
It then results that $[0,x] \in (J\mathbf{G}_{T*})^\perp$, and since $J^2 = -I$, the equation above
shows that

$$(J\mathbf{G}_{T*})^\perp = [JJ(\mathbf{G}_T)^\perp]^\perp = (\mathbf{G}_T)^{\perp\perp}.$$

Since $(\mathbf{G}_T)^{\perp\perp}$ is the closed linear subspace of $\mathbf{H} \oplus \mathbf{H}$ generated by $\mathbf{G}_T$, it
follows that

$$(\mathbf{G}_T)^{\perp\perp} \subset \mathbf{G}_{T'},$$

and hence that $[0,x] \in \mathbf{G}_{T'}$. Because T' is a linear operator, $x = 0$. Thus
any vector x orthogonal to $\mathbf{D}_{T*}$ is necessarily 0, and T^* is densely defined.

Conversely, suppose T^* is densely defined. Then T^{**} exists, and as above,

$$\mathbf{G}_{T**} = (J\mathbf{G}_{T*})^\perp = [JJ(\mathbf{G}_T)^\perp]^\perp = (\mathbf{G}_T)^{\perp\perp}.$$

Since $(\mathbf{G}_T)^{\perp\perp}$ is the closed subspace generated by $\mathbf{G}_T$, this shows that T^{**}
is a closed necessarily linear extension of T. If S is any other closed linear
extension, then $T \subset S$, $S^* \subset T^*$, and $T^{**} \subset S^{**}$. But

$$\mathbf{G}_{S**} = (\mathbf{G}_S)^{\perp\perp} = \mathbf{G}_S;$$

that is, $S^{**} = S$, and therefore $T^{**} \subset S$.

Example 10.3.1   Let $\mathbf{H} = L_2(M)$ for some measure space M, and let k
be a complex-valued locally measurable function on M. The domain of the
multiplication operator M_k is then defined as the set of all f in $L_2(M)$ such
that $kf \in L_2(M)$, and on this domain M_k is defined by

$$M_k(f) = kf.$$

Note that the domain $\mathbf{D}$ of M_k is the same as the domain of $M_{\bar{k}}$, that $\mathbf{D}$ is a
linear subset of $L_2(M)$, and that M_k is a linear operator. Furthermore, $\mathbf{D}$ is
dense in $\mathbf{H}$. To see this, let c_n be the characteristic function of the set of
points x such that $|k(x)| \leq n$. Then c_n is a locally measurable function, and
$c_n f \in \mathbf{D}$ for all f in $L_2(M)$. Since $\|c_n f - f\|_2 \to 0$ as $n \to \infty$, it follows that
D is dense.

Next we shall show that M_k is a closed operator. For this suppose that
$f_n \in D$, that $f_n \to f$, and that $kf_n \to g$ in L_2. By taking an appropriate

subsequence, we may assume $f_n(x) \to f(x)$ a.e.   Then since $kf_n \to g$ in L_2, there is a subsequence f_{n_i} such that

$$k(x)f_{n_i}(x) \to g(x) \qquad \text{a.e.}$$

But since $k(x)f_{n_i}(x) \to k(x)f(x)$ a.e., it results that $kf = g$ a.e.   Thus $f \in \mathbf{D}$ and $M_k f = g$.

The adjoint of M_k will now be shown to equal $M_{\bar{k}}$.   First observe that if $g \in \mathbf{D}$, then

$$\int kf\bar{g} = \int f\overline{\bar{k}g}$$

for all $f \in \mathbf{D}$; so, evidently,

$$M_{\bar{k}} \subset M_k{}^*.$$

On the other hand, if g is in the domain of $M_k{}^*$, then there is a function h in L_2 such that

$$\int kf\bar{g} = \int fh$$

for all f in $\mathbf{D}$.   If c_n is the characteristic function defined above, then for all $f \in L_2$, $c_n f \in \mathbf{D}$, and it follows that

$$\int f(kc_n\bar{g}) = \int f(c_n h)$$

for all f in L_2.   Hence $kc_n g = c_n h$ a.e.   Moreover, $c_n g \to g$, $c_n h \to h$, and $M_{\bar{k}}$ is a closed operator; so $g \in \mathbf{D}$, $\bar{k}g = h$ a.e., and $M_k{}^* \subset M_{\bar{k}}$.   Thus the equation $T = T^{**}$ for a closed densely defined operator is very simply exemplified here.

Now let T be the operator whose domain consists of all functions in L_2 that vanish outside some set on which k is bounded and which agrees with M_k on this domain.   Then, in general, T is not closed.   [Consider, for example, the simple but illuminating case of the operation of multiplication by x acting on functions $f(x)$ in $L_2(-\infty, \infty)$.]   But it is easily seen that the closure of T, that is, T^{**}, is the original operator M_k, whereas

$$T^* = M_{\bar{k}}.$$

Many other operators extended by T will have M_k for their closures.   For example, the restrictions of multiplication by x to the domain of all continuous functions of compact support, or the domain of all hermite functions, or the domain of all infinitely differentiable functions of compact support, are such operators.   The proofs of these facts are nice real variable exercises, simple but relatively typical of what is involved in some of the more concrete applications of operator theory, and are left as such.

In the specific analytical theories in which operator theory is involved, one of the most common problems is that of "diagonalizing" an operator; this is particularly the case in quantum mechanics, but also in the theories of differential and integral equations, and elsewhere.   As mentioned earlier, a normal operator is one for which this is possible, through transformation by a suitable unitary operator.   By this we mean the following

DEFINITION   An operator T in a Hilbert space $\mathbf{H}$ is said to be *normal* if there exists a unitary transformation U of $\mathbf{H}$ onto $L_2(M)$ for some measure space M such that UTU^{-1} is the operation of multiplication by a complex-valued locally measurable function k.   A *self-adjoint* operator is a normal operator T for which the corresponding function k is real-valued.

From Example 10.3.1 it is evident that a normal operator T is densely defined and closed and has the properties

$$TT^* = T^*T, \quad D_T = D_{T^*}, \quad \|Tx\| = \|T^*x\| \quad \text{if} \quad x \in \mathbf{D}_T.$$

For a self-adjoint operator, $T = T^*$.   Conversely, such simple intrinsic properties are sufficient to imply normality or self-adjointness, and the availability of such characterizations is one of the major benefits of working in Hilbert space rather than in more general Banach spaces.   At this point we should mention that a self-adjoint operator is more commonly defined as a densely defined operator which is equal to its adjoint.   This is etymologically unexceptionable, but fails to reveal the diagonalization property which is theoretically crucial.

Since the basic ideas are well indicated and simpler in the self-adjoint case, and since this is also the most important one for applications, our treatment of diagonalizability will be limited to the self-adjoint case.

THEOREM 10.4   *A densely defined operator T in a complex Hilbert space is unitarily equivalent to the operation of multiplication by a real measurable function on L_2 over a measure space if and only if $T = T^*$.*

Proof   As mentioned above, the "only if" part is an immediate consequence of the results given in Example 10.3.1.

Suppose, then, that T is a densely defined operator in the complex Hilbert space $\mathbf{H}$ such that $T = T^*$.   Let $\mathbf{D}$ be the domain of T, and U the operator in $\mathbf{H}$ whose domain consists of all vectors of the form $Tx + ix$, with x in $\mathbf{D}$, and which acts as follows:

$$U: Tx + ix \to Tx - ix.$$

The indicated map U is single-valued, and in fact isometric, since

$$\|Tx + ix\|^2 = \|Tx\|^2 + (Tx, ix) + (ix, Tx) + \|x\|^2$$

$$= \|Tx\|^2 + \|x\|^2 = \|Tx - ix\|^2.$$

From these equations it also follows that if the sequence $\{Tx_n + ix_n\}$ is convergent for some sequence $\{x_n\}$ in $\mathbf{D}$, then the sequences $\{x_n\}$ and $\{Tx_n\}$

are also convergent.   Now T is closed, since it is the adjoint of an operator, and so it would follow under the above conditions that $x_n \to x$ for some x in **D**.   This shows that the domain of U is closed, and it is actually all of **H**. For if y is any vector orthogonal to the domain of U, then

$$(Tx + ix, y) = 0$$

for all x in **D**.   In other words,

$$(Tx, y) = (x, iy)$$

for all x in **D**.   Since T is self-adjoint, it follows that $y \in$ **D** and that $Ty = iy$. This is possible only when $y = 0$, since it implies that

$$(Ty, y) = i(y, y) = (y, Ty) = \overline{(Ty, y)}.$$

Thus U is defined on all of **H**.   By the same argument, with $T + iI$ replaced by $T - iI$, we see that the range of U is also all of **H**, which shows that U is unitary.   The unitary operator U is called the *Cayley transform* of T.

Note now that T is determined by U.   For if $y = Tx + ix$ with x in D, then $Uy = Tx - ix$.   Hence

$$x = \frac{1}{2i}(I - U)y$$

and

$$Tx = \tfrac{1}{2}(I + U)y.$$

Also observe that this implies that $I - U$ is a nonsingular linear operator on **H**.

Since U is unitary, and hence normal, it may be diagonalized by a unitary transformation; that is, U is unitarily equivalent to a multiplication operator. Thus it is no essential loss of generality to assume that **H** is $L_2(M)$ for some measure space M and that U is the operation M_u of multiplication by a locally measurable function u, with $u(t)$ having absolute value 1 l.a.e.   Since $I - U$ is nonsingular, $u(t)$ can equal 1 only on a local null set.   By redefining $u(t)$ as -1 on this local null set, we may in fact assume that it is empty. Then the domain **D** of T consists of all elements of the form

$$x = \frac{1}{2i}(1 - u)y, \qquad y \in L_2,$$

$y = 2i(1 - u)^{-1}x$, and

$$Tx = \tfrac{1}{2}(1 + u)y = i\left(\frac{1 + u}{1 - u}\right)x.$$

Since $|u(t)| = 1$ for all t, the locally measurable function

$$k(t) = i\left[\frac{1 + u(t)}{1 - u(t)}\right]$$

is real-valued, and hence $M_k = M_k^*$.   Moreover, $T \subset M_k$, so that $M_k \subset T^* = T$.   This shows that $T = M_k$, and completes the proof.

We turn now to the problem of extending the "operational calculus" for a bounded normal operator, outlined in Exercise 1, Sec. 9.2, to the unbounded case.   The desired extension is twofold.   We wish to define not only bounded functions of unbounded operators, but unbounded functions as well.

For simplicity, we shall again restrict our attention to the self-adjoint case.   Our problem then is to define $f(T)$ for any self-adjoint operator T and any complex Baire function f on the reals.   In the case that T is a multiplication operator, say, $T = M_h$ on L_2 over some measure space, with h a real locally measurable function, it is natural to make the definition

$$f(M_h) = M_{f \circ h}.$$

Moreover, this definition extends in an invariant fashion to the case of an arbitrary self-adjoint operator.

THEOREM 10.5   *Let T be a self-adjoint operator in a complex Hilbert space* **H**.   *Then for any complex Baire function f, there exists a unique normal operator $f(T)$ in* **H** *with the following property: Whenever U is a unitary transformation such that $UTU^{-1} = M_h$ on L_2 over some measure space, with h a real locally measurable function, then*

$$f(T) = U^{-1} M_{f \circ h} U.$$

Proof   Suppose T is a self-adjoint operator in the complex Hilbert space **H**.   Then there is a measure space M and a unitary transformation U of **H** onto $L_2(M)$ such that $UTU^{-1} = M_h$, where h is a real-valued locally measurable function on M.   Now suppose N is another measure space and that V is a unitary map of **H** onto $L_2(N)$ such that $VTV^{-1} = M_k$, where k is a real-valued locally measurable function on N.   For the existence and uniqueness of $f(T)$ it then suffices to prove that

$$U^{-1} M_{f \circ h} U = V^{-1} M_{f \circ k} V$$

for any complex Baire function f on the reals.   For this it is evidently enough to show that

$$W M_{f \circ h} W^{-1} = M_{f \circ k},$$

where $W = VU^{-1}$.   Now all we know a priori is that $W M_h W^{-1} = M_k$.   However, the Cayley transform of M_h is, clearly, M_u on $L_2(M)$, where

$$u(x) = \frac{h(x) - i}{h(x) + i},$$

and the Cayley transform of M_k is M_v on $L_2(N)$, where v is defined similarly in terms of k.   Moreover, $W M_u W^{-1} = M_v$, since a unitary equivalence between self-adjoint operators is also a unitary equivalence between their Cayley transforms. The functions u and v have their values in the unit circle $|z| = 1$, and for any function p on the unit circle of the form

$$p(z) = \sum_n c_n z^n,$$

where c_n is complex and the sum is extended over a finite set of integers, it is readily verified that

$$WM_{p \, \mathrm{o} \, u} W^{-1} = M_{p \, \mathrm{o} \, v}.$$

These functions p form an algebra **A** which is uniformly dense in the algebra of all continuous functions on the unit circle S. On the other hand, the continuous functions are dense in the algebra **B** of all bounded Baire functions on S, relative to the topology of bounded sequential convergence. Thus **A** is dense in **B** in this topology. Since the spectral integrals $b \to WM_{b \, \mathrm{o} \, u} W^{-1}$ and $b \to M_{b \, \mathrm{o} \, v}$ $(b \in B)$ are continuous on **B** and agree on **A**, it follows that

$$WM_{b \, \mathrm{o} \, u} W^{-1} = M_{b \, \mathrm{o} \, v}$$

for all b in **B**.   For any complex Baire function g on S, there exists a sequence b_i of elements in **B** such that $|b_i| \le |g|$ and $b_i(z) \to g(z)$ for all z.   If ϕ is any element in the domain of $M_{g \, \mathrm{o} \, v}$, it follows from the dominated convergence theorem that

$$\lim_i \int_N |g[v(y)] - b_i[v(y)]|^2 \, |\phi(y)|^2 \, dn(y) = 0.$$

For the same reason, $M_{b_i \, \mathrm{o} \, u}(W^{-1}\phi) \to M_{g \, \mathrm{o} \, u}(W^{-1}\phi)$.   Hence $WM_{b_i \, \mathrm{o} \, u} W^{-1}\phi$ $\to WM_{g \, \mathrm{o} \, u} W^{-1}\phi$, but since $WM_{b_i \, \mathrm{o} \, u} W^{-1} = M_{b_i \, \mathrm{o} \, v}$, it results that

$$WM_{g \, \mathrm{o} \, u} W^{-1}\phi = M_{g \, \mathrm{o} \, v}\phi.$$

Thus $WM_{g \, \mathrm{o} \, u} W^{-1} = M_{g \, \mathrm{o} \, v}$ for any complex Baire function g.   Now let f be a complex Baire function on the reals.   Then the equation

$$g(z) = f\left(i \left(\frac{1+z}{1-z} \right) \right)$$

defines a complex Baire function g on the unit circle.   Moreover, as is seen by elementary algebra, $g \circ u = f \circ h$ and $g \circ v = f \circ k$.   This shows that

$$WM_{f \, \mathrm{o} \, h} W^{-1} = M_{f \, \mathrm{o} \, k}$$

and completes the proof.

Example 10.3.2   Let M be a measure space and h a real locally measurable function on M.   Then the equation

$$U(t) = e^{itM_h}$$

means that $U(\cdot)$ is the continuous one-parameter unitary group on $L_2(M)$, with the property that for any g in $L_2(M)$

$$U(t):g(x) \to e^{ith(x)}g(x).$$

As already indicated in the treatment of bounded operators, the diagonalization, or spectral resolution of a self-adjoint operator, and the operational calculus can in part be given alternatively in terms of spectral measures. The only difference between the cases of bounded and unbounded operators is that, in the latter, the spectral measure is not supported by a bounded set. In the present context, the spectral-measure formulation is useful mainly for understanding the connection between the development given here and that presented in the older literature.

COROLLARY 10.3.1 *For any self-adjoint operator T in a complex Hilbert space* **H**, *there exists a unique spectral measure E on the Baire subsets of the reals to the projections on* **H** *such that*

$$f(T) = \int_{-\infty}^{\infty} f(\lambda)\, dE(\lambda)$$

for all bounded Baire functions f.

Proof Let U be a unitary transformation such that $UTU^{-1} = M_h$ on L_2 over some measure space. Let F be the spectral integral on the bounded Baire functions defined by

$$F(f) = U^{-1}M_{f \circ h}U,$$

and let E be the spectral measure defined by F. Then we have

$$f(T) = \int_{-\infty}^{\infty} f(\lambda)\, dE(\lambda).$$

The uniqueness of E is trivial. For if Q is another spectral measure with the same property, then

$$\varphi_B(T) = E(B) = Q(B)$$

for all Baire sets B.

DEFINITION A sequence $\{T_n\}$ of partially defined operators in a Hilbert space *converges* to another such operator T in case $T_n x \to Tx$ for all vectors x in the domain of T and if in addition this domain is maximal with respect to this property.

COROLLARY 10.3.2 *For any spectral measure E on the reals to the projections on a complex Hilbert space* **H**, *there is a unique map F from the complex Baire functions to the normal operators in* **H** *which extends the integral*

$$f \to \int_{-\infty}^{\infty} f(\lambda)\, dE(\lambda)$$

on the bounded Baire functions, and has the property that if $f_n \to f$ and $|f_n| \leq |f|$, then $F(f_n) \to F(f)$.

Proof   The uniqueness of such an F follows quite generally by consideration of the class of functions on which any two such maps agree, which includes the bounded Baire functions and is closed under dominated sequential convergence, and so includes all Baire functions.

The existence of such an F may be proved in a variety of ways. For our purposes it is most useful and illustrative to reduce to the case in which, for any Baire set B, $E(B)$, is the operation of multiplication by $\varphi_{k^{-1}(B)}$ on $L_2(M)$ for some measure space M and real-valued locally measurable function k. In this case it follows at once from the dominated convergence theorem that

$$F: f \to M_{f \circ k}$$

has the indicated properties.

It therefore only remains to show that every spectral measure E on the linear Baire sets is unitarily equivalent to one of the indicated type. For this purpose, observe that the $E(B)$ form a commutative self-adjoint set of bounded linear operators on **H**.   Hence, by Scholium 9.4, there is a standard measure space M and a unitary map U of **H** onto $L_2(M)$ such that $UE(B)U^{-1}$ is multiplication by a uniquely determined continuous function, say, h_B. Since $h_B h_B = h_B$, the set $S(B)$, where h_B assumes the value 1, is both open and closed.   The sets $S_{mn} = S((m/2^n, (m+1)/2^n])$, where m is an arbitrary integer and n a positive integer, form a countable disjoint family whose union over m is the entire space.   Let k_n be the locally measurable function, in fact continuous function, on M, which assumes the value $m/2^n$ on S_{mn}, provided this set is nonempty.   Then

$$0 \leq k_{n+1}(x) - k_n(x) \leq 2^{-n-1}$$

for all x, and $\sum_n [k_{n+1}(x) - k_n(x)]$ is uniformly convergent, so that k_n converges uniformly to a continuous function k.   Furthermore, if x is a point at which $k_n(x) = m/2^n$, it follows that $k(x) \leq (m+1)/2^n$.   Thus k maps S_{mn} into the interval

$$B_{mn} = (m/2^n, (m+1)/2^n].$$

On the other hand, if x is a point such that $k(x) \in B_{mn}$, there exists an integer j such that $k_j(x) \in B_{mn}$.   But since $k_j(x) = i/2^j$ for some i, we see that $B_{ij} \subset B_{mn}$ and that

$$x \in S_{ij} \subset S_{mn}.$$

Thus $k^{-1}(B_{mn}) = S_{mn}$, and if E' is the spectral measure such that $E'(B)$ is multiplication by $\varphi_{k^{-1}(B)}$, it results that E' agrees with the spectral measure

$$B \to UE(B)U^{-1}$$

on the sets B_{mn}. Now the collection of all subsets on which two spectral measures agree is evidently a σ-ring, and since the linear Baire sets are generated by those of the form B_{mn}, it follows that

$$UE(B)U^{-1} = E'(B)$$

for all Baire sets B.

The integral notation used for the case of a bounded Baire function may now be extended to the case of an arbitrary Baire function f according to the rule

$$\int_{-\infty}^{\infty} f(\lambda)\, dE(\lambda) = F(f).$$

It is then evident that the proof of Corollary 10.3.2 also establishes the following result:

COROLLARY 10.3.3   *For any spectral measure E on the linear Baire sets to the projections on a complex Hilbert space* **H**, *the operator*

$$T = \int_{-\infty}^{\infty} \lambda\, dE(\lambda)$$

is self-adjoint and such that

$$f(T) = \int_{-\infty}^{\infty} f(\lambda)\, dE(\lambda)$$

for any complex Baire function f.

Our final result follows from the foregoing corollary and Corollary 10.3.1.

COROLLARY 10.3.4   *For any self-adjoint operator T in a complex Hilbert space* **H**, *there exists a unique spectral measure E on the Baire subsets of the reals to the projections on* **H** *such that*

$$T = \int_{-\infty}^{\infty} \lambda\, dE(\lambda).$$

Proof Let E be the spectral measure on the reals such that

$$f(T) = \int_{-\infty}^{\infty} f(\lambda)\, dE(\lambda)$$

for all bounded Baire functions f. Then, by diagonalizing T, we see that

$$T = \int_{-\infty}^{\infty} \lambda\, dE(\lambda)$$

and hence that

$$f(T) = \int_{-\infty}^{\infty} f(\lambda)\, dE(\lambda)$$

for all complex Baire functions f.   If Q were another spectral measure such that

$$T = \int_{-\infty}^{\infty} \lambda \, dQ(\lambda),$$

we should then also have $f(T) = \int_{-\infty}^{\infty} f(\lambda) \, dQ(\lambda)$.   Thus Q would in fact equal E.

EXERCISES

1   Let T denote the operation $f(x) \to xf(x)$ on $L_2(-\infty, \infty)$.   Show that T is the closure of its restriction to each of the following domains:

    a.   All continuous functions of compact support.
    b.   All infinitely differentiable functions of compact support.
    c.   All step functions.
    *d.**   All finite linear combinations of hermite functions.

2   An operator T in a Hilbert space is called *hermitian* in case it is densely defined and has the property that $(Tx,y) = (x,Ty)$ for all vectors x and y in its domain.   Show that a given operator T is hermitian if and only if T^* exists and extends T.

3   Show that any hermitian operator has a closure, and that this closure is again hermitian.

4   Concoct an example of a densely defined operator which has no closure. [Compute the adjoint of the map in $L_2(-\infty, \infty)$, $f \to (\int f)v$, where v is a nonzero fixed vector in $L_2(-\infty, \infty)$ and the domain consists of all integrable elements of this space.]

5   Let A denote the self-adjoint generator of the one-parameter unitary group on $L_2(-\infty, \infty)$, $U(t): f(x) \to f(x + t)$ [i.e., the unique self-adjoint operator A such that $U(t) = e^{itA}$].   Show that its domain consists precisely of those functions f in $L_2(-\infty, \infty)$ which are equivalent to absolutely continuous functions whose derivatives are again in L_2.

6   Show that the operator A of Exercise 5 is the closure of its restriction to each of the following domains:

    a.   All infinitely differentiable functions of compact support.
    *b.**   All finite linear combinations of hermite functions.

7   An operator A in a Hilbert space is called *essentially self-adjoint* if it has a closure which is self-adjoint.   Show that A is essentially self-adjoint if and only if A^* and A^{**} both exist and are equal.

8   *a.*   Show that an operator in a Hilbert space which is essentially self-adjoint has a unique self-adjoint extension.
    *b.**   Prove the converse.

9 Show that the operator $f(x) \to k(x)f(x)$ in $L_2(M)$ for a measure space M, k being a given complex-valued measurable function, is the closure of its restriction to the domain of all simple functions on whose support k is bounded.

10 Show that if f and g are any complex-valued measurable functions on a measure space, with corresponding multiplication operators A and B (defined as in Exercise 9), then $A + B$ and AB have closures which are, respectively, the operations of multiplication by $f + g$ and fg.

11 Show that if $U(t) = e^{itA}$ with A self-adjoint, and if f is a continuously differentiable function of compact support on the reals, then for any vector x, $y = \int U(t)xf(t)\,dt$ is in the domain of A, and that $Ay = i\int U(t)xf'(t)\,dt$.

12 Show that a hermitian operator A is essentially self-adjoint if and only if either one of the following conditions holds:
  a. For some real constant $c \neq 0$, $A + icI$ and $A - icI$ have dense ranges.
  b. For every real constant $c \neq 0$, the conclusion of (a) holds.

13* Show that any self-adjoint operator is essentially self-adjoint on any domain which is dense and invariant under the one-parameter unitary group which it generates.

14* Let $\mathbf{H} = L_2\,(0,1)$, and let A denote the self-adjoint generator of the one-parameter unitary group $f(x) \to f(x + t)$, where the addition is reduced modulo 1. Let V denote the unitary operation $f(x) \to e^{ix}f(x)$, and set $A' = VAV^{-1}$.

  a. Show that, on the domain $\mathbf{D}$ of all infinitely differentiable functions vanishing near 0 and 1, A' and A agree within an additive constant, and so commute, on this dense domain.
  b. Show that, nevertheless, the one-parameter unitary groups $U(\cdot)$ and $VU(\cdot)V^{-1}$ do not commute. (Determine the spectral resolutions of A and A' through the use of the theory of Fourier series. Show also that A has a so-called "simple spectrum," i.e., that the set of all bounded functions of A is maximal Abelian in the algebra of all bounded linear operators.)

15 Show that the direct sum of closed operators is closed, defining the direct sum of any set (finite or infinite) of closed operators appropriately.

16 Show that if T is a closed and densely defined linear operator in a Hilbert space, then T^*T is self-adjoint.

17 Show that for any nonnegative self-adjoint operator A (i.e., one such that $A \geq 0$), there exists a unique nonnegative self-adjoint operator B such that $B^2 = A$.

18 Show that two self-adjoint operators in a Hilbert space can be different although they agree on a dense set, but that a self-adjoint operator which vanishes on a dense set must be 0.

19 A *partial isometry* on a Hilbert space $\mathbf{H}$ is an everywhere-defined linear transformation which is an isometry on some subspace $\mathbf{M}$ of $\mathbf{H}$ and annihilates the orthocomplement of $\mathbf{M}$. Show that if T is a closed densely defined operator in the Hilbert space $\mathbf{H}$, and if B is the nonnegative self-adjoint square root of T^*T, then there exists a unique partial isometry V on $\mathbf{H}$ which annihilates the orthocomplement of the range of T and which carries Tx into Bx for every vector x in the domain of T.

20   Show that, in Exercise 19, $B = VT$, $T = V^*B$, and $T^* = BV$.   Show also that B and V commute with every unitary transformation which commutes with both T and T^*.   (The adjoint V^* of any partial isometry V is itself one, and the decomposition $T = V^*B$ is sometimes called the "polar decomposition" of T. On an infinite-dimensional space there will not always exist a decomposition of the form $T = UB$ for some unitary U, and even when this is the case the unitary operator U need not commute with all the indicated unitaries.)

21   An "analytic vector" for a continuous one-parameter unitary group $U(\cdot)$ on a Hilbert space $\mathbf{H}$ is a vector v such that

$$U(t)v = \sum_n \frac{(itA)^n}{n!}\, v,$$

A being the self-adjoint generator of the group, the series being absolutely convergent for all sufficiently small t.   Show that there is always a dense domain of analytic vectors and that the restriction of A to this domain is essentially self-adjoint.

22   Show that if $A_1, \ldots, A_n$ is a finite set of strictly commuting self-adjoint operators in a Hilbert space, i.e., such that their spectral projections all commute in the ordinary sense (or equivalently, the one-parameter groups they generate do so), then $A_1^2 + \cdots + A_n^2$ is self-adjoint, and that any vector analytic for it is also analytic for each of the A_i.

23   Show that any linear partial differential operator with constant coefficients acting in the domain of all infinitely differentiable functions of compact support in $L_2(R^n)$ has a closure which is a normal operator, commuting with all translations $f(x) \to f(x + y)$.

24*   If $t \to T(t)$ is a strongly continuous one-parameter group of bounded linear operators on a Hilbert space $\mathbf{H}$, its generator A is the operator

$$\lim_{\epsilon \to 0} \epsilon^{-1}[T(\epsilon) - I].$$

Show that the operator A is the closure of its restriction to any linear domain which is dense in $\mathbf{H}$ and invariant under all the $T(t)$.

25*   Extend the result of Exercise 24 to the case of an n-parameter group (i.e., representation of the additive group of an n-dimensional vector space) in such a way as to yield the strengthening of Exercise 23, in which the domain is replaced by an arbitrary dense one that is translation-invariant, and invariant under all powers of all the infinitesimal generators of one-parameter subgroups of the given group.

10.4   ABELIAN HARMONIC ANALYSIS

One of the central theorems of Abelian harmonic analysis is the von Neumann uniqueness theorem, which originated in the problem of showing the essential identity of the Heisenberg and Schrödinger formulations of quantum mechanics. From a mathematical viewpoint it is now recognized to be essentially a theorem giving the structure of the unitary representations of certain transformation groups, where we make the

DEFINITION Let $\Gamma = (G,S,F)$ be a topological transformation group such that S is locally compact. A *unitary representation* of Γ is a system $(U,\phi,\mathbf{H})$ such that U is a continuous unitary representation of G on the complex Hilbert space $\mathbf{H}$, and ϕ is a self-adjoint representation of the algebra $\mathbf{C}_0(S)$ on $\mathbf{H}$, for which the following relation holds:

$$(1) \qquad\qquad U(a)\phi(k)U(a)^{-1} = \phi(k_a),$$

where for any function k on S, $k_a(p) = k(a^{-1}p)$ [the group action $F(a,p)$ being denoted simply as ap].

Example 10.4.1 Suppose, additionally, that Γ has the structure of a regular group of measure-preserving transformations; i.e., a regular measure m is given on S which is invariant under the action of G. Let $\mathbf{H}$ denote the Hilbert space $\tilde{L}_2(S,m)$, and let $U(a)$ and $\phi(k)$ be defined as follows:

$$U(a):f \to f_a, \qquad \phi(k):f \to kf,$$

where f is arbitrary in $\mathbf{H}$. It is straightforward to verify that $(U,\phi,\mathbf{H})$ is a unitary representation of Γ; it may appropriately be called the *regular representation* of the given regular group of measure-preserving transformations.

In the simple, but particularly important, case in which G is locally compact, $S = G$, and in which G acts on S via left translations, $F(a,b) = ab$, the general unitary representation is a "multiple," or direct sum, of a certain number of copies of the regular representation. In the special case in which G is a vector group, this result is essentially the cited uniqueness theorem of von Neumann; the general case, developed by Mackey and Loomis, is relevant to and indicative of the theory of "induced" group representations.

THEOREM 10.6 *Let $(U,\phi,\mathbf{H})$ be a unitary representation of the transformation group consisting of the locally compact group G acting on itself by left translation and having the property that if $\phi(f)x = 0$ for all f, then $x = 0$. Then $\mathbf{H}$ is the direct sum of subspaces invariant under U and ϕ, relative to each of which the representation is unitarily equivalent to the regular representation.*

In other terms, this means that $\mathbf{H}$ is the direct sum of subspaces $\mathbf{H}_\lambda$, each of which has the property that

$$U(a)\mathbf{H}_\lambda \subset \mathbf{H}_\lambda, \qquad \phi(k)\mathbf{H}_\lambda \subset \mathbf{H}_\lambda,$$

for all $a \in G$ and $k \in \mathbf{C}_0(G)$; and that there exist operators T_λ such that T_λ is unitary from $\tilde{L}_2(G)$ onto $\mathbf{H}_\lambda$ and satisfies the relations

$$T_\lambda^{-1}\phi(k)T_\lambda = \phi_0(k), \qquad T_\lambda^{-1}U(a)T_\lambda = U_0(a),$$

where $\phi_0(k)$ and $U_0(a)$ act on $\tilde{L}_2(G)$ as follows:

$$U_0(a): f(b) \to f(a^{-1}b); \qquad \phi_0(k): f(b) \to k(b)f(b).$$

Proof of the theorem   Let h and k be arbitrary in $\mathbf{C}_0(G)$, and consider the operator A defined by the weak integral

$$A = \int \phi(h)U(a)\phi(k)\, da;$$

this integral exists, since the integrand is a bounded weakly continuous function of compact support.   Indeed, if x and y are arbitrary in $\mathbf{H}$, then

$$(\phi(h)U(a)\phi(k)x,y) = (U(a)\phi(k)x,\phi(h)^*y),$$

showing that the integrand is weakly continuous; and writing $U(a)\phi(k) = \phi(k_a)U(a)$, the integrand becomes $\phi(h)\phi(k_a)U(a) = \phi(hk_a)U(a)$; if C is a compact set outside of which h and k vanish, then $(hk_a)(b) = h(b)k(a^{-1}b)$, which vanishes unless $b \in C$ and $a^{-1}b \in C$, and hence vanishes unless a is in the compact set CC^{-1}; the boundedness of the integrand is evident.

It suffices to show that A cannot be identically zero for all choices of h and k.   For assuming this has been done, it follows that there exists a vector $y \in H$ and a k such that $A(h,k)y \neq 0$ for some h; let T_0 then denote the map $h \to A(h,k)y$; then T_0 satisfies the following relations, similar to those required of T_λ above:

$$(2) \qquad \phi(k)T_0 = T_0\phi_0(k), \quad U(a)T_0 = T_0U_0(a), \quad k \text{ arbitrary in } \mathbf{C}_0(G),$$

by entirely straightforward verifications.   In addition, on setting $\beta(g,h) = (T_0g,T_0h)$, it is straightforward to verify that the form β satisfies the relation

$$\beta(gh,k) = \beta(g,\bar{h}k)$$

for arbitrary g, h, and k in $\mathbf{C}_0(G)$, in addition to being linear in g and anti-linear in h.   On setting $\Lambda(k) = \beta(g,h)$ when $k = g\bar{h}$, it follows that Λ is well defined, in the sense that if also $k = g'\bar{h}'$, then $\beta(g,h) = \beta(g',h')$; for on introducing an element $f \in \mathbf{C}_0(G)$ which has the value unity on the supports of h and h', $\Lambda(k)$ may be written $\beta(g,fh) = \beta(g\bar{h},f) = \beta(g'\bar{h}',f) = \beta(g',fh')$ $= \beta(g',h')$.   It follows in a similar fashion that Λ is linear, and it is also positive, for $\Lambda(k\bar{k}) = \|T_0k\|^2$, k being arbitrary.

From the relation $U(a)T_0 = T_0U_0(a)$ it follows that Λ is invariant under G.   By the uniqueness of Haar measure, Λ must be proportional to the Haar integral, say $\Lambda_0: \Lambda = c\Lambda_0$.   Now defining $T_1 = c^{-1/2}T_0$, T_1 is isometric, and it satisfies the same relations (2) as does T_0.   It therefore extends uniquely to a unitary transformation T from $\tilde{L}_2(G)$ to a closed linear submanifold of $\mathbf{H}$, say, $\mathbf{H}_1$, on which the required "intertwining" relations (2) hold.   The orthocomplement in $\mathbf{H}$ of $\mathbf{H}_1$ is likewise invariant under the $U(a)$ and the $\phi(k)$, and on applying transfinite induction, it follows that $\mathbf{H}$ is a direct sum of invariant subspaces similar to $\mathbf{H}_1$.

It remains only to show that $A(h,k)$ is not identically zero.   This will be done by defining a closely related operator $A'(f)$ for arbitrary $f \in \mathbf{C}_0(G \times G)$, which coincides with $A(h,k)$ when f has the form $f(a,b) = h(a)k(b)$, and depends linearly and continuously on f; the vanishing of all the $A(h,k)$ will then imply, via the Stone-Weierstrass theorem, the vanishing of $A'(f)$ for all f, which leads immediately to a contradiction.   Specifically, let f^a be defined by the equation

$$f^a(b) = f(b,a^{-1}b);$$

that is, $f^a(b)$, as a function of a and b, is the composition of the function f with the mapping $(a,b) \to (b,a^{-1}b)$.   The integral

$$A'(f) = \int \phi(f^a)U(a)\,da$$

will now be shown to exist (weakly) and to satisfy the inequality

$$\|A'(f)\| \le \lambda(C)\,\|f\|,$$

where $\lambda(C)$ is a constant, depending on the compact set C, C being any compact subset of G such that f vanishes outside of $C \times C$, and $\|f\|$ denotes the supremum of $|f(a,b)|$, for a, b, $\in G$.

To this end, note that f^a vanishes outside C, and that $f^a = 0$ unless $a \in CC^{-1}$ by a simple computation; thus, for any vectors x and y, the function of a: $(\phi(f^a)U(a)x,y)$ is well defined and has compact support.   Now, as noted in Chap. 7, a continuous function of compact support on a locally compact group is uniformly continuous (relative to either the right or left uniform structure on the group), and it follows that the mapping $a \to f^a$ is continuous from G to the functions on G, in the topology defined by the indicated norm $\|f\|$.   Observe next that ϕ is bounded relative to the elements of $\mathbf{C}_0(G)$ supported by any fixed compact set, say, C'.   For if $g \in \mathbf{C}_0(G)$ and $g \ge 0$, then $\phi(g) \ge 0$, since g has the form $g = h\bar{h}$ for some element $h \subset \mathbf{C}_0(G)$, which implies that $\phi(g) = \phi(h)\phi(h)^* \ge 0$.   It follows that if $g' \in \mathbf{C}_0(G)$ and $g' \ge g$, then $\phi(g') \ge \phi(g)$.   Now if f is any element of $\mathbf{C}_0(G)$ which is supported by C', and if h is any fixed nonnegative function in $\mathbf{C}_0(G)$ which has the value 1 on C', then

$$f\bar{f} \le \|f\|^2\, h,$$

which implies that

$$\phi(f)\phi(f)^* \le \|f\|^2\, \phi(h),$$

leading in turn to the inequality

$$\|\phi(f)\phi(f)^*\| = \|\phi(f)\|^2 \le \lambda'(C')^2 \|f\|^2$$

for some constant $\lambda'(C')$.   In particular, $\|\phi(f^a) - \phi(f^{a'})\| \le \lambda'(C) \|f^a - f^{a'}\|$, showing that $\phi(f^a)$ is a uniformly continuous function of a.   It follows easily that $(\phi(f^a)U(a)x,y)$ is a continuous function of a, and is bounded by $\lambda'(C) \|f\| \|x\| \|y\|$.   This shows, finally, that

$$\left| \int (\phi(f^a)U(a)x,y) \, da \right| \le \lambda(C) \|f\| \|x\| \|y\|,$$

the integral existing since the integrand is continuous and has compact support.   Thus the indicated weak integral exists and has the indicated bound.

Now suppose that $A(h,k)$ is zero for all values of h and k.   This means that $A'(f) = 0$ for all f of the form $f(a,b) = h(a)k(b)$.   Since $A'(f)$ depends linearly on f, it follows that $A'(f) = 0$ whenever f is a finite linear combination of functions of the form just indicated.   Now taking C as an arbitrary but fixed compact set in G, the set of all such finite linear combinations is uniformly dense in the set of all continuous functions on $G \times G$ which are supported by $C \times C$, by the variant of the Stone-Weierstrass theorem which applies to functions on a locally compact space, here taken as the interior of the set $C \times C$ (see, Corollary 5.1.1).   In view of the continuity property of A' derived in the preceding paragraph, it follows that A' vanishes.   Now taking f to have the form

$$f(a,b) = h(a)k(ab^{-1}),$$

f^a has the form

$$f^a(b) = h(b)k(a),$$

and it results that

$$\int \phi(h)U(a)k(a) \, da = 0$$

for all choices of h and k in $C_0(G)$.   If N denotes an arbitrary neighborhood of e in G, and k is a nonnegative function supported by N and of total integral 1 on G, then

$$\int \phi(h)U(a)k(a) \, da \to \phi(h) \qquad \text{(relative to the net of all } N \text{ ordered by reverse inclusion)}$$

by the strong continuity of the representation U and a simple argument employed in Chap. 7.   Thus the contradiction $\phi(h) = 0$ for all h has been arrived at, showing that $A(h,k)$ cannot be zero identically in h and k.

An important consequence of the foregoing theorem is the von Neumann uniqueness theorem in the generalized form due originally to Mackey (for the separable case).

COROLLARY 10.4.1 *Let G be a locally compact Abelian group, and let G^* denote its topological character group. Let U and V be continuous representations of G and G^* (respectively) on a Hilbert space $\mathbf{H}$ such that*

$$U(a)V(b^*) = b^*(a)V(b^*)U(a)$$

for all $a \in G$ and $b^* \in G^*$. Then $\mathbf{H}$ is the direct sum of invariant subspaces under U and V, relative to each of which the pair (U,V) is unitarily equivalent to the pair (U_0, V_0) on the space $L_2(G^*)$, where

$$U_0(a): f(x^*) \to x^*(a)f(x^*); \qquad V_0(b^*): f(x^*) \to f(x^* - b^*).$$

Proof By the generalized Stone theorem, there exists a unique spectral measure E on the Baire subsets of G^* such that

$$U(a) = \int b^*(a)\, dE(b^*).$$

Now defining $\phi(f) = \int f(b^*)\, dE(b^*)$ for $f \in \mathbf{C}_0(G^*)$, it is a consequence of the general properties of spectral measures derived earlier that ϕ is a *-homomorphism of $\mathbf{C}_0(G^*)$ into the bounded linear operators on $\mathbf{H}$, and that if $\phi(f)x = 0$ for all f, then $x = 0$. Assuming that $V(c^*)U(a)V(c^*)^{-1} = c^*(a)^{-1}U(a)$ for arbitrary $a \in G$ and $c^* \in G^*$, it follows that

$$V(c^*)\int b^*(a)\, dE(b^*)V(c^*)^{-1} = \int (b^* - c^*)(a)\, dE(b^*),$$

and that

$$V(c^*)\phi(f)V(c^*)^{-1} = \phi(f_{c^*}).$$

Thus $(V,\phi,\mathbf{H})$ is a representation of G^* as a transformation group acting on G^* by translation. The theorem implies, therefore, that $\mathbf{H}$ is the direct sum of invariant subspaces, on each of which (V,ϕ) is unitarily equivalent to the pair (V_0,ϕ_0) acting on the space $L_2(G^*)$, where V_0 is as given in the conclusion of the corollary, and $\phi_0(k)$ is for any $k \in \mathbf{C}_0(G^*)$ the operation of multiplication by k.

It remains only to deduce from the form obtained for ϕ the form given for U in the conclusion of the theorem. As seen earlier, any spectral measure E on a locally compact space S is determined by the corresponding integral, $f \to \int f\, dE, f \in \mathbf{C}_0(S)$. On the other hand, in the case of the spectral measure on G^* associated by Stone's theorem with a given continuous unitary representation of a locally compact Abelian group G, the measure is determined by the representation. It is evident that ϕ_0 is the integral for the spectral measure corresponding to U_0, and it follows that the unitary transformation whose induced action on operators carries ϕ (as restricted to the subspace in question) into ϕ_0 also carries U (restricted to the same subspace) into U_0.

From the generalized von Neumann theorem there follows in turn one of the most useful theorems in Abelian harmonic analysis, namely, that of Plancherel, in the generalized form first established by Weil.

COROLLARY 10.4.2   *Let G be a locally compact Abelian group. If $f \in L_1(G) \cap L_2(G)$, its Fourier transform $\hat{f}$,*

$$\hat{f}(x^*) = \int x^*(a)f(a)\, da,$$

is in $L_2(G^*)$, and the mapping $f \to \hat{f}$ extends uniquely to a unitary transformation from $\tilde{L}_2(G)$ onto $\tilde{L}_2(G^*)$ (with suitable normalization of the Haar measure on G^*).

Proof   Let U_1 and V_1 be the representations of G and G^* (respectively) on $L_2(G)$, defined by the equations

$$U_1(a):f(x) \to f(x - a); \qquad V_1(b^*):f(x) \to \overline{b^*(x)}f(x).$$

It is clear that U_1 and V_1 are continuous unitary representations of G, and straightforward to verify that

$$U_1(a)V_1(b^*) = b^*(a)V_1(b^*)U_1(a)$$

for arbitrary a and b^*.   In addition, the pair (U_1, V_1) is irreducible; i.e., there exists no nontrivial closed linear subspace which is invariant under all the $U_1(a)$ and $V_1(b^*)$.   To see this, note first that the weakly closed algebra of operators generated by the $V_1(b^*)$ contains the full multiplication algebra of G as a measure space relative to Haar measure.   For otherwise there would exist a continuous linear functional on this algebra which vanished on all $V_1(b^*)$, but not on all multiplication operators.   As seen earlier, a linear functional on the multiplication algebra of a localizable measure space—in particular, a direct sum of finite measure spaces such as a locally compact group relative to Haar measure (the group is the union of the cosets of a σ-compact invariant subgroup, namely, that generated by any compact neighborhood of the unit)—which is continuous in the weak-operator topology has the form $M_k \to \int kh$, where h is a fixed integrable function. The vanishing of this linear functional on the $V_1(b^*)$ means the vanishing of $\int b^*(a)h(a)\, da$ for all b^*, that is, the vanishing of the Fourier transform of h, which entails the vanishing of h.   Now if P denotes the projection on a closed linear submanifold of $L_2(G)$ which is invariant under the $U_1(a)$ and $V_1(b^*)$, it follows that P commutes with M_k for every bounded locally measurable functions k on G; since this multiplication algebra is maximal Abelian, it follows that P is itself a multiplication operator, say, $P = M_p$. The invariance of the submanifold in question under the $U_1(a)$ means that P commutes with the $U_1(a)$, which in turn implies that the locally measurable function p on G is invariant under the transformations $p \to p_a$.

By the ergodicity of the action of a locally compact group on itself by left translations, the only such functions p are constants, showing that either $P = I$ or $P = 0$, establishing the irreducibility in question.

It follows now from the generalized von Neumann theorem that there exists a unitary transformation T from $L_2(G)$ onto $L_2(G^*)$ having the property that $T^{-1}U_0(a)T = U_1(a)$, $T^{-1}V_0(b^*)T = V_1(b^*)$, for all $a \in G$, $b^* \in G^*$. If now f and g are arbitrary in $L_1(G) \cap L_2(G)$, then $f * g \in L_2(G)$ and

$$T(f * g) = T \int U_1(a)fg(a)\, da = \int TU_1(a)fg(a)\, da = \int U_0(a)Tfg(a)\, da,$$

which on reference to the form of $U_0(a)$ shows that

$$T(f * g) = Tf\hat{g}.$$

Now choosing a net $\{f_\mu\}$ such that $f_\mu * g \to g$ in $L_2(G)$ for all g, then $T(f_\mu * g) \to Tg$. It follows that $\{Tf_\mu \hat{g}\}$ is convergent in $L_2(G^*)$, and hence that $\{Tf_\mu\}$ is locally convergent in $L_2(G^*)$ in the neighborhood of any point x_0^* for which $\hat{g}(x_0^*) \neq 0$. Now for any point x_0^* in G^* there exists a g for which $\hat{g}(x_0^*) \neq 0$, and it follows from the localizability of a locally compact group as a measure space (or without the use of the concept of localizability, from the fact noted earlier, that such a group is the union of σ-compact open subsets) that there exists a locally measurable function h on G^* to which the net Tf_μ is convergent in L_2 relative to any compact subset. It follows that

$$Tg = h\hat{g}$$

for all $g \in L_1(G) \cap L_2(G)$.

Having made use in the argument just completed of the relation between U_0 and U_1, consider now that between V_0 and V_1: $TV_1(b^*)g = V_0(b^*)Tg = V_0(b^*)(h\hat{g})$; on the other hand, $TV_1(b^*)g = T(\overline{b^*(x)}g(x))$ [where x is a dummy variable] $= h(x^*)\hat{g}(x^* - b^*)$ [where x^* is a dummy variable]; thus

$$h(x^* - b^*)\hat{g}(x^* - b^*) = h(x^*)\hat{g}(x^* - b^*).$$

Translating through b^*, it follows that $h(x^*)\hat{g}(x^*) = h(x^* + b^*)\hat{g}(x^*)$ a.e. on G^*. Since g can be chosen so that $\hat{g}(x^*) \neq 0$ in the neighborhood of any given point, it follows that each point x_0^* in G^* has a neighborhood in which $h(x^*) = h(x^* + b^*)$ l.a.e. (b^* being held fixed). It follows, as earlier, that $h(x^*) = h(x^* + b^*)$ a.e. on G^*. Since this is true for every $b^* \in G^*$, it follows from the ergodicity of translations on G^* that h is a constant, say, c. Now defining $T' = c^{-1}T$, and replacing Haar measure on G^* by its multiple by $|c|^{-\frac{1}{2}}$, T' is unitary from $L_2(G)$ to $L_2(G^*)$, with the modified measure, and coincides with the Fourier transform on $L_1(G) \cap L_2(G)$.

REMARK   If G is a real vector group, and if $\langle x, y \rangle$ is a given real non-degenerate symmetric form on G, there is a canonical isomorphism of G^* with G which carries any element $f \in G^*$ into the unique vector $y \in G$ such that $f(x) = \langle x, y \rangle$ for all $x \in G$.   Relative to the given symmetric form, there is then an identification of G^* with G, relative to which, in turn, the Fourier transform becomes a unique operator on $L_2(G)$ (into itself).   The Fourier transform on euclidean space is commonly regarded in this light; thus if $f \in L_2(E_n)$, its Fourier transform is given by the equation

$$\hat{f}(x) = a \int \exp\,[ib\langle x, y \rangle] f(y)\,dy,$$

where dy denotes the element of Lebesgue measure in the euclidean space E_n, and a and b are positive constants, which are subject to some variation in the literature.   The most common values for b are, in order of relative frequency, $\pm b = 1$, 2π, and $\tfrac{1}{2}$.   For a the values are $a = 1$, in which case the property $(f * g)^\wedge = \hat{f}\hat{g}$ is valid, or a value dependent on b which renders valid the equation $\|f\|_2 = \|\hat{f}\|_2$ (when it is desirable to distinguish between these two slightly different operations, the first may be called the *Fourier*, and the second the *Plancherel*, transform).

The value of a as a function of b for the Plancherel transform may be determined by the substitution of any fixed function for which the integrals in question may be explicitly determined; a generally useful case is that of the function $\exp\,(-x^2/2)$.   Note, first, that

$$\int_{-\infty}^{\infty} \exp\,(-x^2/2) = \sqrt{2\pi}, \text{ since, setting } c \text{ for this integral,}$$

$$c^2 = \left[\int_{-\infty}^{\infty} \exp\left(-\frac{x^2}{2}\right) dx\right]\left[\int_{-\infty}^{\infty} \exp\left(-\frac{y^2}{2}\right) dy\right]$$

$$= \int_{-\infty}^{\infty}\int_{-\infty}^{\infty} \exp\left[-\frac{x^2 + y^2}{2}\right] dx\,dy;$$

on transforming to polar coordinates, it results that

$$c^2 = \int_{0}^{2}\int_{0}^{\infty} \exp\left(-\frac{r^2}{2}\right) r\,dr\,d\theta = 2\pi.$$

Now consider

$$I = \int_{-\infty}^{\infty} e^{itxb}\, e^{-x^2/2}\,dx;$$

completing the square in the exponent,

$$I = \int_{-\infty}^{\infty} \exp\left[-\frac{(x - itb^2)}{2}\right] \exp\left(-\frac{t^2b^2}{2}\right) dx = \exp\left(-\frac{t^2b^2}{2}\right) J(tb),$$

where

$$J(s) = \int_{-\infty}^{\infty} \exp\left[-\frac{(x - is^2)}{2} \right] dx.$$

It is easily seen that $J(s)$ is a continuously differentiable function of the real variable s, and

$$J'(s) = -\int_{-\infty}^{\infty} (x - is) \exp\left[-\frac{(x - is^2)}{2} \right] dx$$

$$= \int_{-\infty}^{\infty} d\left[\exp -\frac{(x - is^2)}{2} \right] = 0,$$

so that $J(s) = \text{const} = J(0) = \sqrt{2\pi}$. Thus $I = \sqrt{2\pi} \exp(-t^2 b^2/2)$, and it follows that $a = (b/2\pi)^{\frac{1}{2}}$ for $n = 1$; for general n it is easily deduced that $a = (b/2\pi)^{n/2}$, through consideration of the transform of $\exp\left[-\left(\sum_i x_i^2/2 \right) \right]$. It follows that the values $b = \pm 2\pi$ are the unique ones for which the transform is both multiplicative and unitary.

EXERCISES

1 Let T denote the Plancherel transform on $L_2(E_n)$, $E_n = n$-dimensional euclidean space. Show that $T^{-1} = TJ$, where J is the transformation $f(x) \rightarrow f(-x)$, and deduce that $T^4 = I$.

2 Let T denote the Plancherel transform on $L_2(E_1)$, P denote the infinitesimal self-adjoint generator of the one-parameter group $f(x) \rightarrow f(x + t)$, and Q denote the same for the group $f(x) \rightarrow e^{izt}f(x)$. Show that f is in the domain of P if and only if f is in the domain of Q, and that in this event $TPf = QTf$. Obtain an analogous result in which the roles of P and Q are interchanged.

3 With the notation of Exercise 2, show that the Plancherel transform is within a constant factor of absolute value 1, the unique unitary transformation T, such that $T^{-1}PT = Q$, $T^{-1}QT = -P$.

4 Show that there exists no unitary transformation U on $L_2(-\infty, \infty)$ such that $U^{-1}PU = Q$ and $U^{-1}QU = P$.

5 *a.* Extend Exercise 2 to the case of functions of several variables.

b. Let the laplacian Δ on $L_2(E_n)$ be defined as $\sum_k P_k^2$, where P_k denotes the infinitesimal self-adjoint generator of the group $f(x) \rightarrow f(x + te_k)$, and e_k denotes the kth basis vector ($k = 1, \ldots, n$). Show that Δ is self-adjoint, and that $T^{-1}\Delta T$ is the operation $f(x) \rightarrow |x|^2 f(x)$, where $|x|$ denotes the euclidean length.

6 Show that if $f \in L_2(-\infty, \infty)$ is continuously differentiable, and if $f' \in L_2(-\infty, \infty)$, then Pf exists (in the notation of Exercise 2) and equals $-if'$.

(*Hint:* Use Exercise 24, Sec. 10.3, to show that P is essentially self-adjoint on the domain D of all functions of compact support.   Show that if $P_0 = P \mid D$, then $(P_0 g, f) = -(g, if')$ for all g in the domain of P_0.)

7* *a.* Show that if $f \in L_2(-\infty, \infty)$ is of class C^n, and if $f^{(n)} \in L_2$, then f is in the domain of P^n and $P^n f = (-i)^n f^{(n)}$.   (*Hint:* Extend Exercise 24, Sec. 10.3, to the case of powers of the generator; deduce that P^n is essentially self-adjoint on the domain of all C^n functions of compact support, and proceed as in Exercise 6.)

   b. Show that if $f \in L_2(E_n)$ is of class C^2, and if $\sum_k (\partial^2 f / \partial x_k{}^2) \in L_2(E_n)$, then

f is in the domain of Δ and Δf is given by the indicated classical expression.

8   Show that if $f \in L_2(E_3)$ is in the domain of Δ, then $f \in L_\infty(E_3)$.   (*Hint:* Use the Plancherel theorem and the Schwarz inequality of Exercise 9.)

9   *a.* Let f be in the domain of P, in the notation of Exercise 2.   Writing f as the inverse Fourier transform of the product $(1 + |x|)^{-1} \cdot (1 + |x|)\hat{f}(x)$ of two L_2 functions, show that f may be redefined on a null set so as to be continuous, and that the resulting function f_1 is unique.

   b. Show that f_1 is absolutely continuous with derivative iPf.   (Evaluate the integral of Pf over a finite interval in terms of f.)

   c. Deduce that the domain of P consists precisely of those functions f in L_2 which differ on a Lebesgue null set from a (locally) absolutely continuous function f_1 whose derivative is in L_2, and that $Pf = -if_1'$.

10   Show that if T is an invertible homogeneous linear transformation in E_n, and if $f \in L_2(R^n)$, then, setting $f_T(x) = f(Tx)$,

$$(f_T)^\wedge = (\hat{f})_{T^{*-1}} |\det T|^{-1}.$$

where T^* denotes the dual transformation to T.   Deduce that if f is rotationally invariant, so also is $\hat{f}$.

11   Show that if $f \in L_2(E_3)$ and $f(x) = g(|x|)$, where $g(\lambda) = g(-\lambda)$ for arbitrary real λ, then $\hat{f} = cr^{-1}(d/dr)\hat{g}(r)$, where c is a constant (dependent on the normalization of the transform) and the derivative is taken after modification of g on a null set.

12   Show directly from the Plancherel theorem that if $f \in L_1(G^*)$, and if $f^\vee$ is defined by the equation $f^\vee(x) = \int_{G^*} \overline{\lambda(x)} f(\lambda)\, d\lambda$, then $f^\vee = 0$ only if $f = 0$. [*Hint:* If g is arbitrary in $L_1(G^*) \cap L_2(G^*)$, then $h = f * g$ is also in $L_1(G^*) \cap L_2(G^*)$; show that $(f * g)^\vee = f^\vee g^\vee$, so that $h^\vee = 0$; but for an element h of $L_1(G^*) \cap L_2(G^*)$, $h^\vee$ is the inverse Fourier transform.]

13   Show that if the element $x^* \in G^*$ is outside the closed subset C^* of G^*, then there exists an element $f \in L_1(G)$ such that $\hat{f}(x^*) \neq 0$ and $\hat{f}(C^*) = 0$.   (*Hint:* Take x^* to be the unit in G^*.   If N^* is a compact neighborhood of e^* such that $N^* = N^{*-1}$ and N^{*2} is disjoint from C^*, then $f = g^2$, where $\hat{g}$ is the characteristic function of N^*, will have the indicated property.)

14   For any element $x \in G$, let $j(x)$ denote the element $x^{**} \in G^{**}$ given by the equation $x^{**}(\lambda) = \lambda(x)$.   Show that j is an isomorphism of G onto G^{**}.   (This is the Pontrjagin-van Kampen duality theorem.)

[*Hint:* (*1*) Note that if U is the regular representation of G, then the map $a \to U(a)$ is a homomorphism of G into the unitary group by virtue of the local compactness of G.   (*2*) Note that by taking the Fourier transform twice, the regular representation of G is unitarily equivalent to the representation V on $L_2(G^{**})$, $V(a)f(x^{**}) = f[x^{**} + j(a)]$.   (*3*) Conclude from (2) that j is an isomorphism of G *into* G^{**}.   (*4*) Note that $j(G)$ is dense in G^{**} since otherwise Exercises 13 and 14 yield a contradiction.   (*5*) Deduce from the local compactness of G that $j(G)$ covers a neighborhood of the unit in G^{**}, and by translation ultimately all of G^{**}.]

15   A "Weyl system" over a finite-dimensional linear vector space $\mathbf{L}$ is defined as a triple $(U,V,\mathbf{H})$ such that U and V are continuous unitary representations on the complex Hilbert space $\mathbf{H}$ of the additive groups of $\mathbf{L}$ and its dual $\mathbf{L}^*$, respectively, satisfying the "Weyl relation" $V(f)U(x) = e^{if(x)}U(x)V(f)$ for all $x \in \mathbf{L}$ and $f \in \mathbf{L}^*$.   For any vector $z \in \mathbf{L} \oplus \mathbf{L}^*$, of the form $z = x \oplus f$, let $W(z)$ denote the operator $e^{if(x)/2}U(x)V(f)$.   Show for arbitrary z and z' in $\mathbf{L} \oplus \mathbf{L}^*$, the "extended Weyl relations" $W(z)W(z') = e^{(i/2)B(z,z')}W(z + z')$, where $B(z,z') = f(x') - f'(x)$.   Show that these extended relations for a given map W from $\mathbf{L} \oplus \mathbf{L}'$ to the unitary operators on a complex Hilbert space $\mathbf{H}$ imply that if $U = W \mid \mathbf{L}$ and $V = W \mid \mathbf{L}^*$, then $(U,V,\mathbf{H})$ is a Weyl system over $\mathbf{L}$.

16   Show that if T is any symplectic transformation on $\mathbf{L} \oplus \mathbf{L}'$, that is, a homogeneous linear transformation which preserves the form $B(z,z')$, and if W satisfies the extended Weyl relations, then so also does W_T, where $W_T(z) = W(Tz)$.   Use the von Neumann uniqueness theorem to conclude that there exists for any symplectic transformation T a unitary operator $U(T)$ on $L_2(\mathbf{L})$ such that $U(TT') = cU(T)U(T')$, where c is a scalar depending on T and T', and such that if $T = S \oplus S^{*-1}$, then $U(T)$ has the form $f(x) \to f(S^{-1}x)$ $[f \in L_2(\mathbf{L})]$.

17   *a.* If $\mathbf{M}$ is a finite-dimensional complex Hilbert space, show there exists a continuous map W from $\mathbf{M}$ to the unitary operators on a complex Hilbert space $\mathbf{K}$ such that $W(z)W(z') = e^{(i/2)B(z,z')}W(z + z')$ for arbitrary z and z' in $\mathbf{M}$ and such that no nontrivial closed linear subspace of $\mathbf{K}$ is invariant under all of the $W(z)$, where $B(z,z') = \mathrm{Im}\,[(z,z')]$.

 b. * Show there exists a continuous unitary representation Γ of the unitary group on $\mathbf{M}$ such that $\Gamma(U)W(z)\Gamma(U)^{-1} = W(Uz)$ for all $z \in \mathbf{M}$.   [If U_t denotes the one-parameter group $z \to e^{it}z$, then the one-parameter group $\Gamma(U_t)$ is essentially that generated by the harmonic-oscillator hamiltonian in quantum mechanics, with number of degrees of freedom equal to the dimension of $\mathbf{M}$. $\Gamma(i)$ is then the Fourier transform.]

18   *a.* With the notation of Exercise 15, show that if $f \in L_2(\mathbf{L} \oplus \mathbf{L}^*) \cap L_1(\mathbf{L} \oplus \mathbf{L}^*)$, then $\int W(z)f(z)\,dz$ exists as a weak integral (dz is the element of Lebesgue measure) and is a Hilbert-Schmidt operator.

 b. * With the notation of Exercise 15, let $c(z,z') = e^{iB(z,z')}$, and define the "skew convolution" of any two functions f and g on $\mathbf{L} \oplus \mathbf{L}'$ as the function h, given by the equation

$$h(z) = \int c(z,z')f(z - z')g(z')\,dz',$$

where dz' is the element of Lebesgue measure.   Show that this operation is associative but not commutative on L_1 and that $\|h\|_2 \leq \|f\|_2 \|g\|_2$.

19 Show directly from the Plancherel theorem that the Fourier transform of an integrable function on a locally compact Abelian group G vanishes at infinity on G^*. (This is the generalized Riemann-Lebesgue lemma. Write the integrable function f as the product of two square-integrable functions; use the Plancherel theorem and the Schwarz inequality.)

XI

SEMIGROUPS AND PERTURBATION THEORY

11.1 INTRODUCTION

A semigroup may be defined as a subset of a group which contains the unit and is closed under multiplication. But the only one to be considered in this chapter is that of all non-negative reals, as a subset of the group of all real numbers under addition. This is the basic case for extension to Lie semigroups, just as the theory of one-parameter groups is the basic case for extension to the representation theory of Lie groups.

Such semigroups of operators arise for example in the theory of diffusion processes and parabolic differential equations; complex one-parameter semigroups arise as analytic continuations of real one-parameter unitary groups having a semibounded generator, such as frequently occur in physical applications; etc. It is important for some of the applications to treat semigroups in Banach spaces, as well as in Hilbert space; but difficult to go effectively beyond the degree of generality represented by Banach spaces.

11.2 THE HILLE-YOSIDA THEOREM

A natural and useful class of semigroups to consider consists of the *regular semigroups*.

DEFINITIONS   A regular semigroup in a Banach space **B** is a map $t \to V(t)$ from $[0,\infty)$ to the linear operators on **B**, which is strongly continuous, and satisfies the relations $V(0) = I$, $V(t + t') = V(t)V(t')$, $\|V(t)\| \leq e^{ct}(t,t' \in [0,\infty);\ c = \text{constant})$. If $c = 0$, the term *contraction semigroup* is used. The *infinitesimal generator* of a continuous semigroup $V(\cdot)$ in a Banach space **B** is the operator A defined by the equation $Ax = \lim_{t \to 0} t^{-1}(V(t) - I)x$ on the domain of all vectors x for which the indicated limit exists.

The following result, known as the Hille–Yosida theorem, follows from the spectral theorem in the case of a semigroup of normal operators on a Hilbert space. Conversely, a major part of the proof of Stone's theorem, or its extension to one-parameter groups of normal operators, is deducible from it.

THEOREM 11.1   *The generator A of a regular semigroup $U(\cdot)$ on a Banach space **B** is closed and densely defined. Moreover, if z is any complex number such that $\mathrm{Re}\, z > c$, where $\|U(t)\| \leq e^{ct}$, then $zI - A$ has a bounded inverse and*

$$(zI - A)^{-1} = \int_0^\infty e^{-tz} U(t)\, dt, \qquad \|(zI - A)^{-1}\| \leq (\mathrm{Re}\, z - c)^{-1}.$$

*Conversely, if A is a closed densely-defined operator in **B** such that $\|(\lambda I - A)^{-1}\| \leq |\lambda - c|^{-1}$ for real $\lambda > c$, then A is the generator of such a semigroup.*

Proof   We recall that a closed densely defined operator T is said to have a bounded inverse if it is univalent and onto; i.e., $Tx \neq 0$ if $x \neq 0$, and the range of T is all of **B**. T^{-1} is then everywhere defined, and being closed, as is easily seen, is bounded.

It is evidently sufficient to consider the case $c=0$, applicable to the trivial modification of A, $A - cI$. Assuming then $c = 0$, set

$$S(z) = \int_0^\infty e^{-tz} U(t)\, dt,$$

taken in the strong or weak operator topology. It is evident that $S(z)$ exists and immediate that $\|S(z)\| \leq (\mathrm{Re}\, z)^{-1}$. We shall show that $S(z) = (zI - A)^{-1}$ by direct computation, but prove first that A is closed and densely defined.

If x is in $\mathbf{D}(A)$, then an easy computation shows that $U(t)x$ is also in $\mathbf{D}(A)$, and

$$\frac{d}{dt} U(t)x = AU(t)x = U(t)Ax;$$

integrating, one obtains the formula (essentially Duhamel's)

$$(*) \qquad U(t)x = x + \int_0^t U(s)Ax \, ds.$$

Now suppose that $u_n \in \mathbf{D}(A)$, $n = 1, 2, \ldots$, and $u_n \to u$, while $Au_n \to y$. It follows from formula (*), with x replaced by u_n, that

$$U(t)u = u + \int_0^t U(s)y \, ds.$$

Hence

$$t^{-1}[U(t) - I]u = t^{-1}\int_0^t U(s)y \, ds \to y,$$

showing that A is closed. Further, for arbitrary x in $\mathbf{B}$, setting $y = \int_0^\infty U(s)xf(s) \, ds$, f being a fixed C^1 function of compact support on R^1, which vanishes on the negative half-axis, then

$$t^{-1}[U(t) - I]y = t^{-1}\left[\int_0^\infty U(s+t)xf(s) \, ds - \int_0^\infty U(s)xf(s) \, ds\right]$$

$$= \int_0^\infty U(s)x[f(s-t) - f(s)]t^{-1} \, ds \to -\int_0^\infty U(s)xf'(s) \, ds.$$

Thus $y \in \mathbf{D}(A)$; but if $f = f_n$, where $\{f_n\}$ is a sequence such that $f_n(s) \geq 0$, $\int_0^\infty f_n(s) \, ds = 1$, and $\int_0^\epsilon f_n(s) \, ds \to 1$ for all $\epsilon > 0$ as $n \to \infty$, then $\int_0^\infty U(s)xf_n(s) \, ds \to x$, so that the set of all such vectors y is dense in $\mathbf{B}$.

Now let ϵ be an arbitrary positive number; then

$$\epsilon^{-1}[U(\epsilon) - I]S(\lambda)x = \epsilon^{-1}\left[\int_0^\infty e^{-\lambda s}\, U(s+\epsilon)x \, ds - \int_0^\infty e^{-\lambda s}\, U(s) \, x ds\right]$$

$$= \epsilon^{-1}\left[\int_\epsilon^\infty e^{-\lambda(s-\epsilon)}\, U(s)x \, ds - \int_0^\infty e^{-\lambda s}\, U(s) \, x ds\right]$$

$$= \epsilon^{-1}\int_0^\epsilon e^{-\lambda s}\, U(s)x \, ds +$$

$$\qquad\qquad + \int_\epsilon^\infty \epsilon^{-1}[e^{-\lambda(s-\epsilon)} - e^{-\lambda s}]U(s)x \, ds$$

$$\to -x + \int_0^\infty \lambda\, e^{-\lambda s}\, U(s) \, ds = -x + \lambda S(\lambda)x.$$

Thus, the range of $S(\lambda)$ is contained in $\mathbf{D}(A)$, and for all $x \in \mathbf{B}$,

$$AS(\lambda)x = [-I + \lambda S(\lambda)]x, \qquad \text{i.e., } (\lambda I - A)S(\lambda) = I.$$

Now suppose $x \in \mathbf{D}(A)$; then

$$S(\lambda)Ax = \int_0^\infty e^{-\lambda s}\, U(s)Ax\, ds = \int_0^\infty e^{-\lambda s}\, AU(s)x\, ds;$$

since A is closed, the latter integral equals in turn

$$A \int_0^\infty e^{-\lambda s}\, U(s)x\, ds = AS(\lambda)x.$$

Thus $S(\lambda)(\lambda I - A)x = x$, showing that $\lambda I - A$ is univalent; together with the preceding paragraph, this shows that $(\lambda I - A)^{-1}$ exists and equals $S(\lambda)$.

Now let A be a given closed operator in $\mathbf{B}$ with dense domain $\mathbf{D}$, such that $\|(\lambda I - A)^{-1}\| \le \lambda^{-1}$ for $\lambda > 0$; it must be shown that A is the generator of a contraction semigroup. To this end, set $V_\lambda(t) = e^{tA_\lambda}$, where $A_\lambda = A_\lambda = \lambda^2(\lambda I - A)^{-1} - \lambda I$; we aim to obtain the desired semigroup as the limit of the semigroup $V_\lambda(\cdot)$ as $\lambda \to \infty$. Note first that if $x \in \mathbf{D}$, then

$$\|\lambda(\lambda I - A)^{-1}x - x\| = \|(\lambda I - A)^{-1}Ax\| \le \lambda^{-1}\|Ax\|,$$

employing the hypothesized bound on $(\lambda I - A)^{-1}$. It follows that $\lambda(\lambda I - A)^{-1}x \to x$ as $\lambda \to \infty$ if $x \in \mathbf{D}$, and since $\lambda(\lambda I - A)^{-1}$ is uniformly bounded, this follows in turn for all $x \in \mathbf{B}$.

Now setting $R(\lambda) = (\lambda I - A)^{-1}$,

$$V_\lambda(t) = e^{-\lambda t}\, e^{t\lambda^2 R(\lambda)} = e^{-\lambda t} \sum_{n=0}^\infty (t\lambda^2)^n (n!)^{-1} R(\lambda)^n$$

which implies that

$$\|V_\lambda(t)\| \le e^{-\lambda t} \sum (t\lambda^2)^n (n!)^{-1} \lambda^{-n} = 1.$$

Writing

$$V_\lambda(t) - V_{\lambda'}(t) = \int_0^t (d/ds)[V_\lambda(s)V_{\lambda'}(t - s)]\, ds,$$

and carrying out the differentiation, it results that

$$V_\lambda(t) - V_{\lambda'}(t) = \int_0^t V_\lambda(s)(A_\lambda - A_{\lambda'})V_{\lambda'}(t - s)\, ds$$

$$= \int_0^t V_\lambda(s)V_{\lambda'}(t - s)(A_\lambda - A_{\lambda'})\, ds,$$

noting that A_λ commutes with $V_{\lambda'}(r)$ for arbitrary λ, λ' and r.

It follows that for arbitrary $x \in \mathbf{B}$,

$$\|(V_\lambda(t) - V_{\lambda'}(t))x\| \le t\|(A_\lambda - A_{\lambda'})x\|.$$

Now if $x \in \mathbf{D}$, then

$$A_\lambda x = (\lambda^2(\lambda I - A)^{-1} - \lambda)x = (\lambda I - A)^{-1}[\lambda^2 I - \lambda(\lambda I - A)]$$
$$= \lambda(\lambda I - A)^{-1}Ax \to Ax,$$

implying that $\|(V_\lambda(t) - V_{\lambda'}(t))x\| \to 0$ uniformly in t, t remaining bounded. From the uniform boundedness of the $V_\lambda(t)$ it follows that the same holds for arbitrary $x \in \mathbf{B}$. It follows straightforwardly in turn that there exists a contraction semigroup $V(\cdot)$ such that $\|(V_\lambda(t) - V(t))x\| \to 0$, uniformly in t on bounded t-intervals.

Recalling that for arbitrary $x \in \mathbf{D}$, $V_\lambda(t)x = x + \int_0^t V_\lambda(s)A_\lambda x \, ds$, it follows on letting $\lambda \to \infty$ that for $t > 0$, $t^{-1}[V(t)x - x] = t^{-1}\int_0^t V(s)Ax \, ds$. Letting $t \to 0$, it follows in turn that $A \subset A'$, where A' denotes the generator of the semigroup $V(\cdot)$. On the other hand, if $\lambda > 0$ then $(\lambda I - A)$ is one-to-one onto $\mathbf{B}$, since $(\lambda I - A)^{-1}$ exists, and $(\lambda I - A')$ can have this property, which it must by the first part of the theorem, only if A' is a trivial extension of A, i.e. $A = A'$.

11.3   CONVERGENCE OF SEMIGROUPS

A central problem in quantum mechanics is that of the establishment of groups or semigroups of operators, representing physically the temporal development, or closely related features, of a physical system. The most natural mathematical models from a physical point of view are frequently relatively singular ones, whose establishment requires approximation by more regular groups or semigroups (which in general will fail to satisfy such physical desiderata as "locality" or "covariance", but will be analytically relatively tractable). In this way the question of modes of approximation of a group or semigroup by another group or semigroup becomes technically important.

DEFINITION   A sequence $\{V_n(\cdot); n = 1, 2, \ldots\}$ of regular semigroups in a Banach space $\mathbf{B}$ (or the corresponding generators) is said to be *convergent* to the regular semigroup $V(\cdot)$ in $\mathbf{B}$ (or the corresponding generator) in case $V_n(t)x \to V(t)x$ for all $x \in \mathbf{B}$ and $t \in [0,\infty)$, and $\|V_n(t)\| \leq e^{ct}$ for some constant c. The basic case is that of contraction semigroups:

THEOREM 11.2   *A sequence of contraction semigroups* $V_n(\cdot); n = 1, 2, \ldots$ *in a Banach space* $\mathbf{B}$ *is convergent to the contraction semigroup* $V(\cdot)$ *in* $\mathbf{B}$ *if and only if one of the following equivalent conditions holds:* (A_n *and* A *denoting the respective generators*)

a. $(\lambda I - A_n)^{-1} \to (\lambda I - A)^{-1}$ *(strongly) for some* λ, $\mathrm{Re}\ \lambda > 0$;

b. *(a) holds for all* λ *such that* $\mathrm{Re}\ \lambda > 0$;

c. for each x in **B**, $V_n(t)x \to V(t)x$ *uniformly on each finite t-interval.*

Proof   We will show that convergence $\to (a) \to (b) \to (c) \to$ convergence. Indeed, (a) follows by bounded convergence from the assumption of convergence, for any value of λ. Now assuming that (a) holds with $\lambda = \lambda_0$, then for any λ with Re $\lambda > 0$

$$(\lambda I - A_n) = (\lambda - \lambda_0)I + (\lambda_0 I + A_n) = (\lambda_0 I - A_n)[I + (\lambda - \lambda_0)(\lambda_0 I - A_n)^{-1}].$$

Using the fact that $(I + B)^{-1}$ exists and is a continuous function of B in the region $[B \colon \|B\| < 1]$, it follows that

$$(\lambda I - A_n)^{-1} \to (\lambda I - A)^{-1} \quad \text{if} \quad |\lambda - \lambda_0| < \text{Re } \lambda_0.$$

Thus (a) holds for all λ in an open set, each point λ_0 of which contains a disk of radius Re λ_0; it follows that (b) holds.

Now assume that (b) holds; this means that

$$\int_0^\infty e^{-\lambda t} V_n(t)x \, dt \to \int_0^\infty e^{-\lambda t} V(t)x \, dt$$

for all $x \in$ **B**. It follows that

$$(*) \qquad \int_0^\infty V_n(t)xf(t) \, dt \to \int_0^\infty V(t)xf(t) \, dt$$

for all f of the form: $f(t) = \sum_{k=1}^N a_k e^{-\lambda_k t}$, $\lambda_k > 0$, N finite. The set of all such f is dense in $L_1(0,\infty)$ (e.g., by the Weierstrass approximation theorem), whence $(*)$ holds for all $f \in L_1(0,\infty)$.

To show that $V_n(t) \to V(t)$ strongly, it suffices to show that $V_n(t)x \to V(t)x$ for all x in a dense set (whether pointwise in t or uniformly on finite t-intervals), by virtue of the uniform boundedness of the $V_n(t)$. Consider in particular the dense set **D** consisting of all vectors x of the form

$$x = \int_0^\infty V(s)zf(s) \, ds$$

for some $z \in$ **B** and $f \in L_1(0,\infty)$. Setting

$$x_n = \int_0^\infty V_n(s)zf(s) \, ds,$$

then

$$V_n(t)x - V(t)x = [V_n(t)x - V_n(t)x_n] + [V_n(t)x_n - V(t)x].$$

Now $\|V_n(t)x - V_n(t)x_n\| \le \|x - x_n\| \to 0$ by what has already been shown. The second term on the right is

$$V_n(t)x_n - V(t)x = \int_0^\infty V_n(t+s)zf(s)\,ds - \int_0^\infty V(t+s)zf(s)\,ds$$

$$= \int_0^\infty V_n(s)zf_t(s)\,ds - \int_0^\infty V(s)zf_t(s)\,ds,$$

where

$$f_t(s) = \begin{cases} f(s-t), & s \geq t \\ 0 & s < t. \end{cases}$$

By what has already been shown, it follows that $V_n(t)x - V(t)x \to 0$ for each t; and noting the continuity of the map $t \to f_t$ from $[0,\infty)$ into $L_1(0,\infty)$ and the estimate: $\left\| \int_0^\infty V(s)g(s)\,ds \right\| \leq \|g\|_1$, it follows by an elementary compactness and approximation argument that the convergence is uniform on each finite t-interval.

COROLLARY 11.3.1   *Suppose that* **D** *is a dense subset of* **B**, *that* A_n *($n = 1, 2, \ldots$) and* A *are generators of contraction semigroups on* **B**, *and that* A *is the closure of its restriction to* **D**. *If* $A_n x \to A x$ *for all* $x \in$ **D**, *then* $A_n \to A$.

Proof Since $\|(\lambda I - A_n)^{-1}\| \leq \lambda^{-1}$ for $\lambda > 0$, and similarly for A, it suffices to show that

$$(\lambda I - A_n)^{-1}u \to (\lambda I + A)^{-1}u$$

for all u in a dense subset of **B**. From the assumption that $A = \overline{A \mid \mathbf{D}}$, it follows that $(\lambda I - A)\mathbf{D}$ is such a subset.

Taking $u \in (\lambda I - A)\mathbf{D}$, say $u = (\lambda I - A)v$, with $v \in$ **D**, then by a simple computation

$$(\lambda I - A_n)^{-1}u - (\lambda I - A)^{-1}u = (\lambda I - A_n)^{-1}(A_n - A)v.$$

It follows that

$$\|(\lambda I - A_n)^{-1}u - (\lambda I - A)^{-1}u\| \leq \lambda^{-1}\|(A_n - A)v\| \to 0.$$

Stability of convergence under bounded perturbation is shown by

COROLLARY 11.3.2   *Let* A_n *($n = 1, 2, \ldots$) and* B *be generators of regular semigroups on the Banach space* **B**, *and suppose that* $A_n \to A$. *If* B *is a continuous linear operator on* **B**, *then* $A_n + B \to A + B$.

Proof By Duhamel's formula, for arbitrary $u \in$ **B**,

$$e^{t(A_n + B)}u = e^{tA_n}u + \int_0^t e^{(t-s)A_n} B\, e^{s(A_n + B)}u\,ds,$$

and similarly for $e^{t(A+B)}$. Subtracting the latter equation from the former,

$$e^{t(A_n+B)} u - e^{t(A+B)} u = (e^{tA_n} u - e^{tA} u)$$

$$+ \int_0^\infty [e^{(t-s)A_n} B\, e^{s(A_n+B)} u - e^{(t-s)A} B\, e^{s(A+B)} u]\, ds.$$

Now setting $g_n(t) = \|e^{t(A_n+B)} u - e^{t(A+B)} u\|$, and noting that the foregoing integrand may be written as

$$[e^{(t-s)A_n} - e^{(t-s)A}]B\, e^{s(A+B)} u + e^{(t-s)A_n} B[e^{s(A_n+B)} - e^{s(A+B)}]u,$$

it follows that

$$(*) \qquad\qquad g_n(t) \le a_n(t) + \|B\| \int_0^t g_n(s)\, ds.$$

where

$$a_n(t) = \|e^{tA_n} u - e^{tA} u\| + \int_0^\infty \|(e^{(t-s)A_n} - e^{(t-s)A})B\, e^{s(A+p)} u\|\, ds.$$

By dominated convergence, $a_n(t) \to 0$, evidently boundedly in each finite t-interval. Now recall Gronwell's inequality, according to which equation $(*)$ implies that $\int_0^t g_n(s)\, ds \le e^{\|B\|t} \int_0^t a_n(s)\, e^{-\|B\|s}\, ds$ (this is derived by setting $G_n(t) = \int_0^t g_n(s)\, ds$ and using the integrating factor $e^{-\|B\|t}$ on the inequality: $G_n'(t) = \|B\|G(t) \le a_n(t)$); it follows that $g_n(t) \to 0$ for all t.

The next result can be regarded as a natural and quite useful generalization of the representation of the exponential function e^x as $\lim_{n\to\infty} \left(1 + \dfrac{x}{n}\right)^n$; the basic estimate can be regarded as a generalization of one for the constant c_n in the estimate

$$\left| e^z - \left(1 + \frac{z}{n}\right)^n \right| \le c_n|z| \qquad (|z| \le 1).$$

THEOREM 11.3 (P. CHERNOFF) *Let $t \to U(t)$ be strongly continuous from $[0,\infty)$ to the linear contractions on the Banach space $\mathbf{B}$, and such that $U(0) = I$. Let $\mathbf{D}$ be a dense linear subset of $\mathbf{B}$ on which the generator A of the contraction semigroup is determined (i.e., the closure of its restriction), and suppose that for all $x \in \mathbf{D}$, $\epsilon^{-1}(U(\epsilon) - I)x \to Ax$, as $\epsilon \to 0$.*
Then $U(t/n)^n \to e^{tA}$ (strongly) for all $t \ge 0$.

Lemma 11.3.1 *Let C be a linear contraction on $\mathbf{B}$. Then the map $t \to e^{t(C-I)}$ from $[0,\infty)$ to the linear operators on $\mathbf{B}$ is a contraction semigroup, and for all $u \in \mathbf{B}$ and positive integers n,*

$$\|e^{n(C-I)}u - C^n u\| \le n^{1/2}\|Cu - u\|.$$

Proof Evidently, if $t > 0$,

$$\|e^{t(C-I)}\| = e^{-t}\left\|\sum_{m=0}\frac{t^m C^m}{m!}\right\| \leq e^{-t}\sum_{m\geq 0}\frac{t^m}{m!} = 1.$$

Thus the indicated map is a contraction semigroup.

Now for arbitrary $u \in \mathbf{B}$,

$$\|e^{n(C-I)}u - C^n u\| = \left\|e^{-n}\sum_{m\geq 0}\frac{n^m}{m!}(C^m - C^n)u\right\|$$

$$\leq e^{-n}\sum_{m=0}\frac{n^m}{m!}\|(C^m_r - C^n)u\|$$

$$\leq e^{-n}\sum_{m\geq 0}\frac{n^m}{m!}\|(C^{|m-n|} - I)u\|$$

$$\leq e^{-n}\left[\sum_{m\geq 0}\frac{n^m}{m!}|m-n|\right]\|(C-I)u\|.$$

The constant factor on the right is easily bounded, by Schwarzing and elementary manipulations, by $n^{1/2}$, completing the proof.

Proof of Theorem 11.3 It is no essential loss of generality to take $t = 1$. Set $A_n = n(U(1/n) - I)$; then A_n generates a contraction semigroup, by Lemma 11.3.1. By Corollary 11.3.1, $A_n \to A$; in particular, $e^{A_n} \to e^A$.

Now let v be arbitrary in $\mathbf{D}$; the lemma then implies that

$$\|e^{A_n}v - U(1/n)v\| \leq n^{1/2}\|U(1/n)v - v\| = n^{-1/2}\left\|\frac{(U(1/n) - I)}{1/n}v\right\| \to 0.$$

Thus $U(1/n)^n V \to e^A v$ for all v in $\mathbf{D}$, but as the operators in question are contractions, this follows for all v.

COROLLARY 11.3.3 (TROTTER) *If A and B and the closure of $A + B$ are all generators of contraction semigroups on the Banach space $\mathbf{B}$, then for all $t \geq 0$:*

$$e^{t(A+B)} = \lim_{n\to\infty}(e^{tA/n}e^{tB/n})^n \quad \text{(strongly)}.$$

Proof Set $U(t) = e^{tA}e^{tB}$; it is straightforward to verify that

$$t^{-1}(U(t) - I)x \to (A + B)x$$

for all $x \in \mathbf{D}(A) \cap \mathbf{D}(B)$. Thus the hypotheses of Theorem 11.3 are satisfied, and the corollary follows.

Illustrative of a number of further corollaries (not required later) is

COROLLARY 11.3.4 *If A is the generator of a contraction semigroup in the Banach space $\mathbf{B}$, then for all $t \geq 0$:*

$$e^{tA} = \lim_{n \to \infty} \left(I - \frac{tA}{n}\right)^{-n} \quad \text{(strongly)}.$$

The proof is left as an exercise.

11.4 STRONG CONVERGENCE OF SELF-ADJOINT OPERATORS

The set **S** of all self-adjoint operators in a Hilbert space **K** is not a simple algebraic object. In general, one cannot even add two operators and obtain one which is essentially self-adjoint, even when they admit a common dense domain; and even when one can, this addition is not associative (i.e., the operation $A + B$ = closure of $A + B$), even assuming all sums involved on either side exist and are self-adjoint. It is a convenient feature of **S**, however, that the topology previously introduced for sequences of semigroup generators specializes to one which coincides with some other relevant topologies.

> **THEOREM 11.4**   *The following notions of convergence in a Hilbert space, of a sequence $\{A_n\}$ of self-adjoint operators to a self-adjoint operator A, are all equivalent*:
>
> a. $e^{itA_n} \to e^{itA}$ *(strongly) for all $t \in R^1$.*
> b. *The same as a, uniformly on every finite t-interval.*
> c. *For all bounded and continuous functions f on R^1,*
>
> $$f(A_n) \to f(A) \quad \text{(strongly)}.$$
>
> d. *The same as c, only for those f of compact support.*
> e. *The same as c, for all f which are characteristic functions of intervals, whose endpoint(s) are not in the point spectrum of A.*
> f. *The Cayley transform of $A_n \to$ that of A (strongly).*
> g. *For some $\lambda \neq 0$, $(A_n + i\lambda I)^{-1} \to (A + i\lambda I)^{-1}$.*

A unified approach to these equivalences, which are due to Rellich, Kato and others, can be derived via the spectral theorem from a well-known theorem of E. Helly (1921) on convergences of interval functions on the line, together with equivalences already obtained. In amplified form Helly's theorem is as follows.

Lemma 11.4.1   *Let $a_1, a_2, \ldots$ and a be in the set M of all regular probability measures on R^1. The following conditions are all equivalent*:

(i) $\int f(x)\, da_n(x) \to \int f(x)\, da(x)$ *for all continuous f on R^1 which have compact support;*

(ii) *the same as (i) for all bounded and continuous f;*

(iii) $\hat{a}_n(t) \to \hat{a}(t)$ *for all $t \in R^1$ (where $\hat{a}(t) = \displaystyle\int_{-\infty}^{\infty} e^{itx}\, da(x)$).*

(iv) $a_n(E) \to a(E)$ *for all intervals E whose endpoints have a-measure zero.*

REMARK   Further equivalent conditions are that $\hat{a}_n(t) \to \hat{a}(t)$ uniformly on every finite t-interval; or that $\hat{a}_n(t) \to \hat{a}(t)$ a.e.; but we shall not need these.

Proof   We show (iii) $\to$ (i) $\to$ (iv) $\to$ (ii) $\to$ (iii).

$Ad\,(iii) \to (i)$:   The $\hat{a}_n(t)$ and $\hat{a}(t)$ are uniformly bounded by 1, so that from (iii) it follows that

$$\int g(t)\hat{a}_n(t)\,dt \to \int g(t)\hat{a}(t)\,dt$$

for all $g \in L_1(R^1)$. By Fubini's theorem, this is the same as the assertion

$$(*) \qquad \int \hat{g}(x)\,da_n(x) \to \int \hat{g}(x)\,da(x).$$

But as is well known (or follows directly from the Riemann–Lebesgue lemma together with the Stone–Weierstrass theorem), the set of all such $\hat{g}$ is dense in the space $\mathbf{C}$ of all continuous functions on R^1 which vanish at ∞ (relative to uniform convergence). It follows that equation $(*)$ remains valid when $\hat{g}$ is replaced by an arbitrary function in $\mathbf{C}$.

$Ad(i) \to (iv)$:   It suffices to consider finite intervals, for assuming this case established, then for arbitrary $\lambda' < \lambda$ such that $a(\{\lambda\}) = a(\{\lambda'\}) = 0$ and $a_n(\{\lambda\}) = 0$ for all n,

$$a((\lambda',\lambda)) = \lim_n a_n((\lambda',\lambda)) \leq \liminf_n a_n((-\infty,\lambda)),$$

implying that

$$a((-\infty,\lambda)) \leq \liminf_n a_n((-\infty,\lambda)).$$

Similarly,

$$a((\lambda,\infty)) \leq \liminf_n a_n((\lambda,\infty)).$$

Adding these two inequalities,

$$1 \leq \liminf_n a_n((-\infty,\lambda) + \liminf_n a_n((\lambda,\infty))$$
$$\leq \liminf_n [a_n((-\infty,\lambda) + a_n((\lambda,\infty))] = 1.$$

Thus the inequalities collapse and

$$\lim_n a_n((-\infty,\lambda) = a((-\infty,\lambda)); \qquad \lim_n a_n(\lambda,\infty)) = a(\lambda,\infty).$$

Now let E be a finite interval whose endpoints have a-measure zero. For any $\epsilon > 0$ there exists (by the regularity of the measure a) an element $f \in \mathbf{C}$ such that $f \geq c_E$ ($=$ characteristic function of E) such that

$$\int f(x)\, da(x) \le a(E) + \epsilon.$$

Now $\int f(x)\, da_n(x) \to \int f(x)\, da(x)$, so for sufficiently large n,

$$\int f(x)\, da_n(x) \le a(E) + 2\epsilon.$$

Hence

$$a_n(E) \le a(E) + 2\epsilon$$

for sufficiently large n, whence $\lim_n a_n(E) \le a(E)$.

Similarly, there exists $f \in \mathbf{C}$, $0 \le f \le c_E$, such that

$$\int f(x)\, da(x) \ge a(E) - \epsilon;$$

and paralleling the preceding paragraph, it follows that $\lim_n a_n(E) \ge a(E)$. It results that $\lim_n a_n(E) = a(E)$.

Ad $(iv) \to (ii)$:   Given any $\epsilon > 0$, there exsits a finite interval E on whose endpoints a vanishes, such that $a(R^1 - E) < \epsilon$. Since $a_n(R^1 - E) \to a(R^1 - E)$, $a_n(R^1 - E) < 2\epsilon$ for sufficiently large n.

If now f is, any bounded and continuous function on R^1, it is evident that there exist continuous functions f_1 and f_2 such that $f = f_1 + f_2$, $f_1 \in \mathbf{C}$, f_2 is supported by $R^1 - \bar{E}$, $\|f_i\| \le \|f\|$. Then

$$\int f\, da_n - \int f\, da = \left[\int f_1\, da_n - \int f_1\, da \right] + \left[\int f_2\, da_n - \int f_2\, da \right];$$

$$\left| \int f_2\, da_n - \int f_2\, da \right| \le \|f_2\|[a_n(R^1 - R) + a(R - E)] \le 3\epsilon\|f\|.$$

On the other hand, $\int f_1\, da_n - \int f_1\, da$ can be made arbitrarily small by choosing n sufficiently large, and it follows that $\int f\, da_n \to \int f\, da$.

Ad $(ii) \to (iii)$:   Trivial.

Proof of Theorem 11.4   We shall show that $(a) \to (d) \leftrightarrow (g) \leftrightarrow (f) \to (e) \to (c) \to (b) \to (a)$.

Ad $(a) \to (d)$: If f is arbitrary in $L_1(R^1)$, it follows that for every vector $u \in \mathbf{K}$,

$$\int e^{ut A_n}\, uf(t)\, dt \to \int e^{it A}\, uf(t)\, dt,$$

i.e., $\hat{f}(A_n) \to \hat{f}(A)$. (d) now follows from the density of the absolutely convergent Fourier transforms in $\mathbf{C}$.

Ad (d) → (g): Trivial, noting that the elements of **C** which have compact support are dense in **C**.

Ad (g) → (d): The function of t, say $h(t) = (i\lambda + t)^{-1}$, is in **C**, and (g) asserts that $h(A_n) \to h(A)$. Thus if **A** denotes the subset of **C** consisting of those $g \in$ **C** such that $g(A_n) \to g(A)$, **A** includes h; and is easily seen to be an algebra which is closed under complex conjugation. Since it separates points (h does so) it must be all of **C** by the Stone-Weierstrass theorem.

Ad (d) → (f): The Cayley transform $(A - iI)^{-1}(A + iI)$ differs trivially from $(A - iI)^{-1}$, a function of A which is a uniform limit of those in (d).

Ad (f) → (d): This follows by reversal of the foregoing argument, together with the already proved equivalence of (d) and (g).

Ad (d) → (c): It must be shown that if E is any real interval whose endpoints are not in the point spectrum of A, then

$$c_E(A_n) \to c_E(A),$$

given that $f(A_n) \to f(A)$ for all $f \in$ **C** (as noted earlier, (d) implies this). To show that $c_E(A_n) \to c_E(A)$ it suffices, since these are projections, to show that for all $x \in$ **K**, $(c_E(A_n)x,x) \to (c_E(A)x,x)$; and it is evidently no essential restriction to set $\|x\| = 1$. If a_n and a are the measures

$$a_n(S) = (c_s(A_n)x,x), \qquad a(S) = (c_s(A)x,x),$$

then $\int f\,da_n \to \int f\,da$ by hypothesis, so the required conclusion follows from Lemma 11.4.1.

Ad (e) → (c): An argument parallel to that in the proof that (iv) → (ii) in the lemma shows that it suffices to prove convergence for the case when f has compact support. But every such f is a uniform limit of step functions, and it is easily seen that these may be chosen so that none of the endpoints of the intervals involved is in any given countable set, such as the point spectrum of A. This hypothesis implies convergence for such step functions, and the implication follows.

Ad (c) → (b): This is a special case of Theorem 11.2 (specifically, that convergence implies (c)).

Ad (b) → (a): Trivial.

COROLLARY 11.4.1 *If the A_n $(n = 1, \ldots)$ and A are non-negative self-adjoint operators in a Hilbert space, then $A_n \to A$ if and only if any one of the following equivalent conditions holds:*

  a, $e^{-tA_n} \to e^{-tA}$ for all $t > 0$.
  b. The same as a for one $t > 0$.
  c. $(\lambda I - A_n)^{-1} \to (\lambda I - A)^{-1}$ for all $\lambda > 0$.
  d. $(\lambda I - A_n)^{-1} \to (\lambda I - A)^{-1}$ for some $\lambda > 0$.

Proof It is immediate that if $A_n \to A$, then all of (a), (b), (c), (d) hold (note that since $A \geq 0$, $e^{-tA} = f_t(A)$ if f_t is any bounded continuous function on R^1 such that $f_t(x) = e^{-tx}$ for all $x \geq 0$; and similarly for $(\lambda I - A)^{-1}$). Thus, it suffices to show that (b) implies convergence, and that (d) implies convergence. The argument is essentially the same for these two implications; consider for example the hypothesis (b). As earlier, the set of all continuous functions F such that $F(A_n) \to F(A)$ forms a uniformly closed algebra $\mathbf{A}$ which is closed under complex conjugation. By hypothesis and the observations made at the beginning, $\mathbf{A}$ includes all functions f_t of the indicated form. These evidently separate points on R^1, so that $\mathbf{A}$ includes all of $\mathbf{C}(R^1)$ by the Stone-Weierstrass theorem.

> *Example*   Let M be a measure space, let $\mathbf{K}$ be the Hilbert space $L_2(M)$, and for any measurable function k on M, let M_k denote the operation of multiplication by k, i.e., $M_k f = kf$, the domain of M_k consisting of all $f \in L_2(M)$ such that $kf \in L_2(M)$. It is easy to verify that $M_k^* = M_k$. In particular M_k is self-adjoint when k is real. It is not difficult to show that M_k is bounded if and only if k is essentially bounded, and that $\|M_k\| = \|k\|_\infty$ (the notation $\|k\|_p$ denotes $(\int |k|^p)^{1/p}$).
>
> Now let $\{k_n\}$ be a sequence of real measurable functions on M. If k is a measurable function on M such that $k_n(x) \to k(x)$ a.e., then $M_{k_n} \to M_k$; the proof is left as an exercise.
>
> Since every self-adjoint operator is, within unitary equivalence, of the form M_k, k real, it follows that if $\{f_n\}$ is any sequence of Baire functions on R^1 such that $f_n(\lambda) \to f(\lambda)$ for all $\lambda \in R^1$, then $f_n(A) \to f(A)$ for every self-adjoint operator A. See also Example 10.3.1, p. 279, and Theorem 10.5, p. 283.

11.5   RELLICH-KATO PERTURBATIONS

We now turn to the theorem of Rellich which is central in classical perturbation theory.

> **THEOREM 11.5**   *Let A be a self-adjoint operator in a Hilbert space $\mathbf{K}$; let $\mathbf{C}_b$ denote the set of all hermitian operators B in $\mathbf{K}$ such that*
>
> $(*)$ $$\|Bx\| \leq a\|Ax\| + b\|x\| \qquad (x \in \mathbf{D}(A))$$
>
> *for some constant $a < 1$. Then the map*
>
> $$B \to A + B$$
>
> *is continuous from $\mathbf{C}_b$ to the self-adjoint operators in $\mathbf{K}$, with the topology in $\mathbf{C}_b$: $B_n \overset{\sim}{\to} B$ means that $\|B_n x - Bx\| \leq c_n\|Ax\| + d_n\|x\|$, where $c_n \to 0$ and $d_n \to 0$.*

Proof   Consider first the self-adjointness of $A + B$, which in fact is the content of the original theorem; the following proof is due to Nagy. Suppose $\lambda > 0$; then for arbitrary $x \in \mathbf{K}$ it follows from the inequality $(*)$ that

$$\|B(A + i\lambda I)^{-1}x\| \le a\|A(A + i\lambda I)^{-1}x\| + b\|(A + i\lambda I)^{-1}x\|.$$

It follows, setting $D = B(A + i\lambda I)^{-1}$, that if λ is sufficiently large, $\|D\| < 1$. Hence $(I + D)^{-1}$ exists and is bounded. Setting $E = (A + i\lambda I)^{-1}(I + D)^{-1}$, then E is a bounded linear operator on $\mathbf{K}$ and $(A + B + i\lambda I)E = (A + i\lambda I)E + BE = (I + D)^{-1} + D(I + D)^{-1} = I$. Thus the range of $A + B + i\lambda I$ is all of $\mathbf{K}$, for all sufficiently large λ, which by an elementary criterion implies that $A + B$ is self-adjoint.

To show that $A + B$ is a continuous function of B it suffices, by Theorem 11.4, to show that $(A + B + i\lambda)^{-1}$ is such for some real $\lambda \ne 0$. This follows from the observation that $B(A + i\lambda I)$ is a continuous function of B in the $\mathbf{C}_b$-topology, if λ is sufficiently large, together with the continuity of $(I + X)^{-1}$ as a function of the linear operator X on the set $[X: \|X\| < 1]$.

REMARK   It is evident that if (*) is weakened to the requirement that it holds with $a = 1$, then the conclusion as to self-adjointness is no longer valid. For if A is unbounded and $B = -A$, then this condition holds, but $A + B$ is then defined only on the domain of A as 0, and so is not self-adjoint. It is of course essentially self-adjoint, and R. Wüst (1971) has shown that this conclusion is valid in general for the perturbation of one self-adjoint operator by another.

EXERCISES

1   With the notation and assumptions of the Example, Section 4, show that $M_{k_n} \to M_k$ if $\{k_n\}$ converges to k in measure.

2   With the hypotheses of Theorem 11.5, show that if A is bounded below, the same is true of $A + B$; and that if $b(A)$ denotes the essential infimum of A, that if $A_n \to A$, then $b(A) \le \liminf_n b(A_n)$.

3   Let $V(\cdot)$ be a regular semigroup on a Banach space $\mathbf{B}$, and let $f(\cdot)$ be continuously differentiable from $[0, \infty)$ into $\mathbf{B}$. Show that if u_0 is any given vector in the domain of the generator A of $V(\cdot)$, then the function $u(\cdot)$ from $[0, \infty)$ into $\mathbf{B}$ given by the equation $u(t) = V(t)u_0 + \int_0^t V(t - s)f(s)\, ds$ is the unique continuously differentiable solution of the abstract differential equation, $u'(t) = Au(t) + f(t)$.

4   With the notation of the preceding example, relax the condition on f to continuity. Show that $u(\cdot)$ is then the unique continuous function from $[0, \infty)$ to $\mathbf{B}$ such that if $0 \le s \le t$, then $u(t) = V(t - s)u(s) + \int_s^t V(t - r)f(r)\, dr$. Deduce that there exists a unique function $W(t,s)$ from the subset Δ of $R \times R$ consisting of (t,s) such that $0 \le s \le t$ to the bounded linear operators on $\mathbf{B}$ which is a strongly continuous function on Δ, and satisfies the relations:

a. $W(t,s)W(s,r) = W(t,r)$ if $r \le s \le t$;

b. $W(t,t) = I$ for all $t \ge 0$. (Such a function may be called a propagator.)

5   Let $V(\cdot)$ be a regular semigroup on the Banach space **B**, and let $B(\cdot)$ be a continuous map from $[0,\infty)$ to **B**. Show there exists a unique propagator $W(.,.)$ such that for all $u \in \mathbf{B}$,

$$W(t,s)u = V(t - s)u + \int_s^t V(t - r)B(r)W(r,s)u \, ds.$$

6   Show that if A is a self-adjoint operator in the Hilbert space **H**, then e^{A+B} is an entire analytic function of $B \in \mathbf{B}(\mathbf{H})$.

7   A *nonlinear semigroup* in a Banach space **B** is a strongly continuous function $T(\cdot)$ from $[0,\infty)$ to the nonlinear operators on **B** with the property that $T(t + t') = T(t)T(t')$ for all t, t'. Show that if $V(\cdot)$ is regular semigroup in **B**, and if K is any Lipschitzian operator on **B** (i.e., $\|K(x) - K(y)\| \le c\|x - y\|$ for some constant c and all $x, y \in \mathbf{B}$), then there is a unique nonlinear semigroup such that

$$T(t)u = V(t)u + \int_0^t V(t - s)K(T(s)u) \, ds$$

for all $u \in \mathbf{B}$.

8   Let L be a given self-adjoint operator in a Hilbert space **H** such that $K \ge \epsilon I$ for some $\epsilon > 0$. Let **K** denote the Hilbert space direct sum $[\mathbf{D}(A)] + \mathbf{H}$, where $[\mathbf{D}(A)]$ denotes the domain $\mathbf{D}(A)$ as a Hilbert space relative to the inner product $(x,y) = (Ax,Ay)$. Let $V(t)$ denote the operator on **K** whose matrix relative to the given decomposition is

$$\begin{pmatrix} \cos tB & B^{-1}\sin tB \\ -B\sin tB & \cos tB \end{pmatrix}.$$

Show that $V(\cdot)$ is a continuous one-parameter unitary group on **K**. Show further that $V(t)$ depends continuously on K (with the strong operator topology on $V(t)$). Show finally that if $(u_0,u_1) \in \mathbf{K}$ and $u_0 \in \mathbf{D}(K^2)$, $u_1 \in \mathbf{D}(K)$, and if $u(t)$ denotes the first component of $V(t)(u_0,u_1)$, then $u''(t) + Ku(t) = 0$ while $u(0) = u_0$, $u'(0) = u_1$.

9   Obtain a suitably formulated abstract solution of the differential equation on R^{n+1}, $[(\partial/\partial t)^2 - \Delta + m^2]\phi + p(\phi) = 0$, with given values for $\phi(0,x)$ and $(\partial/\partial t)\phi(0,x)$, where p is a given real Lipschitzian function, by combining the considerations of Exercise 7 with those of Exercise 8.

11.6   PERTURBATIONS IN A CALIBRATED SPACE

The classical theory originated by Rellich and developed by Kato, Birman, and others, of which Theorem 11.5 is typical, has many important applications, and is best possible without additional constraints on the operators involved. However, the restrictions it imposes on the perturbations admitted are much too strong for applicability to problems in nonlinear quantum field theory, or other areas where the perturbation is highly singular relative to the perturbed operator. Consideration of the simplest relativistic cases has led to a theory involving a scale of spaces, rather than a single space, which is effective in dealing with relatively singular perturbations.

In order to indicate the general idea and develop it in a representative case we make the

DEFINITION   A calibrated Hilbert space $\mathbf{K}$ is one in which there is given for each real $p \geq 2$ an extended-norm $\|\cdot\|_p$ (extended in the sense that it may admit infinite values) which is monotone increasing in p, such that the subspace $\mathbf{K}_p$ on which it is finite is a Banach space, to be denoted as $[\mathbf{K}_p]$, with respect to $\|\cdot\|_p$ as norm; with $\|\cdot\|_2$ equal to the given Hilbert norm in $\mathbf{K}$; and having the properties:

a. If $\|u_n - u\|_p \to 0$, then for all q, $\|u\|_q \leq \sup_n \|u_n\|_q$.

b. The subspace $\mathbf{D} = \bigcap_{p \geq 2} \mathbf{K}_p$ is dense in $[\mathbf{K}_p]$ for all p.

Example   Let M be an arbitrary probability measure space, let $\mathbf{K} = L_2(M)$, and $\|\cdot\|_p$ the L_p norm. Other examples include a non-commutative analog of probability spaces (see Chapter XIV), mixed L_p spaces for functions of several variables, and L_p spaces of entire holomorphic functions.

NOTATION   For any operator T in the calibrated space $\mathbf{K}$, $\|T\|_{p,q}$ will denote its bound as an operator from $[\mathbf{K}_p]$ to $[\mathbf{K}_q]$, i.e., $\|T\|_{p,q} = \sup_{\|u\|_p = 1} \|Tu\|_q$.

DEFINITION   A semigroup $V(\cdot)$ in $\mathbf{K}$ is *locally* (resp. *globally*) *of order α* with respect to the given calibration if for each given $p > 2$ and for all $t > 0$ sufficiently close to 0 (resp. all $t > 0$), $\|V(t)\|_{p,pe^{\alpha t}} \leq e^{\alpha t}$, where a is a constant which may depend on p but is bounded on p-bounded sets.

Example   With the example above based on a probability measure space, let U be any real measurable function, and let $V(t)$ denote the operator in $L_2(M)$: $f \to e^{tU}f$. Then by Hölder's inequality,

$$\|V(t)\|_{p,\,pe^{\alpha t}} \leq \|e^{tU}\|_q$$

if $q^{-1} = p^{-1}(e^{\epsilon t} - 1)$, where $\epsilon = -\alpha$, and $q \geq 1$. From the inequality $e^{\epsilon t} - 1 \geq \epsilon t$, it follows that $\|e^{tU}\|_q \leq \|e^U\|_r^t$, if $\epsilon = p/r$. This implies that if $e^U \in L_r$ for all $r < \infty$, then the semigroup $V(\cdot)$ is of order ϵ for every $\epsilon > 0$.

The general idea on which the theory of perturbations in a calibrated space is based is that if A and B are self-adjoint operators which generate semigroups of order α and β, where β may be negative and the semigroup generated by B consists of unbounded operators, and if $\alpha + \beta > 0$, then $A + B$ will exist in an effective sense, in the presence of appropriate secondary conditions which do not greatly limit the singularity of the perturbation B; and will be of, order at least $\alpha + \beta$. The elaboration of this idea depends mainly on the Lie formula and Duhamel's principle, which apply rather generally, even if bounds are given only locally in p and t. For concreteness and simplicity however we treat here only a useful special case involving global conditions on p and t.

> **THEOREM 11.6** *Let M be a probability measure space, $\mathbf{K}$ the Hilbert space $L_2(M)$ calibrated via the L_p norms, $2 \le p < \infty$; H a given self-adjoint operator in $\mathbf{K}$ of global order $\alpha > 0$, and V a real measurable function on M such that V and e^{-V} are both in L_p for all $p < \infty$. Then there exists a self-adjoint operator H' in $\mathbf{K}$ such that for any sequence $\{f_n\}$ of real functions on R^1 for which $f_n(\lambda) \to \lambda$ and $|f_n(\lambda)| \le |\lambda|$ for all $\lambda \in R^1$, $H + V_n \to H'$, where V_n denotes the operation of multiplication by $f_n \circ V$.*

Proof To avoid undue circumlocution, we use the same symbol for a function on M and the corresponding multiplication operator in $L_2(M)$. It suffices to show that $e^{-t(H+V_n)} \to e^{-tH'}$ for some $t \ne 0$, by Theorem 11.4.

To this end, let $\mathbf{C}$ denote the class of all multiplication operators of the hypothesized type, with fixed bounds on $\|V\|_p$ and $\|e^{-V}\|_p$ for all $p < \infty$; let $\mathbf{C}_b$ denote the subset consisting of bounded functions; and set $[\mathbf{D}]$ for the set $\mathbf{D}$ with the topology of convergence in each L_p space; and note the

Lemma 11.6.1   *For all $V \in \mathbf{C}_b$, $u \in \mathbf{D}$, and suitable constants α' and $c(t)$,*

$$\|e^{-t(H+V)}u\|_{2e^{\alpha' t}} \le c(t)\|u\|_2,$$

where $c(t)$ is bounded on bounded t-intervals.

Proof As an operator on L_2, $e^{-t(H+V)} = \operatorname{str} \lim_n (e^{-tV/n} e^{-tH/n})^n$, by Trotter's theorem, and hence by Fatou's lemma

$$\|e^{-t(H+V)}u\|_q \le \limsup_n \|(e^{-tV/n} e^{-tH/n})^n u\|_q$$

for all $q \ge 1$. To prove the lemma, it suffices to show that the inequality

$$\|(e^{-tV/n} e^{-tH/n})^r u\|_{2e^{\alpha tr/n}} \le e^{\alpha' tr/n}\|u\|_2$$

for some constant α' and integer $r < n$ implies the same inequality with r replaced by $r + 1$; for it then follows for $r = n$, etc. from the observation that it is valid for $r = 0$.

Now setting $u_r = (e^{-tV/n} e^{-tH/n})^r u$, by Hölder's inequality and with suitable $a > 1$, setting $u_r = (e^{-tV/n} e^{-tH/n})^r u$,

$$\|e^{-tV/n} e^{-tH/n} u_r\|_{2e^{\alpha t(r+1)/n}} \le \|e^{-tV/n}\|_a \|e^{-tH/n} u_r\|_{2e^{\alpha t(r+2)/n}}.$$

The term $\|e^{-tH/n} u_r\|_{2e^{\alpha t(r+2)/n}}$ is bounded by $\|u\|_{2e^{\alpha tr/n}}$ which is bounded in turn, by the induction hypothesis, by $e^{\alpha' tr/n}\|u\|_2$. It suffices therefore to show that $\|e^{-tV/n}\|_a \leq c(t)$. Computing,

$$a = 2e^{\alpha r(r+1)/n}(1 - e^{-\alpha t/n})^{-1} < 2e^{\alpha t}(1 - e)^{-1} - \alpha t/n;$$

noting that $n(1 - e^{-\alpha t/n}) \geq c\alpha t$ if n is sufficiently large, as is no essential loss of generality to suppose, for some constant $c > 0$, it follows that

$$\|e^{-tV/n}\|_a \leq \left(\int \exp\left[-2e^{\alpha t}V/c\alpha\right]\right)^{c\alpha t/2e^{\alpha t}} = C(t).$$

Proof of Theorem 11.6   By the Duhamel formula for arbitrary V and V' in $\mathbf{C}_b$, and $u \in L_2(M)$,

$$e^{-t(H+V)}u - e^{-t(H+V')}u = \int_0^t e^{-(t-s)(H+V')}(V - V')\, e^{-s(H+V)}\, u\, ds,$$

the operators $e^{-s(H+V)}$ being uniformly bounded on some finite s-interval say $[0,a]$, and for $V \in \mathbf{C}_b$, as operators on $L_q(M)$, for each $q \in [2,\infty)$. Now taking $u \in \mathbf{D}$, it follows, applying the Hölder inequality, that if $q > 2$,

$$\|e^{-t(H+V)}u - e^{-t(H+V')}u\|_2 \leq c'(t)\|V - V'\|_q,$$

where $c'(t)$ is bounded on finite t-intervals, and may depend on u and q.

Now if V is arbitrary in $\mathbf{C}$ and if V and V' in the preceding paragraph are replaced by V_m and V_n it follows that the sequence $\{e^{-t(H+V_n)}u\}$ is convergent in L_2, say to $S_0(t)u$, where $S_0(t)$ is a linear operator from $\mathbf{D}$ to L_2. In view of the uniform bound on $e^{-t(H+V)}$ for $V \in \mathbf{C}_b$ and t in a finite interval, $\|S_0(t)u\|_2 \leq c''(t)\|u\|_2$ where $c''(t)$ is bounded on finite t-intervals and $u \in \mathbf{D}$. Moreover, the uniformity in the convergence in t on finite intervals shows that the map $t \to S_0(t)u$ is continuous from $[0,a]$ to L_2, for any fixed element $u \in \mathbf{D}$. It follows that $S_0(t)$ has for each t a unique bounded linear extension $S(t)$, from the domain $\mathbf{D}$ to all of L_2; and that $S(t)$ as a function of t is continuous in the strong operator topology.

Now for $V \in \mathbf{C}_b$,

$$\|e^{-t(H+V)}u\|_b \leq \tilde{c}\|u\|_b,$$

where $b \in [2,\infty)$, and $\tilde{c}$ may depend on b. On replacing V by V_n with $V \in \mathbf{C}$, and applying Fatou's lemma, it follows that for $V \in \mathbf{C}$,

$$\|S(t)u\|_b \leq \tilde{c}'\|u\|_b.$$

Thus $S(t)$ leaves $\mathbf{D}$ invariant, and acts continuously relative to each L_p norm, $p \in [2,\infty)$. It follows that for arbitrary $t, t' \in [0,\infty)$, $S_0(t + t') = S_0(t)S_0(t')$; from this it follows in turn that $S(t + t') = S(t)S(t')$.

Each $e^{-t(H+V)}$ is self-adjoint for $V \in \mathbf{C}_b$; hence the same is true of $S(t)$. Thus $S(\cdot)$ is a continuous one-parameter semigroup of bounded self-adjoint operators on $\mathbf{K}$. It is not difficult to deduce from Theorem 9.1 that there exists a unique self-adjoint operator H' in L_2 such that $S(t) = e^{-tH'}$.

This concludes the proof and proves also the

COROLLARY 11.6.1   *H' is bounded below by* $-(\alpha/2) \log (\int e^{-2V/\alpha})$.

Although not necessarily the case for arbitrary calibrated perturbations, $H + V$ is already an essentially self-adjoint operator in the case treated here. In fact, the following more precise statement can be made; "entire vector" for a normal operator A is defined as one in the domain of e^{zA} for all complex values of z.

COROLLARY 11.6.2   *If w is an entire vector for H', then $w \in \mathbf{D}$, $w \in \mathbf{D}(H)$, and $H'w = Hw + Vw$.*

Proof   By Lemma 11.6.1, $\|e^{-tH'}u\|_{2e^{\alpha't}} \leq c(t)\|u\|_2$, where $C(t)$ is bounded on bounded t-intervals. Taking u as $e^{tH'}w$, it follows that w is in all L_p, $p < \infty$, i.e., $w \in \mathbf{D}$.

From the Duhamel formula it follows that

$$e^{-tH}w = e^{-tH'}w + \int_0^t e^{-(t-s)H} V e^{-sH'}w \, ds.$$

(strong integral in any L_p space, $p < \infty$). Hence

$$t^{-1}(e^{-tH} - I)w = t^{-1}(e^{-tH'} - I)w + t^{-1} \int_0^t e^{-(t-s)H} V e^{-sH'}w \, ds.$$

Now $e^{-sH'}w$ is continuous as a function of s, with values in $[\mathbf{D}]$; from Holder's inequality it follows that the same is true of $V e^{-sH'}w$; it follows in turn that $e^{-(t-s)H} V e^{-sH'}w$ is continuous as a function of s and t, in the range $0 \leq s \leq t$, with values in $\mathbf{K}$. Letting $t \to 0$ in the preceding equality, it follows that the right side converges to $H'w - Vw$, which in view of the character of the right side shows that w is in the domain of H and that $Hw = H'w - Vw$. Finally, since the restriction of a self-adjoint generator to a dense domain invariant under the one-parameter group it generates is essentially self-adjoint (see Exercise 21 or 24, page 290; or use Exercise 12, page 289) the restriction of H' to the domain of all its entire vectors is such, and a fortiori $H + V$ is such.

EXERCISES

1   Show that in Theorem 11.1, the semigroup generated by $-H'$ is of order α' for all $\alpha' < \alpha$.

2   With the notation of Theorem 11.1, show the following:

a. $H' - V_n \to H$.

b. H' is a continuous function of $V \in \mathbf{C}$, in the relative topology on $\mathbf{C}$ as a subset of $\mathbf{D}$.

c. If $a > 1$, there exists $b \in R^1$, uniformly bounded on **C**, such that $H \leq aH' + bI$. (If A and B are non-negative self-adjoint operators in a Hilbert space, $A \leq B$ means that if $x \in \mathbf{D}(B^{1/2})$, then $x \in \mathbf{D}(A^{1/2})$, and $\|A^{1/2}\|x \leq \|B^{1/2}x\|$.)

d. H' is affiliated with the W^*-algebra determined by H and V.

3   Let $\mathbf{H} = L_2(R^1, g)$, where $dg = (2\pi)^{-1/2} e^{-x^2/2} dx$, and let H denote the harmonic oscillator hamiltonian in **H**. (This may be defined as the operator $h_n \to nh_n$, where $\{h_n\}$ is the orthonormal set obtained by subtracting from x^n its projection on the subspace spanned by $1, x, \ldots, x^{n-1}$, and then normalizing the result.) It is known that the semigroup generated by $-H$ is globally of positive order with respect to the L_p calibration. Show that the operator aq, where q is the operation of multiplication by x, and a is sufficiently small, is a Kato perturbation of H (i.e. satisfies the hypothesis of Theorem 11.5); but that there is no nonzero value of a such that aq^2 is such. Show, however, that $V = aq^n$ satisfies the conditions of Theorem 11.6 for all $n > 0$ and $a > 0$.

4   Let $U(\cdot)$ and $V(\cdot)$ be given semigroups on a topological linear space. Their *abelian composite* may be defined as the one-parameter family $W(\cdot)$, where $W(t) = \lim_{n} (U(t/n)V(t/n))^n$, when this limit exists for all $t \in [0,\infty)$. Show that the abelian composite of semigroups of global non-negative orders α and β, on a calibrated Hilbert space, if a semigroup, is of global order $\alpha + \beta$.

XII

OPERATOR RINGS
AND SPECTRAL
MULTIPLICITY

12.1 INTRODUCTION

Operator algebra in Hilbert space has developed extensively from a variety of motivations and in several directions, beginning with the double-commutor theorem of von Neumann (1930). Von Neumann himself, in part in collaboration with F. J. Murray, founded during the ensuing two decades the theory of what he called simply "rings", but which are known more specifically as "W^*-algebras" or "von Neumann algebras". Here we shall follow von Neumann's usage of the term "ring" when there appears to be no likelihood of confusion, and otherwise employ the term "W^*-algebra", the "W" referring to the weak topology, in which the algebra is closed, and the "$*$" to its self-adjointness.

Later we shall consider "C^*-algebras", the "C" here referring to the uniform closure. Although the difference may appear to be purely technical, there is a real philosophical difference between the aims and origins of the two types of operator algebras, as well as an historical one. The theory of W^*-algebras is concerned primarily with the *spatial* properties and classification of operator algebras; i.e. their mode of action on Hilbert space, and their classification within unitary equivalence. In contrast, the theory of C^*-algebras is concerned mainly with intrinsic algebraic properties which are independent of the mode of action of the operators in the algebra on Hilbert space, or relate to representations of the algebra in alternative Hilbert spaces. The theory evolved from the discovery by Gelfand and Naimark that C^*-algebras could be characterized simply in essentially algebraic terms; and from their suitability for a representation-independent quantum mechanical formalism, for which von Neumann had earlier sought, without especial success, in terms of modifications of the W^*-algebra concept.

This chapter first treats basic properties of general W^*-algebras, and then proceeds to the consideration of the abelian case. The structure of the general abelian W^*-algebra is simply describable in measure-theoretic terms, and provides an extension of the notion of multiplicity (of spectral values, etc.) to the case of Hilbert space. For an individual self-adjoint operator, a complete set of unitary invariants is quite complicated in the infinite-dimensional case, and in general is difficult to determine effectively. However for abelian W^*-algebras—in particular those generated by given commuting operators—they are typically effectively computable, and commonly suffice for theoretical applications involving multiplicity questions.

The structure theory of abelian W^*-algebras also forms the foundation of the theory of "direct integrals" of Hilbert spaces, which in the infinite-dimensional case replaces the concept of direct sum, in connection with the reduction of given operators or operator algebras. This theory is, however, beyond the scope of the present book.

12.2   THE DOUBLE-COMMUTOR THEOREM

A W^*-*algebra*, or "ring", when the context is sufficiently clear, on a Hilbert space **H**, is defined as an algebra of bounded linear operators on **H** which is closed in the weak operator topology, contains with every element A its adjoint A^*, and includes the identity operator I. The ring **B(H)** of all bounded linear operators on **H** is a simple example, and will occur frequently in the following, although it plays a lesser role in the infinite-dimensional case than do the complete matrix algebras in the finite-dimensional cases.

The notion of *commutor* is applicable to an arbitrary set **S** of bounded linear operators on **H**; it is defined as the set of all bounded linear operators T on **H** such that $ST = TS$ for all elements $S \in$ **S**. It is easily seen that the commutor of **S**, which will be denoted in accordance with standard usage as **S'**, is a weakly closed algebra, whatever S may be, and includes I. It is evident that the double commutor **S''** $= ($**S'**$)'$ always contains the original set **S**. In its basic form, the double commutor theorem identifies W^*-algebras with the sets **S** which are identical with their double commutors and self-adjoint.

THEOREM 12.1   *If* **A** *is a* W^*-*algebra on a Hilbert space, then it is identical to its double commutor.*

Note first

Lemma 12.2.1   *A closed linear manifold* **M** *in a Hilbert space* **H** *is invariant under every operator in a self-adjoint set* **S** *of bounded linear operators on* **H** *if and only if the projection with range* **M** *lies in* **S**′.

Proof   If $A \in \mathbf{S}$ and A leaves **M** invariant, then for $x \in \mathbf{M}$, $PAx = Ax = APx$, while for x orthogonal to **M** and $y \in \mathbf{M}$,

$$(PAx, y) = (Ax, Py) = (x, A^*y) = 0,$$

showing that $PAx = 0 = APx$. By linearity, $PAx = APx$ for all x, i.e. $PA = AP$. On the other hand, if $P \in \mathbf{S}'$ and $x \in \mathbf{M}$, then $PAx = APx = Ax$, showing that $Ax \in \mathbf{M}$, i.e. **M** is invariant under **A**.

Proof of Theorem 12.1   It suffices to show that $\mathbf{A}'' \subset \mathbf{A}$, since the reverse inclusion is trivially valid. We shall show that if $T \in \mathbf{A}''$, then every neighborhood N of T, in the strong operator topology, meets **A**; this will show that T is in the strong closure of **A**, a fortiori in the weak closure, and so since **A** is assumed closed, in **A** itself.

Consider first the case of a neighborhood of the form $N = [B \in \mathbf{B}: \|Bx - Tx\| < \epsilon]$ for some given vector $x \in \mathbf{H}$ and $\epsilon > 0$. Let **M** denote the closure of $\mathbf{A}x$; then by simple approximation, **M** is invariant under **A**. Hence the projection P with range **M** lies in $\mathbf{A}'$. This implies that P commutes with $T \in \mathbf{A}''$, which implies in turn that T leaves **M** invariant. In particular, since $x \in \mathbf{M}$, $Tx \in \mathbf{M}$, i.e. Tx lies in the closure of $\mathbf{A}x$, meaning that N meets **A**.

Now consider the general neighborhood N of T, which may be taken in the form $N = [B \in \mathbf{B}: \|Bx_i - Tx_i\| < \epsilon]$, where $x_1, \ldots, x_n$ is an arbitrary fixed finite set of vectors in **H**. Let **K** denote the n-fold direct sum of **H** with itself; i.e. the set of all n-tuples $(y_1, \ldots, y_n)$ with $y_j \in \mathbf{H}$ for all j and the usual linear structure and the inner product:

$$((y_1, \ldots, y_n), (y_1', \ldots, y_n')) = \sum (y_j, y_j').$$

For any operator $B \in \mathbf{B}$, let $\theta(B)$ denote the operator on **K**:

$$\theta(B): (y_1, \ldots, y_n) \to (By_1, \ldots, By_n);$$

then θ is a *-isomorphism of $\mathbf{B(H)}$ into $\mathbf{B(K)}$. To show that N meets **A** is as earlier the same as showing that $\theta(T)x$ is in the closure **M** of the $\theta(\mathbf{A})x$, where $x = (x_1, \ldots, x_n)$. This in turn is equivalent by the earlier argument to showing that $\theta(T)$ commutes with the projection P with range **M**.

Now evidently, $\theta(P) \in \theta(\mathbf{A})'$. The commutor $\theta(\mathbf{A})'$ can be readily determined by straightforward algebra as the set of all $C \in \mathbf{B(K)}$ whose matrix decomposition $(C_{ij}; i, j = 1, \ldots, n; C_{ij} \in \mathbf{B(H)})$ relative to the defining representation of **K** as the *n*-fold direct sum of **H** with itself has the property that $C_{ij} \in \mathbf{A}'$ for all i and j. Thus $P_{ij} \in \mathbf{A}'$ for all i and j, where (P_{ij}) denotes the matrix decomposition of P relative to the cited decomposition of **K**. But the matrix decomposition of $\theta(T)$ is $(T\delta_{ij})$, and it is now obvious that this matrix commutes with that for P.

COROLLARY 12.2.1   *If* **A** *is any self-adjoint subalgebra of* **B(H)** *which annihilates no non-zero vector in* **H**, *then its strong (or weak) closure is identical with* **A**″.

Proof   The foregoing proof shows that if the condition of closure of **A** is relaxed, so that **A** is any self-adjoint subalgebra of **B** containing I, then the conclusion can still be drawn that **A**″ is contained in the strong closure of **A**; and the reverse inclusion is readily derived by approximation. The condition that $I \in \mathbf{A}$ was used only to show that x lies in the closure of **A**x. To establish this on the weaker hypothesis that **A** annihilates no non-zero vector, let $y = x - Px$, where P is the projection whose range is the closure **M** of **A**x. Then $Ay = Ax - APx$; but $PA = AP$, so that $Ax - APx = Ax - PAx = Ax - Ax = 0$, showing that y is annihilated by **A**, whence $y = 0$, i.e. $x \in \mathbf{M}$. That the weak closure of **A** is identical with its strong closure is a general fact concerning linear sets of operators on a Banach space which follows from the Hahn-Banach theorem.

COROLLARY 12.2.2   *A self-adjoint subset of* **B(H)** *is irreducible if and only if its commutor consists only of scalars.*

Note first the

Lemma 12.2.2   *Any ring is generated by the projections it contains.*

Proof   To show that the ring **A** is generated by the projections it contains is to show that it is the minimal ring containing those projections. Now, if A is any self-adjoint element of **A**, and b is any bounded Baire function on R^1, then $b(A) \in \mathbf{A}$, either by approximation or from the observation that $b(A)$ commutes with every operator in **A**′. In particular, every spectral projection of A lies in **A**; but A is a limit of linear combinations of these projections.

Proof of Corollary 12.2.2   If **S** is an irreducible self-adjoint subset of **B** and **M** is an invariant subspace, then the only projections P contained in **S**′ are 0 and I, for otherwise the range of P would be a non-trivial closed **S**-invariant subspace. It follows from the lemma that **S**′ can then consist only of the scalars $[aI, a \in C]$. If conversely **S**′ consists only of scalars, then **S** must be irreducible, for if **M** is any closed invariant subspace, the projection with range **M** must lie in **S**′, and so can be only 0 or I.

The following more precise result is sometimes useful.

COROLLARY 12.2.3   *If* **A** *is a self-adjoint subalgebra of* **B** *which contains* I, *then the unit ball of* **A** *is strongly dense in the unit ball of* **A**″.

Proof   The map $A \to A^*$ from **B** to **B** is clearly continuous in the weak topology, implying that the real-linear subspace of **A** consisting of its self-adjoint elements is dense in the corresponding subspace of $\mathbf{A}''$, Now let A denote any element of $\mathbf{A}''$ such that $A = A^*$ and $\|A\| \leq 1$, and let $\{A_\mu\}$ be a net of self-adjoint elements of **A** such that $A_\mu \to A$ strongly. Then as shown in Chapter XI, $f(A_\mu) \to f(A)$ for all continuous functions f of compact support on R^1. Choosing f to be such a function with the properties that $f(x) = x$ for $|x| \leq 1$ and $|f(x)| \leq 1$ for all x, then $f(A) = A$, and for each μ, $f(A_\mu)$ lies in the unit ball of **A**. Thus every self-adjoint element of the unit ball of $\mathbf{A}''$ is a strong limit of a net in the unit ball of **A**. If A is now an arbitrary element of **A** with $\|A\| \leq 1$, the conclusion follows on applying the result obtained in the self-adjoint case to the self-adjoint element $\begin{pmatrix} 0 & A \\ A^* & 0 \end{pmatrix}$ of the algebra of all 2×2 matrices of the form $\begin{pmatrix} B_{11} & B_{12} \\ B_{21} & B_{22} \end{pmatrix}$ with the B_{ij} in **A**, as operators on the direct sum $\mathbf{H} \oplus \mathbf{H}$ of **H** with itself, whose weak closure consists of all operators of the same general form except that the B_{ij} are in $\mathbf{A}''$.

Although rings consist only of bounded operators, they are useful in the treatment of unbounded operators as well. One makes the

DEFINITION   An operator T on a Hilbert space **H** is said to be *affiliated with* a given W^*-algebra **A** on **H** in case $UTU^{-1} = T$ for all unitaries $U \in A'$; and we write $T \alpha \mathbf{A}$.

COROLLARY 12.2.4   (i)   *If $T \in B(H)$, and **A** is a W^*-algebra on **H**, then $T \alpha \mathbf{A}$ if and only if $T \in \mathbf{A}$.*

(ii)   *If T is a self-adjoint operator on **H**, then $T \alpha \mathbf{H}$ if and only if $b(T) \in \mathbf{A}$ for all bounded Baire functions b on R^1; or equivalently if all spectral projections of T are in **A**; or again equivalently, if $f(T) \in \mathbf{A}$ for all continuous functions f of compact support on R^1.*

(iii)   *If T is closed and densely defined, then $T \alpha \mathbf{A}$ if and only if both the partially isometric and self-adjoint constituents of T in its canonical polar decomposition (either left or right) are affiliated with **A**.*

Proof   Note first that any ring is generated by the unitaries it contains, for if P is any projection in **A**, then $I - 2P$ is a unitary in **A**, and as already seen, the projections in **A** generate **A**. Part (i) follows from this remark. Part (ii) follows from spectral theory (cf. the proof of Lemma 12.2.2). The "if" part of Part (iii) follows from the observation that any product of operators affiliated with a ring is again affiliated with the algebra. The "only if" part follows from the canonicity; $T = UA$, where $A = (T^*T)^{1/2}$, and U is the partial isometry which carries Ax into Tx on the range of A, and vanishes on the ortho-complement of this range. This decomposition is clearly unitarily invariant; in particular, the constituents U and A are invariant under all unitaries which leave T invariant.

EXERCISES

1   The ring generated by a given set of densely-defined closable operators in the Hilbert space **H** is defined as that generated by the partially isometric operators and spectral projections of the self-adjoint operators in the canonical polar decompositions of the closures of the operators. Show that a set of self-adjoint and/or bounded operators is simultaneously diagonalizable (i.e. can simultaneously be represented as multiplications by real measurable functions on an L_2-space) if and only if the W^*-algebra generated by the operators is abelian.

2   Let $\mathbf{H} = L_2(0,2\pi)$, let $A = -i(d/dx)$ in its usual formulation as a self-adjoint operator with periodic boundary conditions (i.e. $f(x) = \sum_n a_n e^{inx} \in \mathbf{D}(A)$ if and only if $\sum_n n^2|a_n|^2 < \infty$, and then $Af = \sum_n na_n e^{inx}$). Let U denote the unitary operator $f(x) \to e^{ix/2}f(x)$ on **H**, and set $B = UAU^*$. Show that if f is any C^∞ function on $(0,2\pi)$ which vanishes near the endpoints of this interval, then $ABf = BAf$; but that A and B do not "commute strongly", in the sense that the W^*-algebra they generate is abelian.

3   Show that a closed linear subspace of a Hilbert space **H** is invariant under all the operators in a self-adjoint subset **S** of **B(H)** if and only if it is invariant under **S**″.

4   Derive the following formal properties of the operation $\mathbf{A} \to \mathbf{A}'$ on the set of all rings on a given Hilbert space.
   a. $\mathbf{A} \subset \mathbf{B}$ if and only if $\mathbf{B}' \subset \mathbf{A}'$.
   b. $\mathbf{A}$ is abelian if and only if $\mathbf{A} \subset \mathbf{A}'$; and is maximal abelian if and only if $\mathbf{A} = \mathbf{A}'$.
   c. For any two rings $\mathbf{A}$ and $\mathbf{B}$, let $\mathbf{A} \vee \mathbf{B}$ denote the W^*-algebra generated by $\mathbf{A}$ and $\mathbf{B}$. Show that $(\mathbf{A} \vee \mathbf{B})' = \mathbf{A}' \cap \mathbf{B}'$ and that $(\mathbf{A} \vee \mathbf{B})' = \mathbf{A}' \cap \mathbf{B}'$.

5   Show that the lattice of all invariant subspaces of a ring is Boolean if and only if $\mathbf{A}'$ is abelian. (A lattice is Boolean if the union and intersection operations of the lattice satisfy the usual axioms for a Boolean algebra, and if each element has a unique complement.)

6   If G is a locally compact topological group, its left regular representation may be defined as the representation $a \to L(a)$, $a \in G$, where $L(a)$ is the operator on $L_2(G)$ with respect to left invariant Haar measure, which carries $f(x)$ into $f(a^{-1}x)$, $f \in L_2(G)$. Show that the ring **L** generated by the $L(a)$ includes all left convolution operators, $f \to m \divideontimes f$, where m is a finite regular measure on G; and that it is identical to the algebra generated by the convolution operators $L_g: f \to g \divideontimes f$, where g is an arbitrary continuous function of compact support on G.

7   Let M be a measure space and **A** an algebra of bounded measurable functions on M. For $k \in$ **A**, let $M(k)$ denote the operator $f \to kf$ on $L_2(M)$. Show that the map $k \to M(k)$ is a homeomorphism with the w^*-topology on $L_\infty(M)$ as the dual of $L_1(M)$ and the weak operator topology on the image.

8   Let m be any finite regular measure of compact support on R^1, and let T denote the operator, $f(x) \to xf(x)$ on the Hilbert space $\mathbf{H} = L_2(R^1,m)$. Show that the ring generated by T consists of all multiplications by bounded Baire functions of x. (Cf. Ex. 7.)

9   Let **A** be a ring on a Hilbert space **H**. A vector $z \in \mathbf{H}$ is called cyclic (resp., separating) for **A** provided Az is dense in **H** (resp., $Az = 0$ and $A \in$ **A** implies that $A = 0$). Show that a vector is cyclic for **A** if and only if it is separating for **A**′.

10   Show that the unit ball in a ring is compact in the weak operator topology.

12.3   THE STRUCTURE OF ABELIAN RINGS

The operators T of Exercises 8 and 9 of the last section have "simple spectrum" or "multiplicity one" for their proper values, from a formal standpoint; they have cyclic vectors (namely the function identically one), which is one classical criterion, and they generate maximal abelian rings which is another. If we consider only measures on [0,1] which are positive on nonempty open sets, the operators have spectrum consisting of the same interval. Yet it is not difficult to show that two such operators are unitarily equivalent if and only if the corresponding measures are mutually absolutely continuous. On the other hand, continuous singular measures whose effective support is the entire interval [0,1] are well known, and in fact there are at least continuum many absolute continuity equivalence classes of such. Thus self-adjoint operators with the same spectrum and "multiplicity one" for the points of this spectrum, in a certain sense, need not be unitarily equivalent.

The researches of Hahn and Hellinger, Nakano and Wecken have indicated the absence of any simple resolution to the problem of determining effectively calculable complete unitary invariants for self-adjoint operators on Hilbert space, and have led to the standpoint that a more useful problem to pose may be that of the structure of the rings generated by the operators in question. This has a general and complete simple solution for arbitrary abelian rings, which may then be applied to the case of algebras generated by particular operators.

To develop this solution, consider first the case of a finite-dimensional Hilbert space. There is no difficulty in establishing that any abelian ring **A** is *algebraically* isomorphic to an algebra of diagonal matrices; but *spatially*, in its action on the Hilbert space **H**, may involve various multiplicities. For example, **A** may be isomorphic to the complex numbers, consisting only of the scalar operators $[aI: a \in C]$; if **H** is n-dimensional, the elements of **A** can then be represented by $r \times n$ diagonal matrices in which each entry is repeated precisely n times; the algebra is then said to be of *multiplicity n*. Or **A** may be algebraically isomorphic to the algebra of all 2×2 diagonal matrices, but spatially (i.e. in the action on **H**) each such matrix may occur with a certain multiplicity. In general it can be seen that there exist unique projections $P_1, P_2, \ldots$ in **A** which are mutually orthogonal and have sum equal to I, such that the restriction of **A** to the range of P_k has precisely multiplicity k. In this form, the structure theory generalizes to the infinite-dimensional case and leads to a simple complete set of unitary invariants for abelian rings. We begin with the

> DEFINITION Let n be a cardinal number (infinite or finite). An *n-fold copy* of a ring **M** on a Hilbert space **H** is a ring **A** on a Hilbert space **K**, which admits mutually orthogonal closed invariant subspaces $\mathbf{H}_j$, of which it is the direct sum, unitary transformations U_j from $\mathbf{H}_j$ onto **H**, and a *-algebraic isomorphism θ from **A** onto **M** such that for arbitrary $x \in \mathbf{H}_j$, and $A \in \mathbf{A}$, $\theta(A)U_j x = U_j A x$, for all j in the index set J. We say also that $\theta(A)$ is an n-fold copy of A.

Alternatively, **K** is unitarily equivalent to the Hilbert space of all functions $j \to x(j)$ from J to **H** for which $\sum_j \|x(j)\|^2 < \infty$, with the usual linear structure and inner product $(x(\cdot), y(\cdot)) = \sum_j (x(j), y(j))$, in such a way that an operator A on **K** is in **A** if and only if it is carried by this equivalence into an operator of the form $x(\cdot) \to M x(\cdot)$, for some fixed $M \in \mathbf{M}$.

The algebra **A** is said to be of *uniform multiplicity n* in case **M** is maximal abelian. It will later be shown that a ring cannot have two different multiplicities.

> THEOREM 12.2 *Let* **A** *be an abelian ring on a Hilbert space* **H**. *For each cardinal number n there exists a unique closed invariant subspace* $\mathbf{H}_n$ *under* **A** *such that the restriction of* **A** *to* $\mathbf{H}_n$ *has uniform multiplicity n, and which is maximal with respect to this property. The subspaces* $\mathbf{H}_n$ *are mutually orthogonal, hvae direct sum* **H**, *and the projection with range* $\mathbf{H}_n$ *lies in* **A**.

Lemma 12.3.1 *The restriction of an abelian ring of uniform multiplicity n to the range of one of its projections is again of uniform multiplicity n.*

Proof Let **A** on **K** be the ring in question, and suppose **A** on **K** is the n-fold copy of the maximal abelian ring **M** on **H**. Any projection P in **A** is then an n-fold copy, $P = \theta(Q)$, of a projection $Q \in \mathbf{M}$; and the range of P is correspondingly an n-fold direct sum of copies of the range of Q. Any operator $A \in \mathbf{A}$ is an n-fold copy of an operator $B \in \mathbf{M}$; consequently, the restriction of A to the range of P is an n-fold copy of the restriction of B to the range of

Q; and it need therefore only be shown that the set of all such restrictions, $[B \mid \text{range } Q: B \in \mathbf{M}]$, is maximal abelian. But if T is a bounded linear operator on range Q commuting with all $B \mid \text{range } Q$, $B \in \mathbf{M}$, then its extension to an operator T' on all of $\mathbf{H}$ by linearity and setting $T'x = 0$ if $Qx = 0$ commutes with all $B = QB + (I - Q)B$. Hence $T' \in \mathbf{M}$, which implies that T, as the restriction of T' to range Q, lies in $\mathbf{M} \mid \text{range } Q$.

Before stating the next lemma, we make explicit the notion of direct sum of rings in the sense that will be used here.

DEFINITION   Let $\mathbf{A}_j$ on $\mathbf{H}_j$, $j \in J$, be a collection of rings on Hilbert spaces. The *direct sum* (more explicitly: Hilbert space direct sum) of these rings is the ring $\mathbf{A}$ on the Hilbert space direct sum $\mathbf{K} = \Sigma_j \in \mathbf{H}_j$ of the $\mathbf{H}_j$ consisting of all operators of the form $\Sigma_j x_j \to \Sigma_j A_j x_j$, with $A_j \in \mathbf{A}_j$, and $\|A_j\|$ bounded as a function of $j \in J$.

Lemma 12.3.2 *The direct sum of maximal abelian rings is again a maximal abelian ring.*

Proof With the notation of the preceding definition, suppose the $\mathbf{A}_j$ are maximal abelian. Suppose $W \in \mathbf{A}'$; it must be shown that $W \in \mathbf{A}$. Let P_j denote the projection of $\mathbf{K}$ with range $\mathbf{H}_j$; then $P_j \in \mathbf{A}$, so W commutes with P_j, implying that W leaves $\mathbf{H}_j$ invariant. Setting $W_j = W \mid \mathbf{H}_j$, then W_j commutes with all $A \mid \mathbf{H}_j$, $A \in \mathbf{A}$, since $\mathbf{A}$ leaves $\mathbf{H}_j$ invariant also. Since $\mathbf{A}_j$ is maximal abelian, $W_j \in \mathbf{A}_j$. But as a restriction of W, $\|W_j\| \leq \|W\|$, so that $W \in \mathbf{A}$.

Lemma 12,3.3 *Let $\mathbf{A}$ be an abelian ring on a Hilbert space $\mathbf{K}$ and for each $\lambda \in \Lambda$ let P_λ denote a projection in $\mathbf{A}$ such that $\mathbf{A} \mid P_\lambda \mathbf{K}$ is of uniform multiplicity n, n being a fixed cardinal number. If P denotes the supremum of the P_λ (in the lattice of projections on $\mathbf{K}$), then $\mathbf{A} \mid P\mathbf{K}$ is again of uniform multiplicity n.*

Proof Note first that the direct sum of abelian rings each of which is of uniform multiplicity n is again of uniform multiplicity n; this is a straightforward deduction from the lemma just proved. Now well-order the set Λ, and orthogonalize the P_λ by defining $Q_\lambda = P_\lambda - \sup_{\lambda' < \lambda} P_{\lambda'}$. The Q_λ are then mutually orthogonal projections in A whose supremum is the same as that of the P_λ; and $\mathbf{A} \mid Q_\lambda \mathbf{K}$ is again of uniform multiplicity n, by Lemma 12.3.1; so the conclusion follows from the case already considered in which the P_λ are mutually orthogonal.

The following concept is useful in the study of rings on not necessarily separable Hilbert spaces.

DEFINITION   A set of projections on a Hilbert space is *countably decomposable* if any set of mutually orthogonal projections in the set is at most countable. A ring is said to be countably decomposable if the set of all of its projections is such.

We also define a *separating vector* for a ring $\mathbf{A}$ on $\mathbf{H}$ as a vector $z \in \mathbf{H}$ such that if $A \in \mathbf{A}$ and $Az = 0$, then $A = 0$. Note that a given vector z in $\mathbf{H}$ is separating for a ring $\mathbf{A}$ if and only if it is separating for the collection $\mathbf{C}$ of all projections in $\mathbf{A}$, in the sense that if $P \in \mathbf{C}$ and $Pz = 0$, then $P = 0$. Only the "if" part of this statement requires proof, which is as follows. Assuming z is separating for $\mathbf{C}$, if $A \in \mathbf{A}$ and $Az = 0$, then $A^*Az = 0$, and it follows readily that $Pz = 0$ for all spectral projections of A^*A, whence $A^*A = 0$, implying that $A = 0$.

Lemma 12.3.4 *An abelian ring of projections on a Hilbert space is countably decomposable if and only if it has a separating vector.*

Proof The "if" part follows directly from Bessel's inequality, independently of the abelian assumption, since if $[P_j : j \in J]$ is any collection of mutually orthogonal projections $\sum_j \|P_j z\|^2 \leq \|z\|^2$, showing that at most countably many of the $P_j z$ can be non-zero, and hence the same for the P_j.

To prove the (non-trivial) "only if" part, note that by Zorn's lemma, there is a collection $\mathbf{C}$ of projections in the ring $\mathbf{S}$ on $\mathbf{H}$ which is maximal with respect to the following properties:

(i) the elements of $\mathbf{C}$ are mutually orthogonal;

(ii) if $P \in \mathbf{S}$, then the set $\mathbf{S}_P = [Q \mid P\mathbf{H} : Q \in \mathbf{S}]$ has a separating vector. By hypothesis, $\mathbf{S}$ is at most countable; let its elements be enumerated as $P_1, P_2, \ldots$, and let z_n be a separating vector for $\mathbf{S}_{P_n}$. There is no essential loss of generality in assuming that $\|z_n\| = 1$; and we set
$$z = \sum_n z_n/2^n, \quad Q = \sum_n P_n.$$

It will be shown first that z is a separating vector for $\mathbf{S}_Q$, and the proof will be concluded by showing that $Q = I$.

To show that z separates $\mathbf{S}_Q$, suppose that $R_Q z = 0$, where $R \in \mathbf{S}$ and $R_Q = E \mid Q\mathbf{H}$. This means in particular that $RQz = 0$, which implies in turn that $\sum_n P_n Rz = 0$, and finally, since the $P_n Rz$ are mutually orthogonal, that $P_n Rz = 0$ for all n. Hence $RP_n z = 0 = Rz_n$, and it follows from the separating character of z_n for $\mathbf{S}_{P_n}$ that $RP_n = 0$ for all n, implying that $RQ = 0$. But this means that z separates $\mathbf{S}_Q$.

To show that $Q = I$, assume the contrary, and let w be a non-zero vector which is orthogonal to the range of Q. Let R denote the supremum of the projections $S \in \mathbf{S}$ such that $Sw = 0$ and $S \leq I - Q$; then $Rw = 0$ and $R \leq I - Q$. Moreover by the maximality of $\mathbf{C}$, there exist non-zero projections S in the foregoing supremum, so that $R \neq 0$. Now let $Z = I - Q - R$; then $w \in Z\mathbf{H}$, for $w = (I - Q)w = (I - Q - R)w = Zw$. In addition, Z is a projection which is orthogonal to all of the P_j. But w is a separating vector for $\mathbf{S}_z$, since if $SZw = 0$ and $S \in \mathbf{S}$, then $SZ \leq R$ by the definition of R, which together with the fact that $SZ \leq Z \leq I - R$ implies that $SZ = 0$. This contradicts the maximality of **C**, showing that $Q = I$ and completing the proof.

The utility of countable decomposability is indicated by the following basic existential result.

Lemma 12.3.5   *If* **A** *is a countably decomposable abelian ring on a Hilbert space, there exists a non-zero projection* $P \in$ **A** *such that* **A** $|$ *P***H** *is of uniform multiplicity.*

Proof   By Zorn's lemma, there exists a collection **C** of separating vectors for **A** such that the cyclic subspaces $\mathbf{A}z_i$ ($i = 1,2$) spanned by any two vectors vectors z_1 and z_2 in **C** are mutually orthogonal (**C** being non-empty by Lemma 12.3.4). For any $z \in$ **C**, let P_z denote the projection whose range is the closure of $\mathbf{A}z$; set $R = \sum_{z \in \mathbf{C}} P_z$; and let Q denote the supremum of the projections $Q' \in$ **A** such that $Q'(I - R) = 0$. The first non-trivial point is that $Q \neq 0$. For assuming the contrary, there exist no projections $P \neq 0$ in **A** such that $P(I - R) = 0$, so that the restriction map $A \to A \mid (I - R)\mathbf{H}$ is an isomorphism of **A** into an algebra of operators on $(I - R)\mathbf{H}$, which is then countably decomposable, and so has a separating vector z'. But then z' is a separating vector for **A** which is such that $\mathbf{A}z' \subseteq (I - R)\mathbf{H}$ so that $\mathbf{A}z'$ is orthogonal to all the $\mathbf{A}z$, for $z \in$ **C**, contradicting the maximality of **C**.

Observe next that $Q = \sum_{z \in \mathbf{C}} QP_z$; and setting $P'_z = QP_z$, $A \mid P'_z\mathbf{H}$ has the separating vector Qz, and is algebraically isomorphic to **A** $\mid$ Q**H**, via the restriction map $A \mid Q\mathbf{H} \to A \mid QP_z\mathbf{H}$. By Section 9.2, $A \mid P'_z\mathbf{H}$ is maximal abelian, having a cyclic vector. According to an exercise in the same section, two maximal abelian rings which are algebraically isomorphic are unitarily equivalent. It follows that there exists a maximal abelian ring **M** on a Hilbert space **K** such that for each $z \in$ **C**, $\mathbf{A} \mid P'_z\mathbf{H}$ is unitarily equivalent to **M** on **K**. Observe finally that the direct sum of the $P'_z\mathbf{H}$ is $Q\mathbf{H}$; it follows by an argument previously employed that **A** $\mid$ Q**H** is unitarily equivalent to an n-fold copy of **M** on **K**, where n is the cardinal number of **C**.

Lemma 12.3.6   *Any commutative ring on a Hilbert space contains a collection* **C** *of mutually orthogonal projections whose union is* I *and such that for any* $P \in$ **C**, *the ring* **A** $\mid$ *P***H** *is countably decomposable.*

Proof   Let **C** be a collection of projections in **A** which is maximal with respect to the properties:

(1) its elements are mutually orthogonal;

(2) if $P \in$ **C**, then **A** $\mid$ *P***H** is countably decomposable.

We show that the union Q of the projections in **C** is I. For otherwise there exists a non-zero vector z in **H** which is orthogonal to Q**H**. Setting R for the supremum of the projections $R' \in$ **A** such that $R' \leq I - Q$ and $R'z = 0$, then $I - Q - R$ is orthogonal to all the elements of **C**, and is not zero since

$(I - Q - R)z = z \neq 0$. Since z is a separating vector for $\mathbf{A} \mid (I - Q - R)\mathbf{H}$, this ring is countably decomposable, contradicting the maximality of $\mathbf{C}$.

Lemma 12.3.7 *Let $\mathbf{A}$ be an abelian ring on a Hilbert space $\mathbf{H}$, and for each cardinal number n, let R_n be the supremum of the projections $P \in \mathbf{A}$ such that $\mathbf{A} \mid P\mathbf{H}$ has uniform multiplicity n. Then the supremum of the R_n is I.*

Proof Let Q denote the supremum of the R_n, and as the basis for an indirect proof, assume $Q \neq I$, Let $\mathbf{A} \mid (I - Q)\mathbf{H}$ be denoted as $\mathbf{A}_1$, and let P_1 be a non-zero projection in $\mathbf{A}_1$ such that $\mathbf{A}_1 \mid P_1(I - Q)\mathbf{H}$, which will be denoted as $\mathbf{A}_2$, is countably decomposable. Let P denote a projection in $\mathbf{A}$ such that $P \mid (I - Q)\mathbf{H} = P_1$; such a projection evidently exists, and $P_1(I - Q)\mathbf{H} = P(I - Q)\mathbf{H}$. By Lemma 12.3.5, there exists a non-zero projection N_2 in $\mathbf{A}_2$ such that $\mathbf{A}_2 \mid N_2P(I - Q)\mathbf{H}$ has uniform multiplicity, say n, Now if N_1 is a projection in $\mathbf{A}_1$ such that $N_1 \mid P(I - Q)\mathbf{H} = N_2$, and if N is a projection in $\mathbf{A}$ such that $N \mid (I - Q)\mathbf{H} = N_1$ (as earlier, such projections clearly exist), then $\mathbf{A}_2 \mid N_2P(I - Q)\mathbf{H} = \mathbf{A} \mid NP(I - Q)\mathbf{H}$. It follows that $NP(I - Q) \leq R_n$ or $NP(I - Q)R_n = NP(I - Q)$, which by the definitions of R_n and Q implies that $NP(I - Q) = 0$, a contradiction.

Lemma 12.3.8 *Let A be an abelian ring which is of uniform multiplicity both m and n. Then $m\aleph_0 = n\aleph_0$.*

Proof Using Lemmas 12.3.1 and 12.3.4, it suffices to treat the case in which $\mathbf{A}$ is countably decomposable. Let $\mathbf{A}$ on $\mathbf{H}$ be unitarily equivalent on the one hand to an m-fold copy of the maximal abelian ring $\mathbf{M}$ on $\mathbf{K}$ and on the other to an n-fold copy of the maximal abelian ring $\mathbf{N}$ on $\mathbf{L}$. Then both $\mathbf{M}$ and $\mathbf{N}$ are countably decomposable, and so have separating vectors u and v in $\mathbf{K}$ and $\mathbf{L}$, which by virtue of the maximal abelian character of each ring are also respectively cyclic.

According to the definition of uniform multiplicity, $\mathbf{H}$ is the direct sum of m mutually orthogonal subspaces $\mathbf{K}_j$, $j \in J$, each of which is invariant under $\mathbf{A}$ and such that $\mathbf{A} \mid \mathbf{K}_j$ is unitarily equivalent to $\mathbf{M}$ on $\mathbf{K}$. Similarly, it is the direct sum of n mutually orthogonal subspaces $\mathbf{L}_k$, $k \in K$, each of which is invariant under $\mathbf{A}$ and such that $A \mid \mathbf{L}_n$ is unitarily equivalent to $\mathbf{N}$ on $\mathbf{L}$. By the preceding paragraph, there exist vectors $x_j \in \mathbf{K}_j$ and $y_n \in \mathbf{L}_k$ which are cyclic for $\mathbf{A} \mid \mathbf{K}_j$ and $\mathbf{A} \mid \mathbf{L}_k$ respectively.

Now the projection x_{jk} of x_j onto $\mathbf{L}_k$ vanishes except for countably many k, so that the set of all indices k for which $x_{jk} \neq 0$ for some $j \in J$ has cardinality at most $\aleph_0 m$. But for every $k \in K$ there exists a $j \in J$ such that $x_{jk} \neq 0$, for suppose this is not the case; let k' be an index in K such that $x_{jk'} = 0$ for all $j \in J$. Thus x_j is orthogonal to $\mathbf{L}_k$, for all j; in particular, $(x_j, Ay_{k'}) = 0$ for all $A \in \mathbf{A}$ and $j \in J$, or $(A^*x_j, y_{k'}) = 0$ in these cases. But the A^*x_j span $\mathbf{K}_j$ as A ranges over $\mathbf{A}$, so it follows that $y_{k'}$ is orthogonal to $\mathbf{K}_j$ for all $j \in J$, implying that $y_{k'} = 0$, a contradiction. It follows that the cardinality of K

is at most $\aleph_0 m$, i.e. $n \leq \aleph_0 m$. By symmetry, the inequality $m \leq \aleph_0 n$ is also valid, and the lemma follows.

Lemma 12.3.9 *Let $\mathbf{M}$ be a countably decomposable maximal abelian ring on the Hilbert space $\mathbf{K}$. Let x be a cyclic vector for $\mathbf{M}$ and let $y_1, y_2, \ldots$ be an arbitrary sequence of vectors in $\mathbf{K}$. Then there exists a non-zero projection P in $\mathbf{M}$ such that $Py_i \in \mathbf{M}x$ for $i = 1, 2, \ldots$.*

Proof By Chapter VIII, it may be assumed that $\mathbf{M}$ on $\mathbf{K}$ is the multiplication algebra of a finite measure space $M = (R,\mathbf{R},r)$. It follows from the cyclicity of x for $\mathbf{M}$ that $x(p) \neq 0$ a.e. on M. Now setting $E_{ij} = [p \colon |y_i(p)| < j|x(p)|]$, it follows that $\bigcup_j E_{ij}$ differs from R by a null set, and hence for each positive integer i there exists a positive integer j_i such that $r(R - E_{ij_i}) < 2^{-i-1}r(R)$. Now setting $E = \bigcup_i E_{ij_i}$, it follows that

$$r(R - E) \leq \sum_i r(R - E_{ij_i}) < r(R)/2.$$

Defining P as the operation of multiplication by the characteristic function of E, then $P \neq 0$ and $Py_i \in \mathbf{M}x$ for all i, since Py_i is obtainable from x by multiplication by the bounded measurable function which equals $y_i(p)x(p)^{-1}$ for $p \in E$, and vanishes on $R - E$.

Lemma 12.3.10 *If an abelian ring has uniform multiplicities both m and n, where m is finite and $m \leq \aleph_0$, then $m = n$.*

Proof As in the proof of Lemma 12.3.8, it may be assumed that the ring $\mathbf{A}$ on the Hilbert space $\mathbf{H}$ is countably decomposable; and there exist corresponding structures, described here with the same notation as in the cited proof. Now let U_j be a unitary transformation from $\mathbf{K}$ onto $\mathbf{K}_j$ which implements the unitary equivalence of $\mathbf{M}$ on $\mathbf{K}$ with $\mathbf{A} \mid \mathbf{K}_j$, and let x be a cyclic vector for $\mathbf{M}$ on $\mathbf{K}$; it is then no essential loss of generality to suppose that $x_j = U_j x$.

Now setting y_{kj} for the projection of y_k onto $\mathbf{K}_j$, we have $y_k = \sum_j y_{kj}$. Defining y'_{kj} as $U_j^{-1} y_{kj}$, it follows from the preceding lemma that there exists a non-zero projection $P' \in \mathbf{M}$ such that $P'y'_{kj} \in \mathbf{M}x$. Now let P denote the projection in $\mathbf{A}$ which is unitarily equivalent via the given transformation (defining the m-fold uniform multiplicity of $\mathbf{A}$ on $\mathbf{H}$) to the n-fold copy of P'; i.e., $P \mid \mathbf{K}_j = U_j P' U_j^{-1}$. Then $P \neq 0$, and it follows from the relation $P'y'_{kj} \in \mathbf{M}x$ that $Py_{kj} \in \mathbf{A}x_j$ for all k and j, say $Py_{kj} = T_{kj}x_j$. Multiplication of the last equation by P shows that it may be supposed that $T_{kj} = PT_{kj}$.

It follows that $Py_k = \sum_j T_{kj}x_j$ for all $k \in K$. Now we assume that $n > m$ and derive a contradiction. We use the known algebraic fact that in an r-dimensional module over a commutative ring with unit, any $r + 1$ elements are linearly dependent over the ring. We apply this to the module over PA of all ordered n-tuples of elements of PA, and in particular to the $n + 1$ n-tuples $(T_{k1}, T_{k2}, \ldots, T_{km})$ $(k = 1, 2, \ldots, m + 1)$. It results that there exist elements $S_1, S_2, \ldots, S_{m+1}$ of PA which are not all zero, and such that $\sum_{k=1}^{m+1} T_{kj} S_k = 0$. Hence $\sum_{k=1}^{m+1} S_k P y_k = 0$, or $\sum_{k=1}^{m+1} S_k y_k = 0$, but as the L_k are mutually orthogonal, it follows that $S_k y_k = 0$ for all k. But since y_k is separating for $\mathbf{A}$, so all of the S_k must vanish, a contradiction which completes the proof.

Proof of Theorem 12.2   With the notation of Lemma 12.3.9, the R_n are mutually orthogonal. For suppose $R_n R_m \neq 0$. Then $A \mid R_m R_n \mathbf{H}$ is of uniform multiplicity both m and n. By Lemma 12.3.8, either $m = n$, or else one of m and n is finite and the other does not exceed $\aleph_0$, in which latter case Lemma 12.3.10 applies to show that $m = n$.

EXERCISES

1   An operator on a Hilbert space is said to be of uniform multiplicity n if the ring it generates is such. Show that the operation of multiplication by x, on functions $f(x)$ in $L_2(0,1)$, is of uniform multiplicity 1. Show further that the operation of multiplication by x_1, on functions $f(x_1, \ldots, x_r)$ on the unit cube in R^r, is of uniform multiplicity $\aleph_0$.

2   Recall that if A is a given operator in a Hilbert space $\mathbf{H}$, then the *point spectrum* of A consists of all $\lambda \in C$ for which there exists a nonzero vector $x \in \mathbf{H}$ such that $Ax = \lambda x$; and for any such λ, the linear manifold $[x \in \mathbf{H}: Ax = \lambda x]$ is called the eigen-(proper) space corresponding to the eigen-(proper) value λ. Show that if A is a bounded self-adjoint operator in $\mathbf{H}$, then

> *a.* the projection onto the closed linear subspace $\mathbf{H}_0$ of $\mathbf{H}$ spanned by the eigenspaces corresponding to points of the point spectrum of A is contained both in the ring generated by A and its commutor;
>
> *b.* that if $\mathbf{H}_1$ denotes the orthocomplement of $\mathbf{H}_0$, and if $A \sim \int \lambda \, dE_\lambda$ is the spectral resolution of A, then for all $x \in \mathbf{H}_1$, the function of λ, $(E_\lambda x, x)$ is continuous;
>
> *c.* if $E(\cdot)$ is the spectral measure corresponding to A, then for all real λ,
> $$E(\{\lambda\}) \text{ is the strong } \lim_{T \to \infty} (2T)^{-1} \int_T^{-T} e^{is(A - \lambda I)} \, ds.$$

3   Show that a bounded operator with simple spectrum on a Hilbert space has *pure point spectrum*, i.e. $\mathbf{H} = \mathbf{H}_0$ in Exercise 2, if and only if it is unitarily equivalent to the operation of multiplication by x on L_2 over a bounded real interval, with respect to a purely discrete measure.

4   A bounded self-adjoint operator A in Hilbert space is said to have *pure continuous spectrum* in case its point spectrum is empty, i.e. $\mathbf{H} = \mathbf{H}_1$ in Exercise 2.

Show that if m is any regular measure on R^1, then the set of all vectors $x \in \mathbf{H}$ such that the measure m_x on R^1 given by the equation $m_x(B) = (E(B)x,x)$ is absolutely continuous with respect to m, is a closed linear subspace $\mathbf{H}_m$ which is invariant under A; and that the projection with range $\mathbf{H}_m$ lies in the ring generated by A. Show also that if $y \in \mathbf{H}$ is orthogonal to $\mathbf{H}_m$, then the measure m_y is purely singular with respect to m, i.e., is supported by a set of m-measure zero.

5   With the notation of Exercise 4, A is said to have *absolutely continuous spectrum* in case $\mathbf{H} = \mathbf{H}_m$ when m is Lebesgue measure. Show that two self-adjoint operators having the same spectrum and same uniform multiplicity are unitarily equivalent provided they both have absolutely continuous spectrum.

6   Show that a self-adjoint operator on Hilbert space has simple spectrum if and if it is unitarily equivalent to multiplication by x acting on L_2 on a bounded interval relative to some regular measure on the interval.

7   Let the *weighted spectrum* of a bounded self-adjoint operator on a Hilbert space $\mathbf{H}$ be defined as the totality of all measures m on R^1 of the form $m(B) = (E(B)x,x)$, where $E(\cdot)$ is the spectral measure for A. Show that two bounded self-adjoint operators on Hilbert space having simple spectrum are unitarily equivalent if and only if their weighted spectra are identical.

8   Prove that for any abelian ring, there is a maximal abelian ring, unique within unitary equivalence, to which it is algebraically *-isomorphic. Show further that:

    a. the cited isomorphism preserves the operational calculus for bounded Baire functions;

    b. it is bicontinuous in the weak topology.

9   Recall that the *measure ring* of a measure space is the Boolean ring of all locally measurable sets modulo the ideal of local null sets; and that the measure space is called *localizable* if its measure ring is complete as a partially ordered set (with the definition $a \leq b$ if and only if $ab = a$). Show that the Boolean ring of all projections in an abelian ring on a Hilbert space is algebraically isomorphic to the measure ring of a localizable measure space; and show conversely that the measure ring of such a space is isomorphic with the Boolean ring of all projections in its multiplication algebra.

10   Show that two abelian rings on Hilbert space of the same uniform multiplicity are unitarily equivalent if and only if the Boolean rings of projections which they contain are algebraically isomorphic.

11   A finite measure space is called *homogeneous* if it contains no atoms (i.e. minimal elements in its measure ring), and if for every measurable set E of positive measure, the subset of $L_2(M)$ which is supported by E has the same dimensionality as $L_2(M)$. Using the theorem of Maharam to the effect that every non-atomic finite measure space is a direct sum of homogeneous spaces, and that if M is homogeneous and $L_2(M)$ has dimension n, then the measure ring of M is isomorphic to that of $[0,1]^n$ (with Lebesgue measure on $[0,1]$), show that the most general abelian ring of uniform multiplicity m is the direct sum of rings of the following types:

a. all scalars operating on an arbitrary Hilbert space of dimension m;

b. all multiplications by bounded measurable functions on M, operating on all measurable functions $f(\cdot)$ from M to a fixed Hilbert space of dimension m, such that $\|f(\cdot)\|$ is square-integrable on M, M being finite and homogeneous.

12 Show that if A is a bounded self-adjoint operator on a separable Hilbert space, then the ring it generates is identical with the totality of its bounded Baire functions; and show further that every abelian ring on a separable space may be obtained in this way. (This result is due to von Neumann. Hint: Use Exercises 8 and 11.)

13 Obtain a necessary and sufficient condition for measure rings of the spaces considered in Exercise 11 to be ring-isomorphic, and obtain thereby a complete set of cardinal number-valued invariants for abelian W^*-algebras.

XIII

C^*-ALGEBRAS AND APPLICATIONS

13.1 INTRODUCTION

The theory of C^*-algebras dates from the discovery by Gelfand and Naimark that uniformly closed self-adjoint operator algebras on Hilbert space—unlike the rings studied by von Neumann and Murray—could be characterized in simple intrinsic algebraic terms, independently of their action on Hilbert space. This opened up the study of the algebraic isomorphism classes of such algebras, in the sense of emphasizing its cogency. It was soon found that C^*-algebras have certain applications in quantum mechanics, and especially in quantum field theory, in parts of group representation theory, and some other areas, in which W^*-algebras could not be substituted.

In addition the intrinsic algebraic structure of the algebra is in some respect much simpler than the classification of the representations of the algebra on Hilbert space, which provides an additional motivation for the development of the general theory of C^*-algebras. This is already quite evident in the case of abelian algebras; the general abelian C^*-algebra, *as an algebra* is in one-to-one correspondence with the general compact Hausdorff space (the spectrum of the algebra); while spatially, i.e., *in its action on Hilbert space*, the general

340

abelian C^*-algebra is at least as complicated as the general abelian W^*-algebra, which involves the relatively complicated multiplicity theory in Chapter XII.

The theory of C^*-algebras contrasts with that of W^*-algebras in the fundamental role played by positive linear functionals on C^*-algebras; such abstract positivity considerations were emphasized at an early date in the work of M. G. Krein, as well as in more special form in the theory of positive definite functions. It developed that normalized positive linear functionals on C^*-algebras provided a model for the concept of "expectation value in a state" in quantum mechanics which was so suggestive that they came to be known as "states", a usage that will be followed here in our purely mathematical treatment.

The utility of C^*-algebras arise in part from the greater smoothness of representations of suitably chosen C^*-algebras in comparison with that of W^*-algebras in a similar analytic context. For example, an irreducible representation of the C^*-algebra generated by convolutions by L_1 functions on the line corresponds in a natural way to a continuous character of the additive group of the line; while a similar representation of the W^*-algebra generated by the same operator corresponds to a character which in general is not only discontinuous but non-measurable. The completeness of the irreducible representations of C^*-algebras can be considered as a generalization of the discovery by Gelfand and Raikov of the completeness of the continuous irreducible unitary representations of arbitrary locally compact topological groups, a result which was a starting point for the now extensive theory of infinite-dimensional representations of non-compact Lie groups.

This chapter treats the basic general theory of C^*-algebras, and develops some illustrative applications, involving both states and representations.

13.2   REPRESENTATIONS AND STATES

We begin with these concepts and then treat their relations.

DEFINITION   A *state* of a C^*-algebra **A** is a self-adjoint complex-valued linear functional E on **A** which is non-negative on non-negative self-adjoint operators, and normalized by the constraint:

$$\sup_{\|A\| \le 1} E(A^*A) = 1.$$

A *pure state* is one which is not a linear combination with positive coefficients of two other states; a state which is not pure is called *mixed*. A set of states is called *complete* if the only element of **A** which vanishes in every state of the set is zero.

Example   Let x be an arbitrary unit vector in the Hilbert space **H**, and for arbitrary $A \in \mathbf{B(H)}$, set $E(A) = (Ax,x)$. Then E is easily seen to be a state of **B**; and will later be shown to be a pure state. When **H** is finite-dimensional, all pure states have this form, but this is not the case when **H** is infinite-dimensional. Note that if $I \in \mathbf{A}$, then the normalization condition is equivalent to the condition that $E(I) = 1$.

DEFINITION   A (*-, or "self-adjoint") *representation of a C^*-algebra* **A** *on a Hilbert space* **K** *is a mapping ϕ from* **A** *into* **B(K)** *with the properties that for arbitrary $A, B \in$* **B(H)**, *and complex a,*

$$\phi(A + B) = \phi(A) + \phi(B), \qquad \phi(AB) = \phi(A)\phi(B),$$
$$\phi(A^*) = \phi(A)^*, \qquad \phi(aA) = a\phi(A).$$

(Thus correspondence of adjoints is included in the term "representation", and the term "*-representation" is used on occasion only to emphasize this feature.) A *cyclic vector* for such a representation is a vector $z \in$ **K** such that $\phi(\mathbf{A})z$ is dense in **K**; and a representation possessing a cyclic vector is called *cyclic*. The representation ϕ is called *irreducible* in case there is no closed linear subspace of **K** which is invariant under the $\phi(A)$ for all $A \in$ **A**, except for **K** itself and $\{0\}$.

Two representations ϕ_i ($i = 1,2$) of **A** on Hilbert spaces $\mathbf{K}_i$ are *unitarily equivalent* via the unitary transformation T from $\mathbf{K}_1$ onto $\mathbf{K}_2$ in case $T\phi_1(A)T^{-1} = \phi_2(A)$, $A \in$ **A**.

A cyclic representation $(\mathbf{K},\phi)$ is said to be *clothed* if a particular cyclic vector z of unit norm in **K** is designated; and the system $(\mathbf{K},\phi,z)$ is called a *clothed cyclic representation*. Unitary equivalence of clothed representations is defined as inclusive of the correspondence of the designated cyclic vectors.

A set of representations of **A** is called *complete* if no nonzero element of **A** is annihilated by every representation in the set.

Examples   For an irreducible representation, every nonzero vector is cyclic. The identity representation of **B(H)**, $A \to A$, is irreducible; but its left regular representation ϕ on the Hilbert space **K** of all Hilbert-Schmidt operators on **H**: $\phi(A)$ sends $B \in K$ into AB, is not. If M is a probability measure space, the identity representation of its multiplication algebra is cyclic, any non-vanishing element of $L_2(M)$ of unit norm serving as cyclic vector; and a canonical associated clothed representation is given by $(L_2(M),Id,I)$, where *Id* denotes the identity representation and 1 the function identically 1 on M.

THEOREM 13.1   *For any state E of a C^*-algebra* **A**, *there exists a clothed cyclic representation* $(\mathbf{K},\phi,z)$, *unique within unitary equivalence, such that* $E(A) = (\phi(A)z,z)$ *for all* $A \in$ **A**.

Proof   We must first set up a partial substitute for an identity to deal with cases in which an identity is absent.

DEFINITION   An *approximate identity* in a C^*-algebra **A** is a net $\{B_\lambda\}$ in **A** such that $\|B_\lambda\| \leq 1$, $B_\lambda = B_\lambda^*$, and $B_\lambda A \to A$ with λ for all $A \in$ **A**.

Lemma 13.2.1   *Any C*-algebra possesses an approximate identity.*

Proof   Let λ denote an arbitrary finite subset of **A**, the C^*-algebra in question, the totality Λ of all such being ordered by set inclusion. If $\lambda = \{C_1, \ldots, C_n\}$, let $D = \sum_i C_i C_i^*$, and set $B_\lambda = D(\epsilon + D)^{-1}$, where $\epsilon = n^{-1}$. Then $B_\lambda \in \mathbf{A}$ and

$$(B_\lambda - I)D(B_\lambda - I) = \epsilon^2 D(\epsilon I + D)^{-1} \le \epsilon I/2,$$

i.e. $\sum_{i=1}^{n} [(B_\lambda - I)C_i][(B_\lambda - I)C_i] \le \epsilon I/2$, implying that

$$\|B_\lambda C_i - C_i\| \le (2n)^{-1}$$

for all $\lambda \in \Lambda$ and $i = 1, \ldots, n$. Given any $C \in \mathbf{A}$, for any λ with $C \in \lambda$, it follows that $\|(B_\lambda - I)C\| \le (2n)^{-1}$, showing that $B_\lambda C \to C$ with λ.

Lemma 13.2.2   *Any representation* $(\mathbf{K}, \phi)$ *of a* C^*-*algebra* **A** *is a contraction:* $\|\phi(A)\| \le \|A\|$, $A \in \mathbf{A}$.

Proof   Let $\mathbf{A}'$ be the algebra generated by **A** and the identity operator I; thus $\mathbf{A}' = [aI + A : a \in C, A \in \mathbf{A}]$; as a one-dimensional extension of **A**, $\mathbf{A}'$ is closed. Defining ϕ' on $\mathbf{A}'$ by the equation $\phi'(aI + A) = aI' + \phi(A)$, where I' is the identity operator on **K**, then ϕ' is a representation of $\mathbf{A}'$ on **K**. Now if $A \in \mathbf{A}$ and $\|A\| \le 1$, then $I - A^*A$ is positive semi-definite, and therefore of the form B^*B for some $B \in \mathbf{A}'$. Since $\phi'(I - A^*A) = \phi(B)^*\phi(B) \ge 0$, it results that $\phi(A^*A) \le I$ so that $\|\phi(A^*A)\| \le 1$, but $\|\phi(A^*A)\| = \|\phi(A)^*\phi(A)\| = \|\phi(A)\|^2$, and it follows that $\|\phi(A)\| \le 1$, completing the proof.

Lemma 13.2.3   *If* $(\mathbf{K}, \phi, z)$ *is a clothed representation of the* C^*-*algebra* **A**, *then the functional* E *on* A *given by the equation* $E(A) = (\phi(A)z, z)$ *is a state of* **A**.

Proof   The conclusion is immediate from the definition except for the normalization condition. To establish this, note that by Lemma 13.2.2, $\sup_{\|A\| \le 1} E(A^*A) \le 1$. To show that the supremum is precisely 1, let $\{B_\lambda\}$ be an approximate identity for **A**, and note that since $B_\lambda A \to A$ for all $A \in \mathbf{A}$, then $\phi(B_\lambda)\phi(A)z \to \phi(A)z$ for all $A \in \mathbf{A}$; but since $\phi(\mathbf{A})z$ is dense in **K**, it follows that $\phi(B_\lambda)u \to u$ for all $u \in \mathbf{K}$, in particular $\phi(B_\lambda)z \to z$. Hence

$$\sup_{\|A\| \le 1} E(A^*A) = \sup_{\|A\| \le 1} (\phi(A)z, \phi(A)z) \ge \sup_{\lambda} (\phi(B_\lambda)z, \phi(B_\lambda)z) = 1,$$

reversing the inequality and completing the proof.

Proof of Theorem 13.1   Let E be the given state of the C^*-algebra $\mathbf{A}$ on the Hilbert space $\mathbf{H}$. Let $\mathbf{C}$ denote the set of all $A \in \mathbf{A}$ such that $E(BA) = 0$ for all $B \in \mathbf{A}$. Then $\mathbf{C}$ is a left ideal in $\mathbf{A}$. Let θ denote the canonical mapping of $\mathbf{A}$ onto its quotient $\mathbf{A}/\mathbf{C}$ modulo $\mathbf{C}$; thus, if $A \in \mathbf{A}$, then $\theta(A)$ is its residue class modulo $\mathbf{C}$. We now define an inner product in $\mathbf{A}/\mathbf{C}$ by setting $(\theta(A),\theta(B)) = E(B^*A)$, and denote the resulting inner-product space as $\mathbf{K}_0$; to see that the inner product is properly defined, i.e. does not depend on the choice of representative in the residue class, note that if C_1 and C_2 are arbitrary in $\mathbf{C}$, then

$$E((B + C_1)^*(A + C_2)) = E(B^*A) + E(C_1^*A) + E(B^*C_2) + E(C_1^*C_2);$$

and $E(C_1^*A) = \overline{E(A^*C_1)}$, while $E(A^*C_1) = E(B^*C_2) = E(C_1^*C_2) = 0$ since E vanishes on $\mathbf{C}$.

It is straightforward to verify that $(\theta(A),\theta(B))$ has indeed the formal properties of an inner product. In addition, it is evidently positive semidefinite, which implies that Schwarz' inequality is applicable:

$$|(\theta(A),\theta(B))|^2 \leq (\theta(A),\theta(A))(\theta(B),\theta(B)).$$

It follows that if $(\theta(A),\theta(A)) = 0$, then $E(B^*A) = 0$ for all $B \in \mathbf{A}$, which means that $A \in \mathbf{C}$, or that $\theta(A) = 0$. Thus the inner product is positive definite, and $\mathbf{K}_0$ may be completed to a Hilbert space $\mathbf{K}$.

We now define a representation ϕ_0 of $\mathbf{A}$ on the dense subspace $\mathbf{K}_0$ of $\mathbf{K}$ by the equation: $\phi_0(A)\theta(B) = \theta(AB)$; that the definition is proper follows from the fact that $\mathbf{C}$ is an ideal, and it is straightforward to show that ϕ_0 is indeed a *-representation, in the algebraic sense of $\mathbf{A}$ on $\mathbf{K}_0$. We next show that for all $A \in \mathbf{A}$, $\phi_0(A)$ is a bounded operator on $\mathbf{K}_0$, and may therefore be uniquely extended to a bounded operator $\phi(A)$ which is everywhere defined on $\mathbf{K}$. To this end note that for all $A \in \mathbf{A}$, $\|A\|^2 I - A^*A \geq 0$, so that for all $B \in \mathbf{A}$, $B^*(\|A\|^2 I - A^*A)B \geq 0$, which implies that $E[B^*(\|A\|^2 I - A^*A)B] \geq 0$. This in turn implies that $E(B^*A^*AB) \leq \|A\|^2 E(B^*B)$, which means that $(\phi_0(A)\theta(B), \phi_0(A)\theta(B)) \leq \|A\|^2(\theta(B),\theta(B))$, or that $\|\phi_0(A)\| \leq \|A\|$. It is straightforward to conclude that the extension $\phi(A)$ again defines a *-representation of $\mathbf{A}$, on all of $\mathbf{K}$.

To establish the existence of a cyclic vector z in $\mathbf{K}$ with the indicated relation to E, note that since $B_\lambda A \to A$, $(\theta(A),\theta(B_\lambda)) \to E(A)$. Now

$$|(\theta(A),\theta(B_\lambda))| \leq \|\theta(A)\| \, \|\theta(B_\lambda)\| \leq \|\theta(A)\|,$$

showing that $|E(A)| \leq \|\theta(A)\|$, from which it follows that there exists a unique vector $z \in \mathbf{K}$ such that $E(A) = (\theta(A),z)$. The density of $\theta(A)$ in $\mathbf{K}$ further implies that $\theta(B_\lambda) \to z$ weakly, by a simple approximation argument. It follows that $E(AB_\lambda) = (\theta(AB_\lambda),z) = (\phi(A)\theta(B),z) \to (\phi(A)z,z)$.

Consider now the questions of the unicity of the clothed representation $(\mathbf{K},\phi,z)$. Suppose $(\mathbf{K}',\phi',z')$ is another clothed representation of $\mathbf{A}$ with the property that $(\phi'(A)z',z') = E(A)$. Let T_0 denote the operator from $\mathbf{K}_0$ into $\mathbf{K}'$ which carries $\phi(A)z$ into $\phi(A')z'$; this mapping is well-defined, and indeed an isometry, since

$$(\phi(A)z,\phi(A)z) = (\phi(A^*A)z,z) = E(A^*A)$$
$$= (\phi'(A^*A)z',z') = (\phi'(A)z',\phi'(A)z').$$

It follows that T_0 admits a unique extension to a unitary operator T from $\mathbf{K}$ onto $\mathbf{K}'$. Since $T\phi(A)z = \phi'(A)z'$ for all $A \in \mathbf{A}$, the same is true with A replaced by AB, yielding the equation $T\phi(A)\phi(B)z = \phi'(A)\phi'(B)z = \phi'(A)T\phi(B)z$, which implies that $T\phi(A)u = \phi'(A)Tu$ for all $u \in \mathbf{K}$. Since $B_\lambda A \to A$, $\phi'(B_\lambda A)z' \to \phi'(A)z'$, from which it follows that $\phi'(B_\lambda) \to I$ in the strong operator topology. Taking the limit in the equality $T\phi(B_\lambda)z = \phi'(B_\lambda)z'$, it follows that $Tz = z'$, completing the proof that $(\mathbf{K},\phi,z)$ is unitarily equivalent to $(\mathbf{K}',\phi',z')$.

It remains to show the mutual correspondence between pure states and irreducible representations. Suppose that E is a pure state, that $(\mathbf{K},\phi,z)$ is a corresponding clothed representation of $\mathbf{A}$, and that $\mathbf{L}$ is a closed linear subspace of $\mathbf{K}$ which is invariant under $\phi(\mathbf{A})$. Let P denote the projection of $\mathbf{K}$ onto $\mathbf{L}$, which then commutes with all $\phi(A)$, $A \in \mathbf{A}$, and define functionals F and G on $\mathbf{A}$ as follows: $F(A) = (\phi(A)Pz,Pz)$; $G(A) = (\phi(A)Pz,Pz)$. Then it follows that $E = F + G$, and that F and G are non-negative scalar multiples of states. Since E is pure, this can only take place if either F or G vanishes, in which case the invariant subspace $\mathbf{L}$ is trivial, or if F, G, and E are all proportional via constants, say $F(A) = cE(A)$, c being a constant. This implies that

$$(\phi(B^*A)Pz,Pz) = c(\phi(B^*A)z,z)$$

for arbitrary $A, B \in \mathbf{A}$, from which it follows that for arbitrary $u, v \in \mathbf{K}$, $(Pu,v) = c(u,v)$. It follows that either $c = 0$ or $c = 1$, so that $\mathbf{L}$ is again necessarily a trivial subspace.

Suppose now that the representation ϕ is irreducible, and suppose that E has the form $E = aF + bG$, where F and G are states and a and b are positive constants with $a + b = 1$. Defining a sesquilinear form Ψ on $\mathbf{K}_0$ by the equation, $\Psi(\theta(A),\theta(B)) = F(B^*A)$, it follows that Ψ is bounded relative to the inner product in $\mathbf{K}$, and so has the form $\Psi(u,v) = (Su,v)$ for some bounded linear operator S on $\mathbf{K}$, and all $u,v \in \mathbf{K}_0$. But S commutes with all $\phi(A)$ since for arbitrary $B, C \in \mathbf{A}$,

$$(S\phi(A)\theta(B),\theta(C)) = F(C^*AB) = F((A^*C)^*B) =$$
$$= (S\theta(B),\phi(A^*)\theta(C)) = (S\theta(B),\phi(A)^*\theta(C)).$$

By the irreducibility of ϕ, S must in consequence of its commutativity with the $\phi(A)$ be a scalar, say s, but then $F = sE$. Thus E cannot be a non-trivial convex linear combination of two other states, i.e. E is pure.

COROLLARY 13.2.1   *The "vector" states, $E(A) = (Ax,x)$, $\|x\| = 1$, of the algebra $\mathbf{B(H)}$ on a Hilbert space $\mathbf{H}$, are pure.*

Proof   Defining $\phi(A) = A$ for $A \in \mathbf{B}$, the system $(\mathbf{H},\phi,x)$ is a clothed representation of $\mathbf{B}$ which is irreducible, so the corresponding state, having the form given, must be pure.

COROLLARY 13.2.2   *Any C*-algebra has a complete set of irreducible representations.*

Proof   It is sufficient to consider the case in which the given algebra $\mathbf{A}$, say on the Hilbert space $\mathbf{H}$, contains the identity operator I on $\mathbf{H}$. For otherwise the algebra $\mathbf{A_1}$ obtained by adjunction of I to $\mathbf{A}$ has a complete set of irreducible representations, whose restrictions to $\mathbf{A}$ either remain irreducible or vanish identically. It follows that a complete set of irreducible representations of $\mathbf{A_1}$ restricts to a complete set of irreducible representations of $\mathbf{A}$, together with possible zero representations of $\mathbf{A}$.

Now assuming $I \in \mathbf{A}$, the set Σ of all states of $\mathbf{A}$ is compact and convex in the w^*-topology (in the dual to $\mathbf{A}$). Convexity is clear. To show compactness, it suffices, in view of the earlier result that $\|E\| = 1$ for all $E \in \Sigma$, to show that Σ is closed. Now if $\{E_\lambda\}$ is a net in Σ which is w^*-convergent to the continuous linear functional F on $\mathbf{A}$, then $F(I) = 1$, $F(A^*A) \geq 0$, and F is self-adjoint, by continuity, i.e. $F \in \Sigma$, or Σ is closed. Now applying the Krein-Milman theorem (see Exercise 25, page 168), it follows that the set Π of all extreme points of Σ is complete, in the sense that if $\pi(A) = 0$ for all $\pi \in \Pi$ and some $A \in \mathbf{A}$, then $A = 0$. But the extreme points of Σ are the same as the pure states of $\mathbf{A}$; thus the pure states of $\mathbf{A}$ form a complete set of states. Hence the set of all irreducible representations is complete, for an element $A \in \mathbf{A}$ which is carried into 0 by every such representation will vanish in every pure state.

We recall that a set of representations of a group is called complete if the common part of their kernels consists only of the identity of the group.

COROLLARY 13.2.3   (GELFAND-RAIKOV)   *Any locally compact topological group has a set of continuous unitary irreducible representations in Hilbert spaces which is complete.*

Proof   Let G denote the group in question, and set $\mathbf{H} = L_2(G)$ with respect to left-invariant Haar measure on G. Let $\mathbf{A}$ be the uniform closure of the algebra $\mathbf{A_0}$ of all left convolutions by elements of $L_1(G)$. Any irreducible representation of $\mathbf{A}$ on a Hilbert space $\mathbf{K}$ restricts to an irreducible representation of $\mathbf{A_0}$ on $\mathbf{K}$, and by Theorem 10.1 there is a unique continuous unitary representation U of G on $\mathbf{H}$ such that

$$\phi(L_f) = \int U(a)f(a)\, da$$

for all $f \in L_1(G)$, where L_f denotes the operation of left convolution by f. This representation U is again irreducible, for it is immediate from general properties of the integral (or via the Hahn-Banach theorem, employing a weak vector-valued integral) that L_f leaves invariant any subspace of **K** which is invariant under U. The set of all irreducible representations of G thus obtained is complete, for if $a \in G$ and $U(a) = I$ for all such representations U, then $U(a)\phi(L_f) = \phi(L_f)$ for all ϕ and f. But $U(a)\phi(L_f) = \phi(L_{f_a})$, where $f_a(x) = f(a^{-1}x)$, so that it would result from the completeness of the ϕ that $L_{f_a} = L_f$ for all $f \in L_1$. It is not difficult to show that this implies that $a = e$.

DEFINITIONS   The *variance* of a self-adjoint element A of a C^*-algebra **A** in a state E is defined as $E(A^2) - E(A)^2$; A is said to *have the exact value* $E(A)$ *in the state* E in case its variance vanishes in this state. A state of a C^*-algebra in which every self-adjoint element has an exact value is called an *observation* of the algebra.

The following corollary can be regarded as asserting that there exists a "wave function", in the sense of quantum mechanics, corresponding to points of the continuous, as well as of the point spectrum, of any given self-adjoint operators; but that in the case of the continuous spectrum, the wave function must be taken in an appropriate representation space of the "algebra of observables" (taken on occasion as **B(H)**, rather than in the original Hilbert space **H** itself.

COROLLARY 13.2.4   *If F is an observation on a C^*-subalgebra* **C** *of a C^*-algebra* **A**, *then there exists a clothed representation* $(\mathbf{K},\phi,z)$ *of* **A** *such that* $\phi(C)z = F(C)z$ *for all* $C \in \mathbf{C}$.

The following lemma has some independent interest, particularly from the standpoint of quantum mechanics. In this connection it implies that the phenomenology of a quantum system—its states, observables, probability distributions of observables in states (cf. the Exercises, below), etc.—are essentially unaltered by the inclusion of the system in a larger system.

Lemma 13.2.4   *If F is a pure state of a C^*-subalgebra* **C** *containing the identity operator of a C^*-algebra* **A**, *then there exists a pure state E of* **A** *such that* $E \mid \mathbf{A} = F$.

Proof   Let p be the real-valued function on the set $\mathbf{C}_0$ of all self-adjoint elements of **C** given by the equation: $p(A) = \inf [t \in R^1 : tI - A \geq 0]$. It is clear that the set over which the infimum is taken is non-empty, and that p has the properties required for the Hahn-Banach theorem, in the form given in Theorem 6.2, page 157. With f taken as F, the condition that $f(x) \leq p(x)$ on the subspace **M**, taken as **C**, is satisfied, by virtue of the positivity of F. It follows that there exists a state G of **A** such that $G \mid \mathbf{C} = F$.

Now let Δ denote the set of all states G of **A** for which $G \mid \mathbf{C} = F$. There is no difficulty in verifying that Δ is a compact convex set in the dual of **A**, and accordingly has an extreme point, say E. Then E is the required pure state of **A**. For if E were not pure, say $E = a_1E_1 + a_2E_2$, where the a_i are positive and the E_j are states of **A**, then $E \mid \mathbf{C} = a_1E_1 \mid \mathbf{C} + a_2E_2 \mid \mathbf{C}$. Since $E \mid \mathbf{C} = F$ and F is pure, it follows that $F = E_1 \mid \mathbf{C} = E_2 \mid C$, which means that the E_j are in Δ. But since E is an extreme point of Δ, this implies that $E = E_1 = E_2$. Thus E can be expressed in the form $a_1E_1 + a_2E_2$ only if the states E, E_1, and E_2 coincide, which means that E is pure.

Lemma 13.2.5   *Any observation on a C*-algebra is pure.*

Proof   Note first that any observation is multiplicative: if F denotes the observation on the C^*-algebra **C**, then $F(AB) = F(A)F(B)$ for $A, B \in \mathbf{A}$. To see this, it suffices to consider the case in which A and B are self-adjoint, by the linearity and self-adjointness of F. In this case, by Schwarz' inequality, $|F(AB)| \leq F(A^2)^{1/2}F(B^2)^{1/2} = |F(A)F(B)|$. On the other hand, from the relation $AB + BA = (1/4)[(A + B)^2 - (A - B)^2]$ it follows that $F(AB) + F(BA) = 2F(A)F(B)$, but $F(BA) = \overline{F(AB)}$, so that it follows in turn that $\mathrm{Re}\,(F(AB)) = F(A)F(B)$. It follows that $F(AB)$ is real, whence $F(AB) = F(BA)$, and it results that $F(AB) = F(A)F(B)$, as required.

To prove the observation F is pure, it suffices to show that in the associated clothed representation $(\mathbf{K}, \phi, z)$, **K** is one-dimensional, and hence irreducible. Let A and B be arbitrary in the C^*-algebra **C**; then

$$F(AB) = (\phi(AB)z, z) = (\phi(B)z, \phi(A^*)z)$$

on the one hand while on the other

$$F(AB) = F(A)F(B) = (\phi(A)z, z)(\phi(B)z, z).$$

By Schwarz' inequality, $|{<}\phi(B)z, \phi(A^*)z{>}|^2 \leq \|\phi(B)z\|^2\|\phi(A^*)z\|^2$

$$= (\phi(B^*B)z, z)(\phi(AA^*)z, z) = F(B^*B)F(AA^*) =$$
$$= |F(AB)|^2 = |(\phi(B)z, \phi(A^*)z)|^2,$$

showing that the inequality is in this case an equality. But this is the case for Schwarz' inequality only when the vectors involved are proportional, which means that $\phi(B)z$ and $\phi(A^*)z$ must be proportional for arbitrary A and B in **C**. From the cyclicity of z it follows that **K** is one-dimensional.

Proof of Corollary 13.2.4   Note to begin with that it is no essential loss of generality to assume that $I \in \mathbf{C}$. For any observation F of **C** will extend to an observation F_1 of the algebra $\mathbf{C}_1$ of all operators of the form $aI + C$ with $C \in \mathbf{C}$, by defining $F_1(aI + A) = a + F(A)$. There will then exist a clothed representation $(\mathbf{K}, \phi, z)$ of the C^*-algebra $\mathbf{A}_1$ consisting of all operators of the form $aI + A$ with $A \in \mathbf{A}$, such that $\phi(B)z = F_1(B)z$ for all $B \in \mathbf{C}_1$. It follows that the clothed representation $(\mathbf{K}, \phi \mid \mathbf{A}, z)$ has the required property.

Let E be a pure state of **A** whose restriction to **C** is the given observation F, and let $(\mathbf{K},\phi,z)$ be the associated clothed representation. If A and B are arbitrary in $\mathbf{C}$, then

$$E[(AB - E(A)B)(AB - E(A)B))^*] = 0,$$

since F is a homomorphism, which means that $\phi(AB - E(A)B)z = 0$, or $\phi(A)u = F(A)u$ with $u = \phi(B)z$. Finally, for some $B \in \mathbf{C}$, $\phi(B)z \neq 0$ since otherwise $F(B^*B) = 0$ for all $B \in \mathbf{C}$, contradicting the assumption that F is a state of $\mathbf{C}$.

EXERCISES

1   Show that with the notation of Theorem 13.1, $\|\phi\| = 1$.

2   Show that the variance of a self-adjoint operator of $A \in \mathbf{A}$ in a state E of $\mathbf{A}$ is identical with the variance in the usual probabilistic sense of the random variable which represents A on the regular compact measure space obtained by imposing on the spectrum of the C^*-algebra $\mathbf{C}$ generated by A the measure representing $E \mid \mathbf{C}$.

3   A B^*-algebra (the term "abstract C^*-algebra" is also used) is an equivalence class of (concrete) C^*-algebras, relative to algebraic $*$-isomorphism as equivalence. Show that if Δ is any compact Hausdorff space, and $\mathbf{H}$ any Hilbert space, the algebra $\mathbf{A}$ of all continuous functions from Δ to $\mathbf{B}(H)$ (taken in its uniform topology), with the norm: $\|f(\cdot)\| = \sup_{\delta \in \Delta} \|f(\delta)\|_\infty$, and the adjunction operation $f^*(\delta) = f(\delta)^*$, is a B^*-algebra. (*Hint:* Use as representation space, $L_2(\Delta)$ relative to a purely discrete measure in Δ.) Show also that if Δ supports a non-singular regular measure, and if $\mathbf{H}$ is separable, then $\mathbf{A}$ has a faithful $*$-representation by a C^*-algebra on a separable Hilbert space.

4   Show that if $\mathbf{H}$ is finite-dimensional, then the general state of $\mathbf{B}(\mathbf{H})$ has the form $E(X) = \operatorname{tr} XD$, for some fixed positive semidefinite operator D of unit trace (known as the "density operator" in statistical mechanics). Conclude that the most general pure state is a vector state. Show further that if $\mathbf{H}$ is infinite-dimensional, neither conclusion is valid.

5   If G is a group of automorphisms of a C^*-algebra $\mathbf{A}$, a *stationary state* (with respect to G) is defined as one with the property that $E(A) = E(g(A))$ for all $A \in \mathbf{A}$ and $g \in \mathbf{R}$. Show that for any given such state there is in addition to the clothed representation $(\mathbf{K},\phi,z)$ cited in Theorem 13.1, a unique unitary representation Γ of G on $\mathbf{K}$ such that $\Gamma(g)\phi(A)\Gamma(g)^{-1} = \phi(g(A))$ and $\Gamma(g)z = z$ for all $g \in G$, $A \in \mathbf{A}$.

6   With the notation of Exercise 5, the state E is called *ergodic* in case it is an extreme point of the set of all stationary states. Show that E is ergodic if and only if the corresponding extended clothed representation $(\mathbf{K},\phi,z,\Gamma)$ is such that $\mathbf{K}$ is irreducible under the combined action of ϕ and Γ. Show also that if $I \in \mathbf{A}$, then every stationary state is a w^*-limit of convex linear combinations of ergodic states.

7   Let Δ be a compact Hausdorff space on which a regular measure m is given, and let G' be a group of homeomorphisms on Δ which leaves m invariant. Let G denote the group of all automorphisms of the algebra $\mathbf{A} = \mathbf{C}(\Delta)$ of all continuous complex-valued functions on Δ, of the form $f(\delta) \to f(g(\delta))$, $g \in G'$, Show that every ergodic state E of $\mathbf{A}$ has the form $E(f) = \int f \, dn$, where n is a G'-ergodic invariant measure on Δ; and conversely, for any such measure, the state E is ergodic.

8   Give examples to show that the set Π of all pure states of a C^*-algebra with identity may be closed, or may fail to be closed. Show that Π is always a Borel subset of the state space of $\mathbf{A}$.

9*   Show that the quotient algebra of a C^*-algebra $\mathbf{A}$ modulo a closed two-sided ideal $\mathbf{C}$, with the quotient norm: $\|A + \mathbf{C}\| = \inf_{C \in \mathbf{C}} \|A + \mathbf{C}\|$, is algebraically *-isomorphic to a C^*-algebra (i.e. is a B^*-algebra). Deduce that an isomorphism of one C^*-algebra into another is necessarily closed.

10*   Extend the correlation between groups and group representations on the one hand and representations of suitable C^*-algebras on the other to the case of a locally compact transformation group of measure-preserving transformations on a locally compact regular measure space.

XIV

THE TRACE AS
A NON-COMMUTATIVE
INTEGRAL

14.1 INTRODUCTION

The philosophy of the theory of integration has evolved considerably in the past several decades. Newer applications have tended to de-emphasize point-wise features in favor of global and/or algebraic ones. This has led in particular to a realization that the theory of the trace on the ring of all bounded operators on a Hilbert space was parallel in a number of ways to abstract Lebesgue integration theory, and to the development of a natural simultaneous generalization of both theories. In these sections we first develop the theory of the trace in **B(H)** from a standpoint that emphasizes the integration-theoretic analogy on the one hand and points towards the generalization to arbitrary rings on the other. We should point out however that the basic results concerning the trace on **B(H)** were originally obtained quite independently of these considerations, because of their intrinsic interest, and that these results are important for a variety of applications.

Following this we give a brief introduction to the subject describable as "non-commutative integration theory", which serves in particular to make explicit the parallel between the Lebesgue integral and the trace in operator

351

rings. Our treatment is based on the concept of a Hilbert algebra, which can be considered an extension of the notion of "integration algebra" earlier treated, and provides a basis for the extension of the Plancherel theory to non-commutative groups.

14.2 ELEMENTARY OPERATORS AND THE TRACE

Let H be a given complex Hilbert space, and B the ring of all bounded linear operators on H. When H is finite-dimensional, there exists a linear functional T on B having properties:

(1) $T(AB) = T(BA)$;
(2) $T(A^*A) \geq 0$, and $= 0$ if and only if $A = 0$; and this functional, the trace, is unique within proportionality via a positive constant.

In the infinite-dimensional case, no such functional exists; and it is clear that the identity operator on H can have no trace in the usual sense. Nevertheless, certain operators on H appear to possess a trace, e.g. those of the form $A = \Sigma_i c_i P_i$, where the P_i are mutually orthogonal one-dimensional projections and the c_i are constants such that $\Sigma_i |c_i| < \infty$. But natural questions such as the linearity of the class of operators having a trace, the additivity of the trace on this class, do not follow directly along the lines used earlier in this book.

We shall treat here analogs of the L_1- and L_2-spaces—the operators possessing a trace forming an analog to an L_1 space—essentially by regarding the problem as one of generalization from a commutative algebra (that of functions on a measure space) to a non-commutative one (that of operators in a ring). A transition between the two situations is provided by the theory of maximal abelian rings, which as already seen, are closely related to function algebras on measure spaces.

To being with, we set up an analog to the class of simple functions, or continuous functions of compact support in the case of a regular measure in a locally compact space: i.e. a class whose integration properties are particularly simple, but which is sufficient to approximate general integrable elements.

DEFINITION   A linear operator A on a Hilbert space H is said to be of *finite rank* in case its range is finite-dimensional.

THEOREM 14.1   *The set F of all finite-rank operators on H is a self-adjoint two-sided ideal in $B(H)$; and there exists a non-vanishing linear functional T on F, unique within proportionality, such that:*

(i) $T(AB) = T(BA)$;
(ii) $T(A^*A) \geq 0$;
(iii) $T(A^*A) = 0$ *only if $A = 0$, for arbitrary A and B in* B.

Proof It is immediate that the sum and product of finite-rank operators is again such, using the fact that the union of two finite-dimensional subspaces is again such. Also, the adjoint of a finite-rank operator is again of finite rank; indeed, every finite-rank operator is a finite sum of operators of the form $x \to (x,e')e$, where e and e' are fixed vectors, and the adjoint of such a "rank one" operator is obtained by interchanging e and e'. If A is of finite rank and B is arbitrary in **B**, then BA is clearly of finite rank, and so is its adjoint, showing that the finite-rank operators form an ideal in **B**.

We say an operator $A \in \mathbf{B(H)}$ is *supported* by a closed linear manifold $\mathbf{M} \subset \mathbf{H}$ or by the projection P, in case $PA = AP = A$, where P is the projection with range **M**. A finite-rank operator is evidently one supported by a finite-dimensional manifold, and the set $\mathbf{B_M}$ of all linear operators supported by **M** is isomorphic to the algebra $\mathbf{B(M)}$. It therefore admits a functional $T_\mathbf{M}$ with the properties indicated; this functional (the trace) is unique within proportionality; and the proportionality factor may be fixed by the constraint that $T_\mathbf{M}(Q) = 1$ for any one-dimensional projection Q. If $\mathbf{M} \subset \mathbf{N}$ where N is also finite-dimensional, then $T_\mathbf{N} \mid \mathbf{B_M}$ has the characteristic properties of $T_\mathbf{M}$, showing that $T_\mathbf{M} \subset T_\mathbf{N}$, and completing the proof.

At this point there are several options as to the direction by which the trace-class operators (analog of integrable functions) could be introduced. One would be to use the order properties, as in Chapter III, first treating non-negative operators and defining $T(A)$ for any such operators as the supremum of the $T(B)$ as B ranges over the elements of **F** bounded by A; the next step would then be to establish additivity. To illustrate an alternative procedure, we elect here to follow a familiar completion method, from which linearity of the resulting trace is immediate while its positivity requires demonstration.

We introduce an analog to the L_1 norm by the

DEFINITION   For any element $A \in \mathbf{F}$, the "L_1-norm" (or "trace norm"), denoted $\|A\|_1$, is defined as the supremum of the $|T(AX)|$, as X ranges over the elements of **F** of bound at most 1.

To avoid possible confusion with the L_1-norm, as well as because of a valid analogy, the operator bound $\|X\|$ will also be denoted as $\|X\|_\infty$. Note that $\|A\|_1$ is finite since by the linearity of $T(AB)$ as a function of A or B together with the non-negativity of $T(A^*A)$, Schwarz' inequality is applicable and shows that $|T(AX)|^2 \leq T(A^*A)T(XX^*)$. In forming the supremum defining $\|A\|_1$ it is no loss of generality to assume that X is supported by any given finite-rank projection P which supports A, as is easily seen, and it follows that $\|A\|_1^2 \leq T(A^*)T(P)$.

COROLLARY 14.2.1   *The L_1 norm on the algebra F has actually the properties of a norm:*

  a. $\|A\|_1 \geq 0$ *and* $\|A\|_1 = 0$ *if and only if* $A = 0$;

     b. $\|A + B\|_1 \leq \|A\|_1 + \|B\|_1$;

     c. $\|aA\|_1 = |a|\,\|A\|_1$ *for any* $a \in C$.

It has in addition the properties:

     d. $\|A^*\|_1 = \|A\|_1 = T(|A|)$, *where* $A = U|A|$ *is the canonical polar decomposition of* A;

     e. *if* $V \in \mathbf{B}$, *then* $|T(AV)| = |T(VA)| \leq \|V\|_\infty \|A\|_1$, *and* $\|VA\|_1 \leq \|V\|_\infty \|A\|_1$, $\|AV\|_1 \leq \|V\|_\infty \|A\|_1$.

Proof Part (*e*) is a simple deduction from the definition of the L_1-norm. Now writing A in canonical polar form $A = US$ with U partially isometric and S self-adjoint, then $\|U^*A\|_1 \leq \|U^*\|_\infty \|A\|_1 \leq \|A\|_1$, but $U^*A = S$, so that $\|S\|_1 \leq \|A\|_1$. The reverse inequality is clear, and it follows that $\|S\|_1 = \|A\|_1$. To show that $\|A\|_1 = T(|A|)$, let X denote any operator in $\mathbf{B}$ with $\|X\|_\infty \leq 1$, and B for the non-negative square root of S; then

$$|T(XS)|^2 = |T(XB^2)|^2 = |T(BXB)|^2 \leq |T(B^2)|^2$$

by the positivity of T as a functional. Taking the supremum over X, it follows that $\|A\|_1 \leq T(|A|)$, and since the reverse inequality is clear from (*e*), it results that $\|A\|_1 = T(|A|)$. The proofs of the remaining properties are entirely straightforward.

Having established the L_1 norm and its properties on $\mathbf{F}$, general 'integrable' operators may now be introduced as limits of sequences of operators in $\mathbf{F}$ which are Cauchy with respect to the L_1 norm, in analogy with a familiar method of approaching the Lebesgue integral. It is convenient to require an additional mode of convergence of the sequence, and we make the

     DEFINITION An element A of $\mathbf{B(H)}$ is said to be of *trace class* (or integrable, or in $L_1(\mathbf{B})$) if there exists a sequence $\{A_n\}$ in $\mathbf{F}$ which is Cauchy in L_1, and convergent uniformly to A.

As in the Lebesgue theory, some types of continuity must be established in order to proceed. It will suffice to show the

Lemma 14.2.1 *If* $\{A_n\}$ *is a Cauchy sequence in* L_1, *and if* $A_n \to 0$ *strongly, then* $T(A_n) \to 0$.

Proof Let ϵ be an arbitrary positive number, and suppose that

$$\|A_m - A_n\|_1 < \epsilon \qquad \text{for } m, n \geq n_1(\epsilon).$$

Setting P for an arbitrary projection in $\mathbf{F}$ and noting that $T(A) = (T(A) - T(AP)) + T(AP)$ for arbitrary $A \in \mathbf{F}$, it follows that

$$T(A_n) - T(A_nP) = T(A_n - A_{n_1}) + T(A_{n_1} - A_{n_1}P) + T(A_{n_1}P - A_nP).$$

It follows in turn that if $n > n_1$, then

$$|T(A_n) - T(A_nP)| \leq 2\epsilon + |T(A_{n_1}) - T(A_{n_1}P)| = 2\epsilon$$

if P is chosen to contain the range of A_{n1}.

To complete the proof it therefore suffices to show that for any fixed projection $P \in \mathbf{F}$, $T(A_n P) \to 0$ as $n \to \infty$. But $T(A_n P) = T(P A_n P)$, and $P A_n P \to 0$ strongly, so that this is a finite-dimensional question, and clear.

COROLLARY 14.2.2 *If A is a trace class operator, and if $\{A_n\}$ and $\{A'_n\}$ are two sequences of operators in $\mathbf{F}$ each of which is Cauchy with respect to the L_1-norm, and strongly convergent to A, then $\lim_n T(A_n) = \lim_n T(A'_n)$.*

Proof This is immediate from the observation that the sequence $\{A_n - A'_n\}$ is again Cauchy in the L_1-norm and convergent strongly to 0.

Hence we may make the

DEFINITION For any trace-class operator A, the *trace* $T(A)$ is defined as $\lim_n T(A_n)$, where $\{A_n\}$ is any sequence in $\mathbf{F}$ which is Cauchy with respect to the L_1-norm, and strongly convergent to A. The set of all trace-class operators is denoted as $L_1(\mathbf{B})$, and the L_1-norm (or "trace norm") $\|A\|_1$ of any $A \in L_1(\mathbf{B})$ is defined as the supremum of $|T(AB)|$ as B ranges over the unit ball in $\mathbf{B}$.

The general properties of the trace on $L_1(\mathbf{B})$ may be summarized as follows.

THEOREM 14.2 *The trace is a continuous self-adjoint linear functional on $L_1(\mathbf{B})$, which is a self-adjoint two-sided ideal in $\mathbf{B}$; and the trace norm on $L_1(\mathbf{B})$ has all of the properties given in Corollary 14.1.1. Moroever,*

a. For any fixed element $A \in L_1(\mathbf{B})$, $T(AB)$ is a continuous function of B with respect to the strong operator topology, relative to the unit ball in $\mathbf{B}$.

b. A self-adjoint bounded linear operator A on $\mathbf{H}$ is in $L_1(\mathbf{B}(\mathbf{H}))$ if and only if it has the form $A = \sum_j a_j P_j$, where the P_j are mutually orthogonal one-dimensional projections, and $\sum_j |a_j| < \infty$.

Proof Apart from parts (a) and (b), this is entirely straightforward. To show (a), approximate to A by elements of $\mathbf{F}$; it results that it suffices to establish (a) for the case in which $A \in \mathbf{F}$. But if $\{B_\lambda\}$ is a net in the unit ball of $\mathbf{B}$ which converges strongly to B, and if P is any projection in $\mathbf{B}$ which supports A, then $T(AB_\lambda) = T(PAPB_\lambda) = T(APB_\lambda P)$, and $PB_\lambda P \to PBP$ strongly. The question thus reduces to a finite-dimensional one, and is clear.

To show (b), note that for $A = A^* \in \mathbf{F}$, $\|A\|_\infty \le \|A\|_1$, from which it follows by approximation that this inequality holds for all self-adjoint $A \in L_1(\mathbf{B})$. It follows that every such A is a uniform limit of finite-rank operators, and so is compact. By the spectral theorem it must then have the form $\sum_j a_j P_j$ for suitable constants a_j and one-dimensional mutually orthogonal projections P_j. In case $\sum_j |a_j| < \infty$, then $A = \lim_n \sum_{j<n} a_j P_j$ both in $\mathbf{B}$ and

in $L_1(\mathbf{B})$, and is trace-class. If $\sum_j |a_j| = \infty$, then recalling that $\|A\|_1 = \sup_{[B \in \mathbf{B}:\ \|B\|_\infty \le 1]} |T(AB)|$, and taking B as a suitable sum of the P_j, it results that $\|A\|_1 = \infty$, i.e. A is not in $L_1(\mathbf{B})$.

COROLLARY 14.2.3   *If $A \in L_1(B)$ and $A = A^* \ge 0$, then $T(A)$ is the supremum of the $T(B)$ as B ranges over the subset of $\mathbf{F}$ bounded above by A.*

It is clear from (b) that $T(A) \ge 0$ if $A \ge 0$, which implies that $T(A) \ge T(A')$ if $A \ge A'$, so that $T(A)$ is not less than the indicated supremum; and (b) shows it is at least that great.

Part of the argument in the proof of the Theorem 14.2 establishes the

COROLLARY 14.2.4 $L_1(\mathbf{B})$ *is complete as a metric space.*

We could not go on and establish analogs of the L_p spaces, but shall treat here only the case $p = 2$.

DEFINITION   An operator A in $\mathbf{B}$ is said to be *Hilbert-Schmidt* or *square-integrable* in case $A^*A \in L_1(\mathbf{B})$; and its Hilbert-Schmidt, or L_2-norm is defined as $T(A^*A)^{1/2}$. The set of all Hilbert-Schmidt operators will be denoted as $L_2(\mathbf{B})$.

COROLLARY 14.2.3 $L_2(\mathbf{B})$ *is a Hilbert space with respect to the inner product: $(A,B) = T(B^*A)$. Moreover adjunction is a conjugation on this space, and for arbitrary $S \in \mathbf{B}$, $\|SA\|_2 \le \|S\|_\infty \|A\|_2$.*

Proof   The only part of the first statement which is not entirely straightforward is the completeness, and this follows by reduction to the self-adjoint case (noting that $T(A^*A) = T(AA^*)$ for all $A \in \mathbf{F}$ and then by approximation for all $A \in L_2(\mathbf{B})$), and the inequality $\|A\|_\infty \le \|A\|_2$ for this case. The rest of the corollary follows by establishing the results on $\mathbf{F}$ and then approximating as earlier.

EXERCISES

1   Let $\mathbf{H} = L_2(M)$ for a measure space M, and for any function $K \in L_2(M \times M)$, let T_K denote the operator $f \to g$ where $g(x) = \int K(x,y)f(y)\,dy$. Show that the map $K \to T_K$ is unitary from $L_2(M \times M)$ onto $L_2(\mathbf{B(H)})$.

Conclude that an operator A on a given Hilbert space $\mathbf{H}$ is Hilbert-Schmidt if and only if for some orthonormal basis e_j, $\sum_{j,k} |(Ae_j,e_k)|^2 < \infty$, and that the indicated sum is independent of the basis.

2   Show that the uniform closure of the algebra of all finite-rank operators on the Hilbert space $\mathbf{H}$ consists of the algebra $\mathbf{C}$ of all compact operators. (*Hint:* Use the polar decomposition.)

3   Let A be an element of $\mathbf{B}(H)$ such that $|T(AX) \le$ const. $\|X\|_\infty$ for all $X \in \mathbf{F}$. Show that A is of trace class. (*Hint:* Show that AB has the same property as A for

all $B \in \mathbf{B}$. Deduce that all spectral projections of A have the same property, and show this implies that they are all finite-dimensional.)

4   For any $A \in L_1(\mathbf{B})$, let F_A denote the linear functional on $\mathbf{C}$ (with the notation of Ex. 2), $F_A(X) = T(AX)$. Show that the mapping $A \to F_A$ is an isometric isomorphism from $L_1(\mathbf{B})$ onto the dual of $\mathbf{C}$. (*Hint:* Only the "onto" part is non-trivial; to prove it, reduce to the case in which F is a given self-adjoint functional on $\mathbf{C}$, and for each $P \in \mathbf{F}$, obtain $A_P \in \mathbf{B}$ supported by P and such that $F(X) = T(A_P X)$ for all X supported by P. Show that $\|A_P\|_1$ is bounded and that $A_P \to A$, $A \in \mathbf{B}$. Then use Ex. 3.)

5   Let $\mathbf{A}$ denote the algebra of all Hilbert-Schmidt operators on the Hilbert space $\mathbf{H}$. For any $A \in \mathbf{A}$, let L_A and R_A denote the operators $B \to AB$ and $B \to BA$ on $\mathbf{A}$. Show that relative to the inner product defined in $\mathbf{A}$, $L_A = L_{A^*}$, $R_A = R_{A^*}$, $J L_A J = R_{A^*}$ where J is the conjugation: $A \to A^*$. Show also that if $\mathbf{L}$ and $\mathbf{R}$ are the rings generated by the L_A and R_A, then $\mathbf{L}' = \mathbf{R}$.

6   For any $B \in \mathbf{B}$, let ϕ_B denote the linear functional on $L_1(\mathbf{B})$, $A \to T(AB)$. Show that ϕ is an isometric isomorphism from $\mathbf{B}$ onto the Banach dual of $L_1(\mathbf{B})$. (*Hint:* Follow the initial lines of the proof of Ex. 4.)

14.3   HILBERT ALGEBRAS

A simple and applicable formulation of abstract integration theory which encompasses both the theory of the trace on $\mathbf{B(H)}$, and the essential elements of the theory of the Lebesgue integral, may be based on a kind of non-commutative variant of the notion of integration algebra treated in Chapter VIII.

DEFINITION   A *pre-Hilbert algebra* is a system $(\mathbf{A},{}^*,(.,.))$ consisting of an associative algebra $\mathbf{A}$ over C, an adjunction operation $*$ on $\mathbf{A}$, and an inner product $(.,.)$ on $\mathbf{A}$, having the following properties:

(i) For arbitrary $a, b, c \in \mathbf{A}$, $(ab,c) = (b,a^*c)$ and $(a^*,b^*) = (b,a)$;

(ii) For any given element $a \in \mathbf{A}$, there is a constant c such that $\|ax\| \leq c\|x\|$ for all $x \in \mathbf{A}$;

(iii) the linear span of the products ab, for a and b arbitrary in $\mathbf{A}$, dense in $\mathbf{A}$ (relative to the norm defined by the inner product).

Example   Let G be a unimodular locally compact group, and let $\mathbf{A}$ denote the set of all continuous functions of compact support on G, as an algebra relative to convolution as multiplication. For $f \in \mathbf{A}$, let f^* be defined by the equation $f^*(x) = \bar{f}(x^{-1})$, and for $f, g \in \mathbf{A}$, let $(f,g) = \int_G f(x)\bar{g}(x)\,dx$, where dx denotes the element of Haar measure on G. There is no difficulty in verifying that $(\mathbf{A},{}^*,(.,.))$ is then a pre-Hilbert algebra. A slight variation of this example is obtained by taking $\mathbf{A}$ as the set of all bounded measurable functions which vanish outside a set of finite measure. As might be expected, these two examples have a close relation, and it will be seen that with an appropriate notion of equivalence they are not essentially different.

Now the example just given, although an important one, on which the generalized Plancherel theorem is based, does not have a clear relation to integration theory, because the direct analog to integration takes place essentially on a dual to the group G, and a kind of generalized Fourier transform is needed to make the analogy explicit. The following examples do have such a clear relation, and may profitably be borne in mind as the theory of Hilbert algebras is developed.

> $a.$ Let M be an arbitrary measure space; let **A** denote the subalgebra of $L_{\infty}(M)$ consisting of elements which vanish outside a set of finite measure define $*$ as complex conjugation, and (f,g) as $\int_M f\bar{g}$.
>
> $b.$ Let **F** denote the algebra of all finite-rank operators on a given Hilbert space, define $*$ as adjunction, and (A,B) as tr B^*A.

This section is devoted to some basic results concerning Hilbert algebras which together convey a more specific notion of how Lebesgue integration theory may be given a useful non-commutative extension.

DEFINITION   With the notation of the preceding definition, for $a \in$ **A** let L'_a (resp. R'_a) denote the operation $x \to ax$ (resp. $x \to xa$), $x \in$ **A**. Let the continuous linear extensions of L'_a and R'_a to the Hilbert space completion **H** of **A** be denoted as L_a and R_a. For arbitrary $x \in$ **H**, let L'_x (resp. R'_x) denote the operator in **H** with domain **A** defined by the equation $L'_x a = R_a x$ (resp. $R'_x a = L_a x$), $a \in$ **A**. In case L'_x (resp. R'_x) is bounded, x is called a left (resp. right) *bounded* element of **A**, and the closure of L'_x (resp. R'_x) is denoted as L_x (resp. R_x). The adjoint of an element $x \in$ **H**, denoted as x^*, is defined by continuity from the notion given in **A**.

THEOREM 14.3   *An element of the completion of a pre-Hilbert algebra is left-bounded if and only if it is right-bounded; and the set* **C** *of all such elements forms a $*$-algebra relative to the multiplication:* $xy = L_x y = R_y x$. *Further the set of all (left or right) bounded elements of the completion of the pre-Hilbert algebra* $(\mathbf{C},*,(.,.))$ *is identical with* **C**.

DEFINITION   A Hilbert algebra is a pre-Hilbert whose algebra of *bounded* elements (as elements which are left- or right-bounded will henceforth be called) is identical with itself; this algebra is called the *bounded algebra* of the system.

Proof of Theorem 14.3   Note first that if $a \in$ **A** and $L_a = 0$, then $a = 0$, for then $ab = 0$ for all $b \in$ **A**, whence $(ab,c) = 0$ for all $c \in$ **A**; but $(ab,c) = (a,cb^*)$ (using the properties of adjunction operations, i.e. $(ab)^* = b^*a^*$ and $a^{**} = a$), so it follows that a is orthogonal to a dense subset of **A**, and hence vanishes. It follows that the mapping $a \to L_a$ is a $*$-isomorphism of **A** into **B(H)**, and that $L_a^* = L_{a^*}$.

Setting J for the operation $x \to x^*$ on **H**, it follows also that $R'_{a^*} \subset JL_a J$ if

$a \in \mathbf{A}$, since this inclusion follows from the equality: $(R'_{a\ast}b,c) = (JL_aJb,c)$ for arbitrary $b, c \in \mathbf{A}$, which equality is straightforward. Thus every element $a \in \mathbf{A}$ is right-bounded, and $R_{a\ast} = JL_aJ$; and similarly regarding left-boundedness.

Observe next that the set $\mathbf{C}_r$ of all right-bounded elements of $\mathbf{H}$ is self-adjoint, and that $R^*_x = R_{x\ast}$ if $x \in \mathbf{C}_r$. It suffices to show that $(R'_x a,b) = (a,R'_{x\ast}b)$ for all $a, b \in \mathbf{A}$. In fact, $(R'_x a,b) = (L_a x,b) = (x,L_{a\ast}b) = (LJ_{a\ast}b,Jx) = (R_a b^*,x^*) = (L_{b\ast}a,x^*) = (a,L_b x^*) = (a,R'_{x\ast}b)$. A similar argument shows that the set $\mathbf{C}_r$ of all left-bounded elements is also-adjoint. But $x \in \mathbf{C}_r$ means that $|(R'_x x,b)| \le c\|a\|\,\|b\|)$ for all $a, b \in \mathbf{A}$ and some constant c; and

$$(R'_x a,b) = (L_a x,b) = (Jb,JL_a Jx^*) = (Jb,R_{a\ast}x^*) = (Jb,L'_{x\ast}a^*),$$

showing that R'_x is bounded if and only if $L'_{x\ast}$ is bounded, i.e. if x^* is left-bounded, but this is equivalent to x itself being left-bounded. Thus $\mathbf{C}_r = \mathbf{C}_l = \mathbf{C}$.

We show next that if $x, y \in \mathbf{C}$, then $L_x y = R_y x$. Since $\mathbf{A}$ is dense in $\mathbf{H}$, it suffices to show that $(L_x y,z) = (R_y x,z)$ for all $z \in \mathbf{A}$. In fact, on the one hand $(L_x y,z) = (y,L_{x\ast}z) = (y,R_z x^*) = (JR_z Jx,y^*) = (L_{z\ast}x,y^*) = (x,L_z y^*)$; while on the other hand,

$$(R_y x,z) = (x,R_{y\ast}z) = (x,L_z y^*).$$

Now note that $\mathbf{C}$ is an algebra relative to the multiplication: $xy = L_x y = R_y x$. It suffices to show that $R'_{xy} \subset R_y R_x$ for arbitrary $x, y \in \mathbf{C}$, for this shows that $xy \in \mathbf{C}$ then and that $R_{xy} = R_y R_x$, implying that the multiplication is associative. To this end, observe first that since R_a and L_b commute for arbitrary a and b in $\mathbf{A}$, the same is true when a and b are in $\mathbf{C}$; for, if p and q are arbitrary in $\mathbf{A}$,

$$(R_a L_b p,q) = (L_b p,R_{a\ast}q) = (R_p b,L_q a^*) = (JL_q Ja,JR_p Jb^*) =$$
$$= (R_{q\ast}a,L_{p\ast}b^*) = (L_p a,R_q b^*) = (R_a p,L_{b\ast}q) = (L_b R_a p,q).$$

It is evident that $(\mathbf{C}, *, (\,.\,,\,.\,))$ is a pre-Hilbert algebra, and it remains only to show that its bounded algebra is $\mathbf{C}$. Now if x is any right-bounded element of this system, the operator R''_x with domain $\mathbf{C}$ given by the equation $R''_x a = L_a x$ is bounded. Hence $R''_x \mid \mathbf{A}$ is bounded, but $R''_x \mid \mathbf{A} = R'_x$, implying that $x \in \mathbf{C}$.

Some simple algebraic features largely derived in the course of the foregoing proof are summarized in

COROLLARY 14.3.1 *In any Hilbert algebra, the mapping $x \to L_x$ (resp. $x \to R_x$) is a *-isomorphism (*anti-isomorphism) of the bounded algebra into $\mathbf{B(H)}$, $\mathbf{H}$ being the Hilbert space of the system. Furthermore, for arbitrary bounded x and y in $\mathbf{H}$,*

$$JL_x J = R_{x\ast}, \qquad L_x R_y = R_y L_x, \qquad JR_x J = L_{x\ast}.$$

Proof Only the one-to-one character of the maps $x \to L_x$ and $x \to R_x$ has not been established earlier. But if $L_x = 0$, then $(L_x a, b) = 0$ for all $a, b \in \mathbf{A}$, from which it follows that x is orthogonal to all $a^* b$, and hence vanishes.

The next result can be regarded as a formal generalization of the maximal abelian character of the multiplication algebra of a localizable measure space.

> COROLLARY 14.3.2 *Let L and R denote the weak closures of* $[L_a : a \in \mathbf{A}]$ *and* $[R_a : a \in \mathbf{A}]$. *Then* $\mathbf{L}' = \mathbf{R}$ *and* $\mathbf{R}' = \mathbf{L}$.

Proof Note first that $\mathbf{L}$ and $\mathbf{R}$ contain I, and so are W^*-algebras. For by Corollary 12.1.1, this follows from the fact that no element of $\mathbf{H}$ is annihilated by all L_a, $a \in \mathbf{A}$ (resp. all R_a).

Now suppose $T \in \mathbf{L}'$. Then $TL_a = L_a T$ for $a \in \mathbf{A}$, and if a and b are in $\mathbf{A}$, then $TR_a b = TL_b a = L_b Ta = R'_{Ta} b$, showing that R'_{Ta} is bounded. Thus $TR_a = R_{Ta}$, and putting $\mathbf{R}_1$ for the weak closure of the totality of the R_a with a bounded, it follows from the observation at the beginning of the proof that $T \in \mathbf{R}_1$. Now $\mathbf{R}_1$ commutes elementwise with the weak closure $\mathbf{L}_1$ of the set of all L_a with a bounded, i.e. $\mathbf{L} \subset \mathbf{R}'_1$, which inclusion implies that $\mathbf{L}_1 \subset \mathbf{L}$. Since it is obvious that $\mathbf{L} \subset \mathbf{L}_1$, it follows that $\mathbf{L} = \mathbf{L}_1$.

The following two corollaries have already been established, in the course of the preceding proofs.

> COROLLARY 14.3.3 *If* $T \in \mathbf{L}$ (resp. $\mathbf{R}$) *and a is a bounded element, then Ta is bounded and* $TL_a = L_{Ta}$ (resp. $TR_a = R_{Ta}$).

> COROLLARY 14.3.3 *The weak closure of the set of all* L_a *with* $a \in \mathbf{A}$ *is the same for all *-subalgebras* $\mathbf{A}$ *which are dense in* $\mathbf{H}$ (*and similarly for the* R_a).

> DEFINITION The weak closure of the set of all L_a (resp. R_a) as a ranges over any dense subalgebra $\mathbf{A}$ will be called the *left ring* (resp. *right ring*) of the pre-Hilbert (or Hilbert) algebra. The intersection of these two rings is called the *center* of the algebra.

The following corollary will be useful later.

> COROLLARY 14.3.5 *Let K be a closed linear manifold in* $\mathbf{H}$ *that is invariant under the left ring of a Hilbert algebra with Hilbert space* $\mathbf{H}$ *and bounded algebra* $\mathbf{C}$. *Then* $\mathbf{K} \cap \mathbf{C}$ *is dense in* $\mathbf{K}$.

Proof If x is arbitrary in $\mathbf{K}$ and a_n is a sequence in $\mathbf{B}$ that converges to x, then putting P for the projection with range $\mathbf{K}$, $Pa_n \to x$, $Pa_n \in \mathbf{K}$, and as $\mathbf{K}$ is invariant, $P \in \mathbf{L}' = \mathbf{R}$, so that by Corollary 14.3.3, $Pa_n \in \mathbf{C}$.

The next corollary generalizes a feature of multiplication algebras.

> COROLLARY 14.3.6 *For any operator T in the center of a Hilbert algebra, with conjugation* J, $JTJ = T^*$.

Proof It suffices by a simple argument to consider the case in which T is a projection, and the conclusion is then equivalent to the assertion that J leaves invariant any closed linear manifold **K** in **H** that is invariant under **L** and **R**. As $\mathbf{K} \cap \mathbf{C}$ is dense in **K**, it suffices to show that if $a \in \mathbf{K} \cap \mathbf{C}$, then $a^* \in \mathbf{K}$. Now if $a^* \in \mathbf{K}$, there exists an element c in **H** which is orthogonal to **K**, but not orthogonal to a^*. Since the orthocomplement of **K** is likewise invariant, c may be assumed to be bounded. If b is arbitrary in **C**, then $(a^*b,c) = (b,ac) = 0$, for $ac \in \mathbf{K}$ and yet ac is orthogonal to **K**, by the invariance of **K** and its orthocomplement under left and right multiplications by elements of **C**, But since I is a strong limit of the R_b, it follows that $(a^*,c) = 0$, a contradiction.

The following concept isolates properties obtained for Hilbert algebras which define a kind of non-commutative generalization of the maximal abelian rings, and play an analogous role in the multiplicity theory of non-commutative rings.

> DEFINITION A *standard ring* (via the conjugation J) on a Hilbert space **H** is a ring **A** for which there exists a conjugation J such that $JAJ = \mathbf{A}'$ and $JTJ = T^*$ for all $T \in \mathbf{A} \cap \mathbf{A}'$.

In terms of this definition, two of the preceding corollaries assert that the left (or right) ring of a Hilbert algebra is standard.

We show next how an analog to a measure may be defined canonically for an arbitrary Hilbert algebra. In the case of Example (a) above, the analog coincides essentially with the measure; in the case of Example (b), with the dimension; the general notion will be called a *gage*.

> DEFINITION A *gage* on a W^*-algebra **A** is a completely-additive non-negative function m on the projections in **A** which is unitarily invariant: $m(U^*PU) = m(P)$ if P is any projection and U any unitary in **A**; and has the (non-triviality) feature that any nonzero projection in **A** bounds a projection in **A** on which m is finite and positive.

THEOREM 14.4 *Let* **L** *be the left ring of a Hilbert algebra, and for any projection $P \in \mathbf{L}$, define $m(P) = \|x\|^2$ or ∞, according as $P = L_x$ for some bounded x, or not. Then m is a gage on* **L**.

Proof To establish complete additivity, note first that if L_x and L_y are orthogonal projections, x and y being bounded, then x and y are orthogonal, for $L_x L_y = 0 = L_{xy}$, showing that $xy = 0$, so that $(xy,z) = 0$ for all bounded z; but $x = x^*$ so that $(xy,z) = (y,xz) = (y,R_z x) = 0$; and since I is contained in the strong closure of the algebra generated by the R_x (as noted in the proof of the preceding theorem), it results that $(y,x) = 0$. Now let P_λ be an indexed family of mutually orthogonal projections in **L** with sum P, and consider first the case in which $\sum_\lambda m(P_\lambda)$ is finite. Then

$P_\lambda = L_{x_\lambda}$ with $x_\lambda \in \mathbf{C}$, and the x_λ are mutually orthogonal. Since $m(P_\lambda) = \|x_\lambda\|^2$, $\sum_\lambda \|x_\lambda\|^2 < \infty$, so that $\sum_\lambda x_\lambda$ exists; denote this sum as x. Now for arbitrary $a \in \mathbf{C}$, $Pa = \sum_\lambda P_\lambda a = \sum_\lambda R_a x_\lambda = R_a x = L'_x a$. It follows that $P = L_x$, whence $m(P) = \|x\|^2 = \sum_\lambda \|x_\lambda\|^2 = \sum_\lambda m(P_\lambda)$.

Consider now the case in which $\sum_\lambda m(P_\lambda) = \infty$. The conclusion is then valid in case $m(P) = \infty$, so suppose that $m(P) < \infty$. Then $P = L_x$ for some bounded x, and from the relation $P_\lambda = P_\lambda P$ it follows, using the earlier derived result that $L_{Ty} = TL_y$ if $T \in \mathbf{L}$ and y is bounded, that $x_\lambda = P_\lambda x$. Hence $\sum_\lambda \|x_\lambda\|^2 = \sum_\lambda \|P_\lambda x\|^2 = \|Px\|^2 < \infty$, in contradiction with the assumption that $\sum_\lambda m(P_\lambda) = \infty$.

To show unitary invariance, let U be unitary and P be a projection, both in $\mathbf{L}$. If $P = L_x$ for some $x \in \mathbf{C}$, then $L_x U = (U^* L_{x^*})^* = (L_{U^* x^*})^* = L_{JU^* x^*}$, so $U^* PU = L_{U^* JU^* x^*}$. Hence $m(U^* PU) = \|U^* JU^* x^*\|^2 = \|x\|^2 = m(P)$. If on the other hand $m(P) = \infty$, then $m(U^* PU) = \infty$, for otherwise $U^* PU = L_y$ for some bounded y, whence $m(P) < \infty$ by the first part of this paragraph.

It remains only to show that every nonzero projection P in $\mathbf{L}$ bounds a projection in $\mathbf{L}$ on which m is positive and finite. Let P_0 denote the supremum of the projections bounded by P having the latter property; then either $P = P_0$, which is the desired conclusion, or $Q = P - P_0$ bounds no non-vanishing projection on which m is finite. Now $Q\mathbf{H}$ is invariant under $\mathbf{R}$ and hence if $Q \neq 0$, it results from Corollary 14.3.5 that $Q\mathbf{H}$ contains a nonzero self-adjoint bounded element a. Setting $\int_\lambda dE(\lambda)$ for the spectral resolution of L_a and defining $S(\epsilon)$ as $\int_{L(\epsilon)} \lambda^{-1} dE_\lambda$, where $\epsilon > 0$ and $L(\epsilon) = [\lambda : |\lambda| > \epsilon]$, then $S(\epsilon)L_a = L_{S(\epsilon)a} = \int_{L(\epsilon)} dE(\lambda)$. It follows that $S(\epsilon)L_b$ is a projection on which n is finite; and it is bounded by Q, for $S(\epsilon)L_a Q = S(\epsilon)(QL_a)^* = S(\epsilon)L_{Qa}^* = SL_a$. Hence $S(\epsilon)L_a = 0$ for all ϵ, but this yields the contradiction $L_a = 0$.

REMARKS   Given a "gage space", i.e. a system $(\mathbf{H}, \mathbf{A}, m)$ consisting of a Hilbert space $\mathbf{H}$, a W^*-algebra $\mathbf{A}$, and a gage m on $\mathbf{A}$, there is a "trace" defined on a class of operators affiliated with $\mathbf{A}$ which is parallel to the trace class operators in case $\mathbf{A} = \mathbf{B}(\mathbf{H})$ and the usual integrable operators in case $\mathbf{A}$ is the multiplication algebra of a measure space (M, m) with $\mathbf{H} = L_2(M, m)$. It is an accident that in the case of $\mathbf{B}(\mathbf{H})$, this L_1-space contains only bounded operators; in general it will consist of closed not necessarily bounded operators affiliated with $\mathbf{A}$.

All of the major developments in the theory of the abstract Lebesgue integral have analogs in this more general setting. In particular, the Riesz-Fischer theorem, the Radon-Nikodym theorem, the theory of the L_p spaces, including their duality, for $1 \leq p \leq \infty$, have extensions which subsume the classical theorems treated earlier in this book. This "noncommutative integration" or trace theory is however beyond the scope of the present work. We mention only that even measurable functions have analogs, so-called "measurable operators", for which there is an effective calculus despite their unboundedness, unlike the case of general unbounded operators: they may be added, multiplied freely, and the closure of the resulting operator is again measurable, etc.

EXERCISES

1   Let G be a locally compact topological group, and let $\mathbf{H}$ denote $L_2(G,m)$, where m denotes Haar measure. Show that if $\mathbf{K}$ and $\mathbf{L}$ are closed linear manifolds in $\mathbf{H}$ which are invariant under all right translations and all left translations, then the corresponding projections commute.

2   A *probability Hilbert algebra* is one in which the algebra $\mathbf{A}$ contains a unit e, and in which $(e,e) = 1$. Suppose given a net $(\mathbf{A}_\lambda,{}^*_\lambda,(.,.)_\lambda) = \Gamma_\lambda$, $\lambda \in \Gamma$, of probability Hilbert algebras together with isomorphisms $\theta_{\lambda',\lambda}$ from Γ_λ into $\Gamma_{\lambda'}$, for $\lambda \le \lambda'$, where "isomorphism" here means "algebraic *-isomorphism preserving the given inner product". Suppose the given isomorphisms have the property that $\theta_{\lambda'',\lambda'}\theta_{\lambda',\lambda} = \theta_{\lambda'',\lambda}$ whenever $\lambda \le \lambda' \le \lambda$. Show that there exists a direct limit probability Hilbert algebra $(\mathbf{A},{}^*,(.,.)) = \Gamma$, in the sense that there are isomorphisms ϕ_λ from Γ_λ into Γ, such that $\phi_{\lambda'}\theta_{\lambda',\lambda} = \phi_\lambda$ whenever $\lambda \le \lambda'$, and that the union of the ranges of the Γ_λ are dense. Show also that the limit is unique, within unitary equivalence.

3   A Hilbert algebra is called *simple* or *factorial* (after von Neumann's term *factor*, signifying a central simple W^*-algebra) in case the common part of its left and right rings consists only of scalars. Show that a direct limit of factorial probability Hilbert algebras is again factorial.

4   Let Γ_n denote the Hilbert algebra whose algebra $\mathbf{A}$ consists of the complex matrices of order $2^n \times 2^n$, with the usual hermitian conjugate as adjoint operation, and the inner product $(A,B) = 2^{-n} \operatorname{tr} B^*A$; let $\theta_{n+1,n}$ denote the mapping $A \to \begin{pmatrix} A & 0 \\ 0 & A \end{pmatrix}$; and for $m > n + 1$, let $\theta_{m,n}$ be defined so that the condition of Exercise 2 is satisfied. Show that the direct limit of the Γ_n has as its left ring a probability Hilbert algebra whose left ring is a factor, which is not algebraically isomorphic to all bounded operators on a Hilbert space. (*Hint:* Show that the state E defined by the unit is a trace, i.e. $E(ab) = E(ba)$ for arbitrary a and b in $\mathbf{L}$; and that no such function can exist on an infinite-dimensional $\mathbf{B(H)}$. This factor is called the "approximately finite" factor, and arises in the theory of the free fermion quantum field.)

5   Let M be a given measure space, let $\mathbf{A}$ denote the algebra of all measurable functions ϕ from $\mathbf{M}$ to the Hilbert-Schmidt operators on a given separable Hilbert space $\mathbf{L}$, for which $\int_M \|\phi(p)\|_2^2 \, dp < \infty$ and $\sup_{p \in M} \phi\|(p)\| < \infty$. Show that $(\mathbf{A},{}^*,(.,.))$ is a Hilbert algebra with $\phi^*(p) = \phi(p)^*$ and $(\phi,\psi) = \int_M (\phi(p),\psi(p)) \, dp$.

6*   Let x be a self-adjoint element of the Hilbert space of a Hilbert algebra. Show that L_x' is essentially self-adjoint.

7*   Let **M** denote the set of all closed densely defined operators affiliated with the left ring of a probability gage space. Show that if one element of **M** extends another, then the two are identical.

8*   With the notation of Exercise 7, show that if A, $B \in \mathbf{M}$, then the closures of $A + B$ and AB exist and are again in **M**.

9   Show that if **H** is the Hilbert space of a Hilbert algebra, if $x_n \to x$ in **H**, and if $\|L_{x_n}'\|$ is finite and uniformly bounded in n, then x is a bounded element, and that L_{x_n} converges strongly to L_x.

10   Let **M** be a W^*-algebra on which is defined a positive central linear functional E which vanishes on non-zero projections. For A, $B \in \mathbf{M}$, define $(A,B) = E(B^*A)$. Show that $(\mathbf{M},^*,(\ldots))$ is a Hilbert algebra; and that the map $A \to L_A$ is an algebraic $*$-isomorphism of **M** onto the left ring of this Hilbert algebra.

11   Let m be a gage on the W^*-algebra **A** on the Hilbert space **H**. Let **C** denote the set of all operators in **A** whose range is contained in that of a projection in **A** on which m is finite. Show that **C** is a 2-sided ideal **A**.

12   With the notation of Exercise 11, let E be a central positive linear functional on **C**, and for $A \in \mathbf{C}$, define $\|A\|_1$ as the supremum of $|E(AX)|$ as X ranges over the unit ball in **C**. Show that the conclusions of Corollary 14.2.1 remain valid. Show further that the conclusion of Lemma 14.2.1 remains valid, with the added hypothesis that $\|A_n\|_\infty$ remains bounded.

SELECTED
REFERENCES

These references are intended to provide a rounded, rather than complete, indication of the relevant book literature of possible interest, stimulus, and use to the reader.†

Introductory

A sampling of the many books providing suitable background material for the present one:

1 Bourbaki, N.: "Les Structures fondamentales de l'analyse," 4 vols., Hermann & Cie, Paris, 1939–1960.   Lucid, precise, and up to date; slightly dry.

2 Dieudonné, J.: "Foundations of Modern Analysis," Academic Press Inc., New York, 1960.   Easily accessible presentation of solid material.

3 Gleason, A.. "Fundamentals of Abstract Analysis," Addison-Wesley Publishing Company, Inc., Reading, Mass., 1966.   An intensive contemporary treatment of foundations.

4 Hausdorff, F.: "Set Theory" (translation), Chelsea Publishing Company, New York, 1962.   A nourishing classic, dated technically but modern in spirit.

† Much of the relevant journal literature is included in the bibliography of I. Segal, Algebraic Integration Theory, *Bull. Am. Math. Soc.* 71, pp. 419–489, 1965.

365

5 Kelley, J. L.: "General Topology," D. Van Nostrand Company, Inc., Princeton N.J. Quite detailed treatment of the main logical preliminary.

6 Rudin, W.: "Principles of Mathematical Analysis," 2d ed., McGraw-Hill Book Company, New York, 1964. Interpolates effectively between calculus and the present material.

Classics of integration theory

7 Caratheodory, C.: "Reelle Funktionen," B. G. Teubner Verlagsgesellschaft, mbH, Leipzig, 1939. A quite accessible prewar synthesis of real variable and integration theory, including a treatment of "outer measure."

8 Lebesgue, H.: "Leçons sur l'intégration," Gauthier-Villars, Paris, 1928. An authoritative latter-day account of some of the truly germinal work of the era.

9 Saks, S.: "Theory of the Integral" (translation), 2d ed., Hafner Publishing Company, Inc., New York, 1937. A good introduction to abstract integration theory is contained in the first few chapters; the bulk of the book is an account of differentiation theory, written at a relatively mature stage in its development.

10 Weil, A.: "L'Intégration dans les groupes topologiques et ses applications," 2d ed., Hermann & Cie, Paris, 1953. A brilliant and stimulating, though concentrated, synthesis of classical methods, and a contemporary outlook, as of just before the war.

Postwar expositions of integration theory

11 Bourbaki, N.: "Intégration," 4 vols., Hermann & Cie, Paris, 1952–1963. A smooth, solid presentation of the part of the theory which is readily handled in terms of locally compact regular models.

12 Hahn, H., and A. Rosenthal: "Set Functions," The University of New Mexico Press, Albuquerque, N. Mex., 1948. Contains much useful information not readily available elsewhere.

13 Halmos, P. R.: "Measure Theory," D. Van Nostrand Company, Inc., Princeton, N.J., 1950. A detailed account of abstract theory from a set-theoretic viewpoint.

14 McShane, E. J.: "Integration," Princeton University Press, Princeton, N.J., 1961. An updated, clear presentation of quasi-classical theory.

15 Royden, H. L.: "Real Analysis," The Macmillan Company, New York, 1963. Lucid treatment in considerable detail of elementary theory.

Some Further Developments in Analysis and Applications

Functional analysis

16 Hille, E., and R. S. Phillips: "Functional Analysis and Semi-groups," rev. ed., American Mathematical Society, Providence, R.I., 1957.

17 Riesz, F., and B. Sz.-Nagy: "Functional Analysis," Frederick Ungar Publishing Co., New York, 1955.

18 Taylor, A. E.: "Introduction to Functional Analysis," John Wiley & Sons, Inc., New York, 1958.

19 Yosida, K.: "Functional Analysis," Academic Press Inc., New York, 1965.

Generalized functions and partial differential equations

20 Gelfand, I. M. et al.: "Generalized Functions," vols. 1 and 4 (translation), Academic Press Inc., New York, 1964.

21 Hörmander, L.: "Linear Partial Differential Operators," Academic Press Inc., New York, 1963.

22 Lions, J. L.: "Équations differentielles operationelles et problèmes aux limites," Springer-Verlag OHG, Berlin, 1961.

23 Schwartz, L.: "Théorie des distributions," 2 vols., Hermann & Cie, Paris, 1950–1951.

Fourier analysis

24 Bochner, S.: "Lectures on Fourier Integrals," Princeton University Press, Princeton, N.J., 1959.

25 Carleman, T.: "L'Intégrale de Fourier et questions qui s'y rattachent," Almqvist and Wiksells, Uppsala, Sweden, 1944.

26 Mandelbrojt, S.: "Series de Fourier et classes quasi-analytiques de fonctions," Gauthier-Villars, Paris, 1935.

27 Paley, R. E. A. C., and N. Wiener: "The Fourier Integral and Certain of Its Applications," Cambridge University Press, London, 1933.

28 Wiener, N.: "The Fourier Integral and Certain of Its Applications," Cambridge University Press, London, 1933.

29 Zygmund, A.: "Trigonometric Series," rev. ed., 2 vols., Cambridge University Press, London, 1959.

Hilbert space

30 Akhiezer, N. I., and I. M. Glazman: "Theory of Linear Operators in Hilbert Space," 2 vols., Ungar Publishing Co., New York, 1961–1963.

31 Dixmier, J.: "Les Algèbres d'opérateurs dans l'espace Hilbertien," Gauthier-Villars, Paris, 1957.

32 Dixmier, J.: "Les C*-algèbres et leurs représentations," Gauthier-Villars, Paris, 1964.

33 Stone, M. H.: "Linear Transformations in Hilbert Space and Their Applications to Analysis," American Mathematical Society, Providence, R.I., 1932.

Probability theory

34 Doob, J. L.: "Stochastic Processes," John Wiley & Sons, Inc., New York, 1953.

35 Dynkin, E. B.: "Markov Processes," 2 vols. (translation), Academic Press Inc., New York, 1964.

36 Feller, W.: "Introduction to Probability Theory and Its Applications," 2 vols., John Wiley & Sons, Inc., New York, 1957–1966.

37 Loève, M.: "Probability Theory," 3d ed., D. Van Nostrand Company, Inc., Princeton, N.J., 1953.

368 Selected References

Topological algebras, groups, and linear spaces

38 Bourbaki, N.: "Espaces vectoriels topologiques," Hermann & Cie, Paris, 1953–1955.
39 Gelfand, I. M., and M. A. Neumark: "Unitäre darstellungen der klassischen Gruppen" (translation), Berlin, Akademie-Verlag GmbH, Berlin, 1957.
40 Hoffman, K.: "Banach Algebras of Analytic Functions," Prentice-Hall Inc., Englewood Cliffs, N.J., 1962.
41 Kelley, J. L., and I. Namioka et al.: "Linear Topological Spaces," D. Van Nostrand Company Inc., Princeton, N.J., 1963.
42 Loomis, L. H.: "Introduction to Abstract Harmonic Analysis," D. Van Nostrand Company, Inc., Princeton, N.J., 1953.
43 Nachbin, L.: "The Haar Integral" (translation), D. Van Nostrand Company, Inc., Princeton, N.J., 1965.
44 Naimark, M. A.: "Normed Rings," Hafner Publishing Company, Inc., New York, 1964.
45 Rickart, C. E.: "General Theory of Banach Algebras," D. Van Nostrand Company, Inc., Princeton, N.J., 1960.

Applications

46 Courant, R., and D. Hilbert: "Methods of Mathematical Physics," rev. ed., 2 vols., John Wiley & Sons Inc., New York, 1953–1962.
47 Jeffreys, H., and B. S. Jeffreys: "Methods of Mathematical Physics," 3d ed., Cambridge University Press, London, 1956.
48 Jost, R.: "General Theory of Quantized Fields," American Mathematical Society, Providence, R.I., 1965.
49 Kato, T.: "Perturbation Theory for Linear Operators," Springer-Verlag OHG, Berlin, 1966.
50 Mackey, G. W.: "The Mathematical Foundations of Quantum Mechanics," W. A. Benjamin, Inc., New York, 1963.
51 Segal, I.: "Mathematical Problems of Relativistic Physics" (Appendix by G. W. Mackey, Group Representations in Hilbert Space), American Mathematical Society, Providence, R.I., 1963.
52 von Neumann, J.: "Mathematical Foundations of Quantum Mechanics" (translation), Princeton University Press, Princeton, N.J., 1955.

Die Grundlehren der mathematischen Wissenschaften in Einzeldarstellungen mit besonderer Berücksichtigung der Anwendungsgebiete

Eine Auswahl

MIX
Papier aus verantwortungsvollen Quellen
Paper from responsible sources
FSC® C105338

If you have any concerns about our products,
you can contact us on
ProductSafety@springernature.com

In case Publisher is established outside the EU,
the EU authorized representative is:
Springer Nature Customer Service Center GmbH
Europaplatz 3, 69115 Heidelberg, Germany

Printed by Libri Plureos GmbH
in Hamburg, Germany